CRYSTALLIZATION OF POLYMERS, SECOND EDITION

In *Crystallization of Polymers*, second edition, Leo Mandelkern provides a self-contained, comprehensive, and up-to-date treatment of polymer crystallization. All classes of macromolecules are included and the approach is through the basic disciplines of chemistry and physics. The book discusses the thermodynamics and physical properties that accompany the morphological and structural changes that occur when a collection of molecules of very high molecular weight are transformed from one state to another.

The first edition of *Crystallization of Polymers* was published in 1964. It was regarded as the most authoritative book in the field. The first edition was composed of three major portions. However, due to the huge amount of research activity in the field since publication of the first edition (involving new theoretical concepts and new experimental instrumentation), this second edition has grown to three volumes.

Volume 1 is a presentation of the equilibrium concepts that serve as a basis for the subsequent volumes. In this volume the author shows that knowledge of the equilibrium requirements is vital to understanding all aspects of the polymer crystallization process and the final state that eventually evolves.

This book will be an invaluable reference work for all chemists, physicists and materials scientists who work in the area of polymer crystallization.

LEO MANDELKERN was born in New York City in 1922 and received his bachelors degree from Cornell University in 1942. After serving in the armed forces during World War II, he returned to Cornell, receiving his Ph.D. in 1949. He remained at Cornell in a post-doctoral capacity until 1952.

Professor Mandelkern was a staff member of the National Bureau of Standards from 1952 to 1962 where he conducted research in the physics and chemistry of polymers. During that time he received the Arthur S. Fleming Award from the Washington DC Junior Chamber of Commerce "As one of the outstanding ten young men in the Federal Service".

In January 1962 he was appointed Professor of Chemistry and Biophysics at the Florida State University, Tallahassee, Florida, where he is still in residence.

He is author of *Crystallization of Polymers*, first edition, published by McGraw-Hill in 1964. He is also author of *Introduction to Macromolecules*, first edition 1972, second edition 1983, published by Springer-Verlag.

Besides the Arthur S. Fleming Award he has been the recipient of many other awards from different scientific societies including the American Chemical Society and the Society of Polymer Science, Japan.

Professor Mandelkern is the author of over 300 papers in peer reviewed journals and has served on the editorial boards of many journals, inc~~~~~ the *Journal of the American Chemical Society*, the *Journal of Polymer Science* and

CRYSTALLIZATION OF POLYMERS
SECOND EDITION

Volume 1
Equilibrium concepts

LEO MANDELKERN

R. O. Lawton Distinguished Professor of Chemistry, Emeritus
Florida State University

CAMBRIDGE
UNIVERSITY PRESS

CAMBRIDGE UNIVERSITY PRESS
Cambridge, New York, Melbourne, Madrid, Cape Town, Singapore,
São Paulo, Delhi, Dubai, Tokyo, Mexico City

Cambridge University Press
The Edinburgh Building, Cambridge CB2 8RU, UK

Published in the United States of America by Cambridge University Press, New York

www.cambridge.org
Information on this title: www.cambridge.org/9780521020138

First published 1964 by McGraw-Hill, New York
Second edition 2002
First paperback edition 2010

A catalogue record for this publication is available from the British Library

Library of Congress Cataloguing in Publication data
Mandelkern, Leo.
Crystallization of polymers / Leo Mandelkern. – 2nd ed.
p.   cm.
Includes bibliographical references and index.
Contents: v. 1 Equilibrium concepts –
ISBN 0 521 81681 5
1. Polymers.   2. Crystallization.   3. Crystalline polymers.   1. Title.
QD281.P6 M3   2002
547′.84–dc21   2002067656

ISBN 978-0-521-81681-6 Hardback
ISBN 978-0-521-02013-8 Paperback

To Berdie, my wife,
and
to my grand-daughter, Sarah,
whose memory will be everlasting

Contents

Preface to second edition

Since the publication in 1964 of the first edition of *Crystallization of Polymers* there has been a vast amount of scientific activity in the study of crystalline polymers. This abundance that we enjoy has ranged from the synthesis of new classes of crystalline polymers to the application of sophisticated experimental techniques, accompanied by significant theoretical advances. Consequently, a large body of literature has resulted. As might be expected, many divergent opinions have been presented. The central problems in this subject were reviewed at a seminal Discussion of the Faraday Society (vol. 68, 1979). At this meeting different points of view were ardently presented. Since that Discussion, which can be considered to be a turning point in the investigation of the crystallization behavior of polymers, a coherent body of work has evolved. Some problems that were posed have been resolved. The differences in many others have been clarified. It appeared to the author that it was an appropriate time to bring together, in a coherent manner, the present status of the field. This was the motivation for the present work. Some aspects of crystalline polymers can be given a definitive analysis. On the other hand, there are still some problems that remain to be resolved. The different points of view will be presented in these cases. A strong effort has been made to present these matters in as an objective and scholarly manner as possible.

There is an extraordinary range of literature dealing with all aspects of the behavior of crystalline polymers. Therefore, no effort has been made here to present an annotated bibliography. Emphasis has been given to the basic, underlying principles that are involved. A considered effort has been made to present as diverse a set of examples as possible, illustrating the principles involved. Some works that should have been included may have been omitted. The author apologizes for this inadvertent error. There is a natural prejudice to select ones own material when appropriate. One hopes that this has not been overdone here. Fundamental principles are emphasized in these volumes. However, it has been the author's experience that these principles can be applied in an effective manner to the control of both microscopic and macroscopic properties of crystalline polymers. Thus, the book should be helpful in understanding and solving many technological problems involving crystalline polymers.

Students and investigators entering this research field for the first time should find a clear and objective perspective of the existing problems, as well as those that are reasonably well understood. For those who have been carrying out research in crystalline polymers, the problems are defined in a manner so as to indicate the directions that need to be taken to achieve resolution.

It was pointed out in the preface to the first edition, that it was composed of three major portions. These three portions have now grown to three volumes. The first of these is concerned with equilibrium concepts. The second deals with the kinetics and mechanisms of crystallization. Morphology, structure and properties of the crystalline state are discussed in the third volume. There is a strong interconnection between these major subjects.

The author is indebted to several generations of students and post-doctoral research associates, whose dedication, enthusiasm and love of research has sustained and contributed greatly to our research effort. It is also a pleasure to acknowledge a great debt to Mrs. Annette Franklin for her expert typing of the manuscript and preparing it in final form.

The permissions granted by *Acta Chimica Hungarica*; Chemical Society; *Colloid and Polymer Science*; *European Polymer Journal*; John Wiley and Sons, Inc.; *Journal of the American Chemical Society*; *Journal of Applied Polymer Science*; *Journal of Materials Science*; *Journal of Molecular Biology*; *Journal of Physical Chemistry*; *Journal of Polymer Science*; *Liquid Crystals*; *Macromolecules*; *Macromolecular Chemistry and Physics*; Marcel Dekker, Inc.; *Polymer*; *Polymer Engineering and Science*; *Polymer Journal*; *Pure and Applied Chemistry*; *Rubber Chemistry and Technology*; and Springer-Verlag to reproduce material appearing in their publications is gratefully acknowledged. Thanks also to Mrs. Emily Flory, Professor C. Price and Professor J. E. Mark for the permissions that they granted.

Tallahassee, Florida   *Leo Mandelkern*
August, 2001

Preface to first edition

We have been witnessing in recent years an unprecedentedly high degree of scientific activity. A natural consequence of the intensity of this endeavor is an ever-expanding scientific literature, much of which contains information of importance and interest to many diverse disciplines. However, it is a rare scientific investigator indeed who has either the time or the opportunity to digest and analyze critically the abundance we enjoy. Nowhere is this problem more acute than in the studies of the properties and behavior of macromolecular substances. Because of the somewhat belated recognition of its molecular character, this class of substances has been susceptive to quantitative investigations only for the past 30 years. During this period, however, there has developed a very rapidly increasing amount of activity and knowledge, in the realm of pure research as well as in industrial and practical applications. The problems presented have engaged the attention of individuals representing all the major scientific disciplines. In this situation it was inevitable that many subdivisions of polymer science have evolved. It appeared to the author that some of these areas could be subjected to a critical and, in some instances, a definitive analysis. Such endeavors also serve the purpose of acting as connecting links between the different specialities. At the same time they tend to underscore the more general and fundamental aspects of the scientific problems.

The present volume was suggested and stimulated by the aforementioned thoughts. We shall be concerned here with the phenomena and problems associated with the participation of macromolecules in phase transitions. The term crystallization arises from the fact that ordered structures are involved in at least one of the phases. The book is composed of three major portions which, however, are of unequal length. After a deliberately brief introduction into the nature of high polymers, the equilibrium aspects of the subject are treated from the point of view of thermodynamics and statistical mechanics, with recourse to a large amount of experimental observation. The second major topic discussed is the kinetics of crystallization. The treatment is intentionally very formal and allows for the deduction of the general mechanisms that are involved in the process. The equilibrium properties and the kinetic mechanisms must, in principle, govern the morphological characteristics of the crystalline state, which is the subject matter of the last chapter. The latter topic has been under intensive investigation in recent years. Many new concepts have been introduced which are still in a state of continuous revision. Consequently, a very detailed delineation of morphological structure has not been attempted. Instead, the discussion and interpretation have been restricted to the major features, which find their origin in the subject matter of the previous chapters.

Although many of the problems that fall within the scope of this work appear to be in a reasonable state of comprehension, there are some important ones that are not. It is hoped that these have become at least more clearly defined. Although no effort has been made to present a bibliographic compilation of the literature, care has been taken to avoid the neglect of significant work. Primary emphasis has been placed on principles, and this consideration has been the main guide in choosing the illustrative material. In this selection process a natural prejudice exists for material with which one is more familiar. This partiality, which appears to be an occupational hazard, has not been completely overcome in the present work. A great deal of what has been learned from studies of the simpler polymers can be applied to the properties and function of the more complex polymers of biological interest. Consequently, whenever possible, a unified approach has been taken which encompasses all types and classes of macromolecules, their diverse origin and function notwithstanding.

It was the author's pleasure and very distinct privilege to have the opportunity to be associated with Prof. P. J. Flory's laboratory some years ago. The author owes to him a debt, not only for the introduction to the subject at hand, but also for an understanding of the problems of science in general and polymer science in particular. As will be obvious to the reader, this book leans very heavily on his gifted and inspired teachings and research. However, the responsibility for the contents and the interpretations that are presented rests solely with the author.

The generous assistance of many friends and colleagues is gratefully acknowledged. Dr. N. Bekkedahl read and criticized a major portion of the manuscript and rendered invaluable aid to the author. Criticisms and suggestions on various chapters were received from Drs. T. G. Fox, W. Gratzer, G. Holzworth, H. Markowitz, and D. McIntyre. Dr. R. V. Rice and Mr. A. F. Diorio generously contributed electron micrographs and x-ray diffraction patterns for illustrative purposes.

The permission granted by *Annals of the New York Academy of Science*; *Chemical Reviews*; *Die Makromolekulare Chemie*; Faculty of Engineering, Kyushu University; John Wiley & Sons, Inc.; *Journal of the American Chemical Society*; *Journal of Applied Physics*; *Journal of Cellular and Comparative Physiology*; *Journal of Physical Chemistry*; *Journal of Polymer Science*; *Kolloid-Zeitschrift*; *Polymer*;

Proceedings of the National Academy (U.S.); *Proceedings of the Royal Society*; *Review of Modern Physics*; *Rubber Chemistry and Technology*; *Science*; and *Transactions of the Faraday Society* to reproduce material originally appearing in their publications is gratefully acknowledged.

Tallahassee, Florida   *Leo Mandelkern*
May, 1963

1

Introduction

1.1 Background

Polymers of high molecular weight have now been accepted as respectable members of the molecular community. This situation was not always true.(1) It is now recognized, however, that polymer molecules possess the unique structural feature of being composed of a very large number of chain units that are covalently linked together. This property is common to all macromolecules despite their diverse origin, their widely differing chemical and stereochemical structures and uses and function. It is, therefore, possible to study this class of substances from a unified point of view that encompasses the relatively simpler polymers prepared in the laboratory, as well as the more complex ones of nature. The characteristic thermodynamic, hydrodynamic, physical, and mechanical properties possessed by high polymeric substances can be explained, in the main, by their covalent structure and the attendant large size of the individual molecules.

Although one is dealing with molecules that contain thousands of chain bonds, macromolecular systems still retain the ability to exist in different states. This property is common to all substances, high polymers included. Two states of matter that are observed in monomeric substances, the liquid and crystalline states, are also found in polymers. The liquid or amorphous state is characterized by some amount of rotation about the single bonds connecting the chain atoms in the polymer. Hence, in this state a single polymer molecule can assume a large number of spatial conformations. The bonds in a collection of such chain molecules in the liquid state, adopt statistical orientations and their centers of gravity are randomly arranged relative to one another. The structural units of a collection of such molecules in this state are arranged in a random, disoriented array and are essentially uncoordinated with one another. However, under appropriate conditions of either temperature, pressure, stress, or solvent environment, a spontaneous ordering of portions of the chain molecules can take place. This ordering results from the strong preference of the chain bonds to assume a set of highly favored specific orientations or rotational states. Therefore, in contrast to the amorphous or liquid polymer, the individual molecules now exist in a state of conformational order. The individual ordered chains, or more specifically portions of them, can then be organized into a regular three-dimensional array with the chain axes usually being parallel to one another, although a few exceptions have been found. The structure of the individual molecules may be such that they are fully extended, or they could be in a helical conformation, or they may fold back upon one another, as circumstances dictate. The significant factor is that a state of three-dimensional order is developed that in its major aspects closely resembles the crystalline state of monomeric substances. This general structural arrangement of the constituent molecules is termed the crystalline state of polymers. Since in virtually all cases the ordering process is not complete, this state is more properly termed a semi-crystalline one.

It is axiomatic that an individual polymer molecule that possesses a high degree of chemical and structural regularity among its chain elements is capable of undergoing crystallization. Indeed, crystallization has been observed in a wide variety of such polymers. It is found, moreover, that a significant amount of structural irregularity can be tolerated without preventing the crystallization process. However, even for a polymer possessing a highly regular structure, conditions must be found that are kinetically favorable for crystallization to occur in the allotted observation time. For example, poly(isobutylene), a polymer of apparently regular structure, can be easily crystallized by stretching. For a long time this polymer was not thought to be crystallizable without the application of an external stress. However, it has been demonstrated that crystallinity can be induced merely by cooling. Many months must elapse, at the optimum temperature, before the development of the crystalline state can be definitely established. Kinetic factors, therefore, are quite important. It is thus not surprising that some polymers thought to have a regular structure have not as yet been crystallized.

The understanding of the structure and properties of semi-crystalline polymers involves many different experimental techniques, scientific disciplines and theoretical approaches. The totality of the problem, and the interrelation between its different facets, are shown schematically in Fig. 1.1.(2) Essentially, all properties are controlled by the molecular morphology, that in turn is determined by the crystallization mechanisms. Information about mechanisms is obtained from studies of crystallization kinetics. In order to interpret kinetics, the equilibrium requirements need to be established. It has long been recognized that the crystalline state that is actually observed in polymers, more often than not represents one that is not at equilibrium and can be considered to be metastable. However, knowledge of the equilibrium requirements is vital to understanding all aspects of the crystallization process and the final state that eventually evolves. Based on the overview

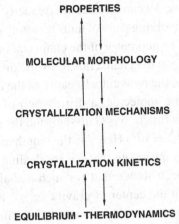

Fig. 1.1 Schematic representation and interrelation of problem areas in the study of crystalline polymers.(2)

given by the schematic of Fig. 1.1, the study of crystalline polymers divides itself naturally into three parts. Therefore, in this work the first of these, Equilibrium Concepts, comprise Volume 1. Volume 2 is concerned with Crystallization Kinetics and Mechanisms. Molecular Morphology and Properties are treated in Volume 3.

Many important properties of polymeric systems reside in the details of the conformation of the individual chains. This is particularly true with regard to their crystallization behavior. Hence it is appropriate that, before embarking on a discussion of the major subject at hand, attention be given to the general principles involved in determining the conformation of individual long chain molecules and the nature of the liquid state.

1.2 Structure of disordered chains and the liquid state

The spatial geometry of a long chain molecule depends on the bond distances between the chain atoms, the valence angles, and the hindrance potentials for internal rotation about single bonds. The conformation of a given chain backbone (fixed bond lengths and valence angles) is completely specified by the rotation angles about each of its single bonds. The large number of conformations available to a given molecule results from the permissible variations in the rotational angles among the skeletal bonds. These conformations differ from one another according to the value of the rotational angle for each individual bond.

As a convenient starting point in developing the statistical methods that are needed to analyze chain conformation, and for the purpose of calculating the dimensions of real molecules, a highly hypothetical model of a chain made up of completely freely rotating single bonds and bond angles is assumed. The geometric properties of such a chain model can be calculated exactly as long as long-range intramolecular interactions involving pairs of units remotely separated along the chain contour are neglected.(3) The geometry of the chain can be conveniently described either by the distance between the chain ends or by the distance of a chain element from the center of gravity of the molecule. Because of the large number of different conformations available to a molecule, a distribution of end-to-end distances is calculated. This distribution function is Gaussian, and the mean-square end-to-end distance is found to be $\langle r^2 \rangle_{0f} = nl^2$. Here l is the length and n the number of links in the chain. The subscripts designate that we are dealing with an isolated, freely jointed chain. It has also been shown that for such a chain the root-mean-square distance of an element from the center of gravity $\langle s^2 \rangle_{0f}^{1/2}$ and $\langle r^2 \rangle_{0f}$ are related by $\langle s^2 \rangle_{0f}^{1/2} = \langle r^2 \rangle_{0f}^{1/2}/6$. For the chain model assumed, these linear dimensions depend on the square root of the number of bonds and hence are many times smaller than the extended length of the macromolecule. The most frequent conformations expected in the liquid state will, therefore, be those that are highly coiled. Calculations of the dimensions of freely rotating chains have also been made for cases where more than one kind of bond and valence angle are present.(3,4) Hence, it is possible for a comparison to be made between the actual dimensions of many real chains and their freely rotating counterparts.

In a real chain, the freedom of internal rotation and thus chain dimensions are tempered by the hindrance potential associated with a given bond, as well as steric interferences and interactions between neighboring substituents attached to the main chain atoms. Also of concern is the question of whether the bond rotations of neighboring bonds are independent or interdependent with one another. The hindrance potentials (for single bonds in polymer chains) are expected to resemble those of similar bonds in monomeric molecules.(5,6,7) For example, a threefold symmetric potential is appropriate to describe the rotational states of ethane. However, the potential for the central bond of butane needs to be modified. Although three minima still exist in the potential function all are not of equal energy. The lowest one is for the planar or trans configuration. The other two minima represent gauche forms, which are obtained by rotations of $\pm 120°$ from the trans position. The two gauche forms are of the same energy and exceed that for the trans form by about 500 to 800 cal mol^{-1}. It has been assumed that a similar potential function is applicable to the hindered rotation of bonds in the long chain polyethylene molecule. Hence, for this polymer the lowest energy form is the planar all-trans configuration which corresponds to the fully extended chain. Although the trans state is energetically favored, gauche states are allowed at favorable temperatures so that it is still possible to generate highly irregular conformations.

For polymers whose chain structures are more complex than that of polyethylene the simple potential function described above needs to be modified. However, the potential functions are still characterized by minima that represent the low energy, highly favored rotational states. Hence, for real chains the angular position of each bond may be considered to occur effectively in one of the available minima. Bond rotations are thus limited to angular values that lie within fairly narrow ranges that can be regarded as discrete states. This approximation has been termed the rotational isomeric state. With this model an elegant mathematical apparatus is available that allows for a quantitative description of the chain conformation and can take into consideration the interdependence of rotational potentials on the states of neighboring bonds.[7,8,9] The partition function of the chain can be calculated using the method of the one-dimensional Ising lattice that was developed for the treatment of ferromagnetism.[7,8,9] From this calculation, the average dimensions of the single, isolated real chain can be deduced as well as the angular position of the energy minima. The chain dimensions are conveniently characterized by their characteristic ratios defined as $C_n = \langle r^2 \rangle_0 / nl^2$, where n represents the number of chain bonds and $\langle r^2 \rangle_0$ the actual mean-square end-to-end distance of an isolated chain unperturbed by long-range intramolecular interactions. The characteristic ratio is a measure of the spatial domain of the chain and will obviously be greater than that of the freely jointed chain. The C_n value can be obtained experimentally by several different physical chemical methods. In the disordered or liquid state individual chains are said to adopt a statistical conformation, since the conformation is governed by the rules of statistical mechanics. A compilation of values for C_∞, characteristic of an infinitely long chain, is given in Table 1.1 for a set of representative polymers. A more complete set of data can be found elsewhere.[9a]

Usually, there is good agreement between the experimentally determined values of C_∞ and the theoretical expectations. For the polymers listed, C_∞ values range from about 2 to 20, significantly greater than what would be calculated for a free rotating or freely jointed chain. Freely jointed, or rotating chains do not give either a good or a universal representation of the spatial characteristics of real chains. There is a certain element of arbitrariness in calculating C_∞ for chains that contain rings in the backbone because of the ambiguity in specifying the required single or virtual bonds.[10] For example, depending on the virtual bond chosen, C_∞ for poly(ethylene terephthalate) is calculated to be either 4.70 or 5.45. To avoid this ambiguity the spatial extent of the chain can also be expressed as $(\langle r^2 \rangle_0 / M)_\infty$. This latter quantity is calculated to be 0.93 Å^2 g mol^{-1} for poly(ethylene terephthalate), a value that is comparable to the experimentally determined one.

Polymers that have C_∞ values in the range listed in Table 1.1 are considered to be "flexible" chains. These values are in marked contrast to another class of polymers, such as the poly(p-phenylene amides) and the corresponding polyesters, where NH is replaced by O. For these polymers C_∞ values are calculated to be in the range of 125–225.[11] Although for sufficiently high molecular weights these polymers can

Table 1.1. *Values of C_∞ for some representative polymers*[a]

Polymer	T °C	C_∞	Reference
Poly(methylene)	138–142	6.6–6.8	a,b,c,d
	140	7.9[b]	e
	25	8.3	f
Poly(tetrafluoroethylene)	327	9.8[b]	g
	325	8 ± 2.5	h
Poly(isobutylene)	24	6.6	i
	24	7.2[b]	j
Poly(oxyethylene)	35–45	4.0[b]	k
	30	4.0–5.5[b]	l
		4.0–5.6	l
Poly(dimethyl siloxane)	20.0, 22.5	6.35–7.7[b]	m
Poly(hexamethylene adipamide)	25	5.9	n
	25	6.10[b]	o
Poly(caproamide)	25	6.08[b]	o
1,4 Poly(isoprene)			
cis	50	3.84[b], 4.55[b], 4.92[b]	p,q
	50	4.7	r,s
trans	50	6.60[b], 6.95[b]	r
	56	6.60, 7.4	s,t,u
Poly(propylene)			
isotactic	140	4.2[b]	v
syndiotactic	140	11[b]	v
atactic	140	5.5[b], 5.3[b]	v,w
Poly(methyl methacrylate)			
isotactic	27.6, 26.5	9.1–10.0, 10.0[b]	x,y,z
syndiotactic	8, 35	6.5, 7.2	y
		7.2[b]	aa
atactic	4–70	6.9 ± 0.5	bb,cc, dd,ee
Poly(styrene)			
isotactic	30	11	ff,gg
syndiotactic	30	15–30[b]	hh
atactic	30	10	gg,ii, jj,kk
Poly(L-proline)	30	14 (water)	ll
		18–20 (organic solvents)	ll
Poly(L-glycine)	30	2[b]	mm
Poly(L-alanine)	30	9[b]	mm

[a] Experimental values are given for C_∞ except when otherwise noted.
[b] Calculated values for C_∞.

Notes to Table 1.1 (*cont.*)
References
  a. Chiang, R., *J. Phys. Chem.*, **70**, 2348 (1966).
  b. Stacy, C. J. and R. L. Arnett, *J. Phys. Chem.*, **69**, 3109 (1965).
  c. Nakajima, A., F. Hamada and S. Hayashi, *J. Polym. Sci.*, **15C**, 285 (1966).
  d. Chiang, R., *J. Phys. Chem.*, **69**, 1645 (1965).
  e. Han, J., R. L. Jaffe and D. Y. Yoon, *Macromolecules*, **30**, 7425 (1997).
  f. Fetters, L. J., W. W. Graessley, R. Krishnamoorti and D. J. Lohse, *Macromolecules*, **30**, 4973 (1997).
  g. Smith, G. D., R. L. Jaffe and D. Y. Yoon, *Macromolecules*, **27**, 3166 (1994).
  h. Chu, B., C. Wu and W. Beck, *Macromolecules*, **22**, 831 (1989).
  i. Fox, T. G., Jr., and P. J. Flory, *J. Am. Chem. Soc.*, **73**, 1909 (1951).
  j. Suter, U. W., E. Saiz and P. J. Flory, *Macromolecules*, **16**, 1317 (1983).
  k. Mark, J. E. and P. J. Flory, *J. Am. Chem. Soc.*, **87**, 1415 (1965).
  l. Smith, G. D., D. Y. Yoon and R. L. Jaffe, *Macromolecules*, **26**, 5213 (1993).
  m. Crescenzi, V. and P. J. Flory, *J. Am. Chem. Soc.*, **86**, 141 (1964).
  n. Saunders, P. R., *J. Polym. Sci., A*, **2**, 3765 (1964).
  o. Flory, P. J. and A. D. Williams, *J. Polym. Sci., Pt. A-2*, **5**, 399 (1967).
  p. Abe, Y. and P. J. Flory, *Macromolecules*, **4**, 230 (1971).
  q. Tanaka, S. and A. Nakajima, *Polym. J.*, **3**, 500 (1972).
  r. Mark, J. E., *J. Am. Chem. Soc.*, **88**, 4354 (1966).
  s. Wagner, H. and P. J. Flory, *J. Am. Chem. Soc.*, **74**, 195 (1952).
  t. Mark, J. E., *J. Am. Chem. Soc.*, **89**, 829 (1967).
  u. Poddabny, Pa, E. G. Erenburg and M. A. Eryomina, *Vysokomal. Soedin. Ser. A*, **10**, 1381 (1968).
  v. Suter, U. W. and P. J. Flory, *Macromolecules*, **8**, 765 (1975).
  w. Alfonso, G. C., D. Yan and Z. Zhou, *Polymer*, **34**, 2830 (1993).
  x. Krause, S. and E. Cohn-Ginsberg, *J. Phys. Chem.*, **67**, 1479 (1963).
  y. Schulz, G. V., W. Wunderlich and R. Kirste, *Makromol. Chem.*, **75**, 22 (1964).
  z. Jenkins, R. and R. S. Porter, *Polymer*, **23**, 105 (1982).
  aa. Vacatello, M. and P. J. Flory, *Macromolecules*, **19**, 405 (1986).
  bb. Fox, T. G., *Polymer*, **3**, 111 (1962).
  cc. Schultz, G. V. and R. Kirste, *Z. Physik Chem. (Frankfurt)*, **30**, 171 (1961).
  dd. Chinai, S. N. and P. J. Valles, *J. Polym. Sci.*, **39**, 363 (1959).
  ee. Vasudevon, P. and M. Santoppa, *J. Polym. Sci. A-2*, **9**, 483 (1971).
  ff. Krigbaum, W. R., D. K. Carpenter and S. Newman, *J. Phys. Chem.*, **62**, 1586 (1958).
  gg. Kurata, M. and W. H. Stockmayer, *Fortschr. Hochpolym. Forsch.*, **3**, 196 (1963).
  hh. Yoon, D. Y., P. R. Sundararajan and P. J. Flory, *Macromolecules*, **8**, 776 (1975).
  ii. Krigbaum, W. R. and P. J. Flory, *J. Polym. Sci.*, **11**, 37 (1953).
  jj. Altares, T., D. P. Wyman and V. R. Allen, *J. Polym. Sci., A*, **2**, 4533 (1964).
  kk. Orofino, T. A. and J. W. Mickey, Jr., *J. Chem. Phys.*, **38**, 2513 (1963).
  ll. Mattice, W. L. and L. Mandelkern, *J. Am. Chem. Soc.*, **93**, 1769 (1971).
  mm. Brant, D. A., W. G. Miller and P. J. Flory, *J. Mol. Biol.*, **23**, 47 (1967).

be treated as statistical coils, they are in fact highly extended, asymmetric chains. We shall be concerned here primarily with flexible type chains.

The discussion of chains in statistical conformation is based on the properties of individual, isolated chains. Except for crystallization from very dilute solution the crystallization process involves a collection of such chains. The question can then be raised as to the relation between the conformation of an isolated chain and that when present in a collection of such chains in the molten or liquid state. Flory has argued on theoretical grounds (12) that polymers in an undiluted melt should be essentially unperturbed, i.e. the chain dimensions should be the same as the isolated chain devoid of long-range intramolecular interaction and thus correspond to the θ condition. This conclusion is based on the premise that although a molecule in the bulk state, or in concentrated solutions, interferes with itself, it has nothing to gain by expanding. The reason is that the decrease in interaction with itself that would occur is compensated by increased interference with its neighbors. Hence, the chain prefers to remain in the θ condition. This theoretical expectation is borne out by experiment. Small-angle neutron scattering measurements of the radii of gyration of many polymers in the bulk are in close agreement with the values for the isolated, unperturbed chain, as determined under θ conditions.(13–18)

The discussion of the liquid state up to this point has been a fairly idealized one, since only the conformation and spatial extent of the chains have been taken into account. Other factors, not as easily susceptible to calculation, also need to be considered. These factors principally involve a description of topological structures such as chain entanglements, loops and knots being among the possibilities. Such structures can be expected in a collection of random long chain molecules in the liquid state and should play a major role in the crystallization process. Unfortunately, the quantification of such topological defects has been difficult. Only chain entanglements, characterized by the molecular weight between entanglements, M_e, have been given quantitative meaning by indirect measurements. It is assumed the points of entanglement acts as crosslinks. Then elementary rubber elasticity theory can be applied to the measured plateau modulus.(19) Values of M_e for selected polymers are given in Table 1.2.(20,21).

There is a considerable variation in the M_e values among the different polymers. The values range from 830 g mol^{-1} for linear polyethylene to as high as 12 000 g mol^{-1} for poly(dimethyl siloxane). The main factors governing M_e are the flexibility of the chain and the presence of branches. From a topological viewpoint, branches and their length are known to affect the entanglement density.(21) The difference in M_e's between polyethylene and the poly(propylenes) can be explained on this basis. Irrespective of the M_e value, the entanglement density will be significant for high molecular weight chains.

1.3 The ordered polymer chain

Under suitable conditions the allowable rotational states can be restricted. A given bond or sequence of bonds will be limited to rotational angles that correspond to the lowest minima in the potential function describing the hindrance to rotation.

Table 1.2. *Molecular weight between entanglements for selected polymers*[a]

Polymer	T °C	M_e (g mol^{-1})
Polyethylene	140	830
Poly(propylene) atactic	140	4600
	25	3500
atactic[b]	30–240	7050
isotactic[b]	170–220	6900
syndiotactic[b]	170–220	2170
Poly(2-methyl-1,3-pentadiene)	25	4700
Poly(1,4-cis-isoprene)	25	3100
Poly(isobutylene)	140	7300
	25	5700
Poly(dimethyl siloxane)	140	12000
	25	9600
Poly(ethylene oxide)	140	1600
Poly(ethylene terephthalate)	275	1200
Poly(carbonate)	200	1300
Poly(capolactam)	270	2000
Poly(oxymethylene)	200	2100
Poly(phenylene oxide)	220	2700–3600
Poly(tetrafluoroethylene)	380	3700

[a] Data taken from Ref. (20) where a more comprehensive table can be found.
[b] Eckstein, A., J. Sahm, C. Friedrich, R. D. Maier, J. Sassmannshausen, M. Bochmann and R. Mülhaupt, *Macromolecules*, **31**, 135 (1998).

Consequently, a highly ordered chain structure is evolved with the concomitant loss of the conformational versatility that characterizes the disordered chain and the liquid state. For example, the trans state represents the bond orientation with the lowest energy in polyethylene. When successive bonds in the chain assume this orientation, a fully extended planar zigzag conformation results, as is illustrated in Fig. 1.2. From the multitudinous number of conformations available to the chain in the liquid state only one ordered structure survives that is characteristic of the crystal. An extended planar, or nearly planar, ordered conformation is characteristic of many polymers including polyamides, polyesters, cellulose derivatives, polydienes, and one of the low energy forms of the polypeptides.

The rotational states allowed for vinyl polymers derived from monomer units of the type $-CH_2-CH_2R-$ depend on the configurations of the successive

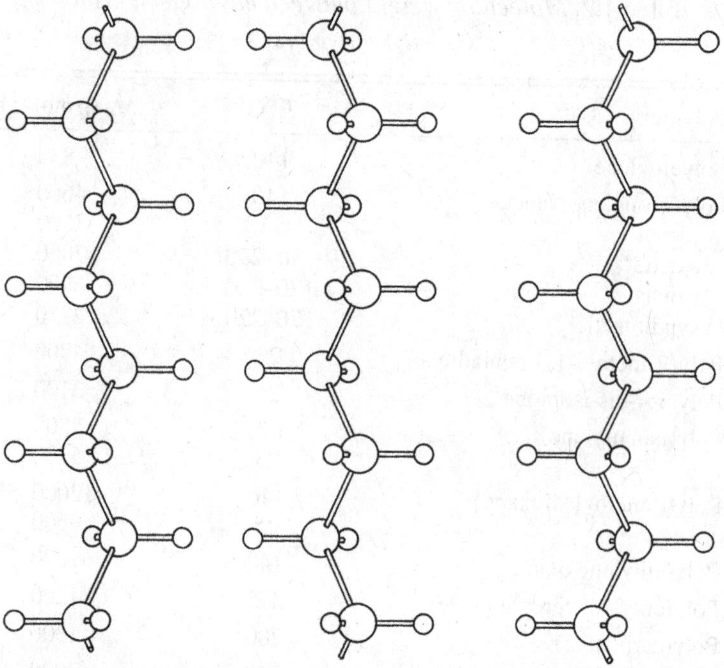

Fig. 1.2 Representation of ordered structure of portions of polyethylene chains. (From Natta and Corradini (22))

asymmetric carbon atoms bearing the substituent. For an isotactic polymer, wherein the substituent bearing carbon possesses the same tetrahedral configuration, the planar zigzag chain is excluded because of the steric interfaces between the neighboring R groups. In the trans state, successive substituent groups are within 2.5 to 2.6 Å of each other. This is not an allowed distance since it results in gross overcrowding. The crowding problem can be alleviated by having alternate bonds assume gauche positions. In this geometric pattern, the substituent groups are adequately separated. If the required rotations are executed in a regular manner so that the sequence of trans–gauche bond orientations is followed, then a helical chain structure is formed. Since there are two gauche positions, if the rotations are always executed in the same direction, either a right-handed or left-handed helix can be generated with the same molecule. If the substituent group is not too bulky, then it is found that the helix contains three chemical repeating units for each geometrical repeating unit. A helix of this type is illustrated in Fig. 1.3a.(22) This helical form allows the substituent groups to be sufficiently far apart. As examples, the nearest distances between nonbonded carbon atoms now become 3.2 Å in isotactic polypropylene and 3.3 Å in isotactic polystyrene.

Polymers containing bulkier side-groups require more space, so that much looser helices are formed. Typical examples of the latter type are illustrated in Fig. 1.3b, c, and d. These structures give rise to larger repeating units. For example,

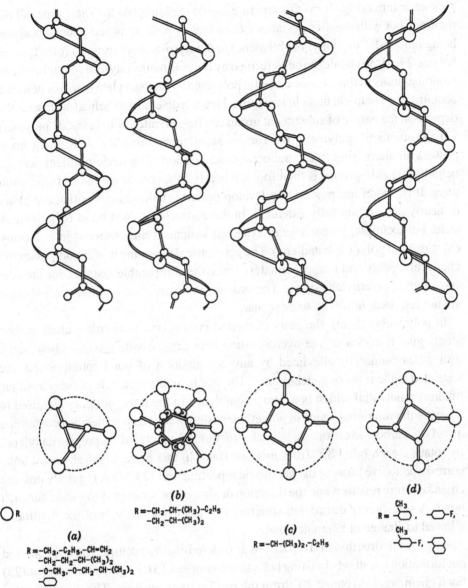

(a)

R =-CH₃,-C₂H₅,-CH=CH₂
  -CH₂-CH₂-CH-(CH₃)₂
  -O-CH₃,-O-CH₂-CH-(CH₃)₂
  ⟨◯⟩

(b)

R =-CH₂-CH-(CH₃)-C₂H₅
  -CH₂-CH-(CH₃)₂

(c)

R =-CH-(CH₃)₂,-C₂H₅

(d)

R = ⟨◯⟩—CH₃ , ⟨◯⟩-F,
    ⟨◯⟩-F, ⟨◯◯⟩

Fig. 1.3 Representation of some typical ordered helical structures for isotactic polymers. (From Natta and Corradini (22))

poly(3-methyl-butene-1) in which the side-group is $CH(CH_3)_2$, has a repeating unit that is composed of four monomer units. This side-group leads to more acute overcrowding so that the angle of the gauche bonds is changed from 120° to about 100°. Instead of having the strict trans position at 0° it is modified to about −26°. For polymers in which the branching occurs at the second atom of the side-chain, as in poly(4-methyl-hexene-1) the helix is comprised of seven monomer units in two geometric turns (Fig. 1.3b). Its structure can be explained by the same type of bond rotations but with smaller deviations from that of pure trans and gauche positions being required. Poly(vinyl naphthalene) and poly(*o*-methyl styrene) form fourfold helices (Fig. 1.3d) while the helix formed by poly(*m*-methyl styrene) contains eleven monomer units in three turns. Isotactic poly(methyl methacrylate) forms a helix that contains five chemical units in two turns. Hence a diversity of helical structures that depend on the nature of substituent group can be generated with isotactic polymers.

In syndiotactic polymers, the carbon atoms containing the substituent group possess an alternating D, L tetrahedral configuration. The steric problem between neighboring side-groups is therefore not nearly as severe as for the isotactic structures. It is, therefore, possible to develop ordered chain structures that are planar or nearly planar, and fully extended. In these structures each bond is in the trans state. For example, planar zigzag extended structures are observed in poly(vinyl chloride) and poly(1,2-butadiene). The geometrical repeating unit encompasses two chemical repeats and is approximately twice the comparable distance for the non-substituted polyethylene chain. The ordered structures for syndiotactic polymers are not required, however, to be planar.

In poly(isobutylene), the pairs of methyl groups on the alternate chain carbon atoms give rise to a severe overcrowding between the side-groups. These steric difficulties cannot be alleviated by any combination of bond rotations that are restricted to the trans or gauche states. The bonds in this molecule possess a unique hindrance potential which bears no resemblance to the threefold potential used to describe the rotational states in polyethylene and other chain molecules containing a carbon–carbon skeleton. A helical structure is generated in poly(isobutylene) by rotating each bond 82° from its trans state. In this helix, eight chemical units correspond to five turns of the geometric repeating unit.(23,24) A regularly ordered chain structure results when the direction or sign of the rotation is the same for each bond. A statistically disordered structure evolves when the sign of the rotation is allowed to change at alternate bonds.

Poly(tetrafluoroethylene) also forms an ordered chain structure. The fully ordered conformation is a slowly twisting helix that comprises 13 CF_2 groups in a repeat.(25) Each chain bond is rotated 20° from the precise trans position. The reason for this distortion is that, if the structure were planar zigzag, the nonbonded fluorine atoms would be uncomfortably close to one another. The rotation about each chain bond again relieves the overcrowding.

Helically ordered chain structures are not limited to molecules containing a carbon–carbon backbone structure. They also manifest themselves in polypeptides, proteins, and nucleic acids. A very important ordered structure of polypeptides is the alpha-helix deduced by Pauling, Corey, and Branson.(26) In this structure (as contrasted with the extended ordered configuration of a polypeptide chain) the maximum number of hydrogen bonds between the carbonyl oxygen and amino

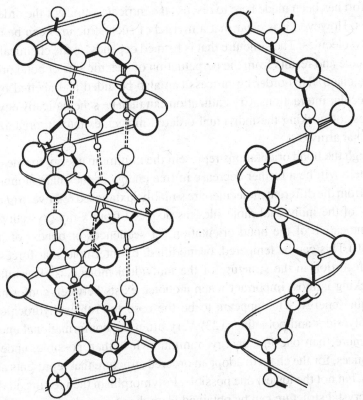

Fig. 1.4 Comparison of the alpha-helix formed by polypeptides (left) with a 3.5 helix generated by an isotactic polymer (right). (From Natta and Corradini (22))

nitrogen are formed intramolecularly. The hydrogen bonding occurs between every third amino acid residue along the chain. A nonintegral helix results which contains 3.6 residues per turn. The peptide group is planar, in analogy to deductions from crystallographic studies of similarly constituted monomeric substances, and each CO and NH group forms a hydrogen bond. A comparison of the structure of an alpha-helically ordered polypeptide chain with a 3.5 helix formed by an isotactic vinyl polymer is shown in Fig. 1.4. In the latter case, the structure is not stabilized by any intramolecular hydrogen bonds. In another example, poly(L-proline), which is a polyimino acid, does not possess the capacity for intramolecular hydrogen bonds. However, because of the influence of steric factors, an ordered helical chain conformation exists, where the imide group is planar and in the trans state.(27)

The ordered structures of nucleic acids involve more than one chain molecule. The structure of deoxyribonucleic acid (DNA), as deduced by Watson and Crick (28) and Wilkins *et al.*,(29) involves two intertwined chains helically woven so as to resemble a twisted ladder. The rungs of the ladder, which render the structure stable, are formed through the hydrogen bonding of complementary purine and pyrimidine bases.

No effort has been made here to discuss the intricate details of the ordered chain structures. However, it is clear that a myriad of such structures can be developed by chain molecules. The structure that is formed depends on the chemical nature of the molecule and results from the perpetuation of specific sets of bond orientations along the chain. This ordering process can also be aided and abetted by specific intermolecular interactions. Crystallization can then be schematically envisaged as the process of packing the individual ordered molecules into an organized three-dimensional array.

Although the bond orientations represent the minimum energy for the chain as a whole, there will be a further decrease in free energy as the chain atoms and substituents from the different molecules are suitably juxtaposed relative to one another. The form of the individual molecules, as deduced from x-ray crystallography, is usually indicative of the bond orientation (or sequences of bonds) of minimum energy.(30) This can be tempered, or modified, by intermolecular forces that can cause a distortion in the structure of the individual molecules. The influence of chain packing is most important when a choice exists between conformations of nearly equal energy. This appears to be the case for rubber hydrochloride and certain polyesters and polyethers.(23) Very often the conformational energy map contains more than one low energy minimum. It is then possible, under certain circumstances, for the chain to adopt an ordered structure that represents an energy minimum, but not the lowest one possible. Polymorphism then results, in that more than one crystal structure can be obtained from the same polymer.

The arrangements of the atoms in the crystalline regions of a polymer can be determined by the conventional methods of x-ray crystallography.(31) The ordered chain conformation and the packing can be established in this manner. Although single crystals are usually not available to polymer crystallographers, many of the characteristics of the unit cell such as the crystal system, dimensions, and positions of the atoms have been deduced for a wide variety of polymers. Normal bond distances, angles, and other elements of structure appear to be the general rule. The role of the chemical repeating unit is analogous to the part played by molecules in crystals of low molecular weight organic compounds. The important realization that the complete molecule does not have to be contained within the unit cell was very influential in establishing the macromolecular hypothesis.(1,32) Unit cells are usually composed of from one to eight chain repeating units. It is also not uncommon to have more than one chain within the unit cell.

1.4 Morphological features

When the structural features of crystalline polymers are examined beyond the level of the unit cell, it is very important that their semi-crystalline character

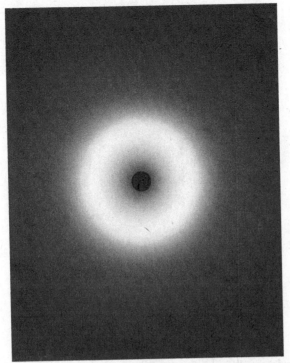

Fig. 1.5 Wide-angle x-ray pattern of noncrystalline natural rubber. (Courtesy of A. F. Diorio)

be recognized. This situation becomes immediately apparent from x-ray diffraction studies. Several different types of wide-angle x-ray patterns can be obtained from polymeric systems. Discrete Bragg reflections do not appear in the pattern when the polymer is noncrystalline. Only a diffuse halo or in some cases two haloes are observed, as is illustrated in Fig. 1.5. This pattern is for noncrystalline natural rubber at 25 °C. A typical pattern obtained when a polymer is crystallized merely by cooling is given in Fig. 1.6 for a linear polyethylene specimen. Discrete Bragg reflections are now observed. These are in the form of a series of concentric circles. The pattern is qualitatively similar to that obtained from powder patterns of crystalline monomeric substances. However, the line widths are not as narrow. The crystallites are randomly arranged. From a macroscopic point of view there is, on an average, no preferred orientation of the crystallographic directions.

Different kinds of preferred orientations can also be developed with crystalline polymers. The native state of many macromolecules of biological interest, such as the fibrous proteins, is characterized by a preferred crystalline orientation. Similar conditions can also be obtained in other polymers by deformation of the specimen

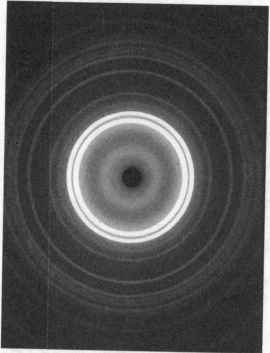

Fig. 1.6 Wide-angle x-ray diffraction pattern of linear polyethylene crystallized by cooling. (Courtesy of A. F. Diorio)

during or subsequent to the crystallization process. Examples of wide-angle x-ray diffraction patterns of three axially oriented crystalline polymers, natural rubber, linear polyethylene, and the naturally occurring fibrous protein collagen, are given in Fig. 1.7. The reflections have now become discrete spots as a result of the preferential orientation of different crystallographic planes. The natural rubber and polyethylene patterns are reminiscent of those obtained from a well-developed single crystal with rotational symmetry about an axis perpendicular to the incident x-ray beam. It should be noted that, despite the close similarity to the conventional single crystal pattern, the persistence of a diffuse halo is still easily discerned. For the same crystallographic structure, i.e. in the absence of polymorphism, the recorded Bragg spacings are identical, whether the specimen is oriented or not. Other types of orientation are also possible, such as biaxial, where the polymer chains tend to lie in a plane. The different types of orientation can be identified and described by wide-angle x-ray diffraction.

There is substantial evidence to indicate, at all the levels of morphology that are amenable to study, well-defined organized structures exist. Small-angle x-ray studies indicate structures having linear dimensions that correspond to hundreds of angstroms.(34,35) A typical low-angle x-ray pattern from a highly axially oriented

(a)

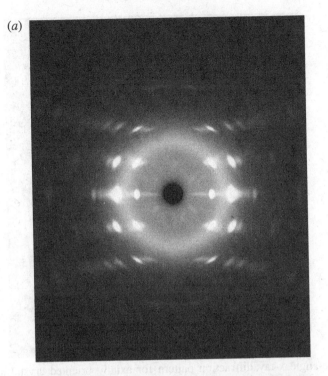

(b)

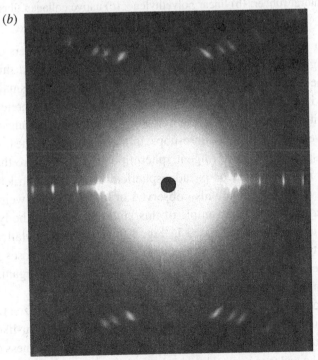

Fig. 1.7 (*cont.*)

(*c*)

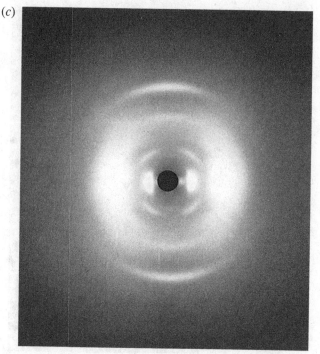

Fig. 1.7 Wide-angle x-ray diffraction pattern for axially oriented crystalline macromolecules. (a) Natural rubber: (b) linear polyethylene: (c) native collagen fiber. (Courtesy of
A. F. Diorio)

fiber of linear polyethylene is shown in Fig. 1.8.(33) Several orders of diffraction,
corresponding to a long period of $410 \pm 20\,\text{Å}$, are resolved in this sample. In
addition to the discrete maxima, diffuse scatter also occurs at the small angles. The
light scattered by thin films of crystalline polymers can be interpreted in terms of
structural entities whose size is in the range of several thousand angstroms.(35)

When viewed under the light microscope, thin films of crystalline homopolymers
very often display highly birefringent spherulitic structures. Here the crystallites
are arranged in a spherical or pseudo spherical array. Such structures are not
unique to polymers as they are also observed in low molecular weight inorganic
and organic compounds. An example of this kind of crystalline body, grown in a
thin polyethylene film, is illustrated in Fig. 1.9.(36) A more detailed discussion
of these structures will be given subsequently. For present purposes it suffices to
note that the existence of spherulites is evidence of structural organization at the
level of several micrometers.

Typical electron microscope studies of homopolymers crystallized from the pure
melt are shown in Figs. 1.10 and 1.11 respectively.(37,38) Lamellar-like crystallites
are the characteristic habit. Detailed studies indicate that the thickness of the lamellae is usually the order of several hundred angstroms, depending on the crystallization conditions. Lateral dimensions on the other hand are the order of a micrometer.

Fig. 1.8 Low-angle x-ray diffraction pattern of an axially oriented crystalline linear poly-ethylene specimen.(33)

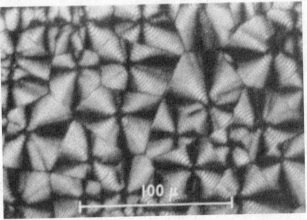

Fig. 1.9 Light micrograph of spherulitic structures grown in crystalline, linear polyethy-lene. (From Price (36))

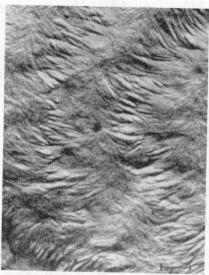

Fig. 1.10 Electron micrograph of melt crystallized linear polyethylene. (From Eppe, Fischer and Stuart (37))

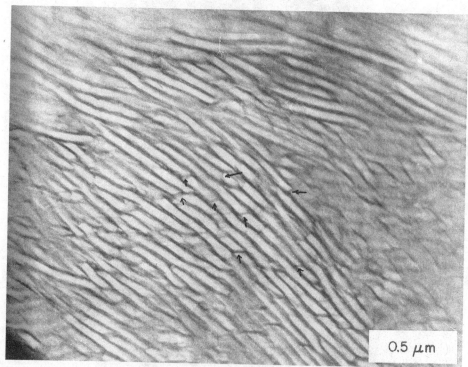

0.5 μm

Fig. 1.11 Transmission electron micrograph of linear polyethylene sample ($M_w = 1.89 \times 10^5$, $M_n = 1.79 \times 10^5$) crystallized isothermally at 131.2 °C. Light areas crystallites; dark areas noncrystalline regions.(38)

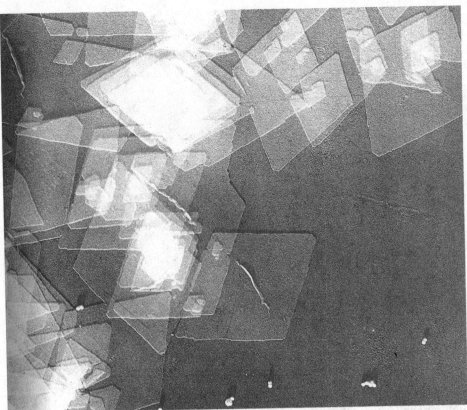

Fig. 1.12 Electron micrograph of linear polyethylene ($M_v = 50\,000$) isothermally crystallized at 89 °C from a dilute tetralin solution. (Courtesy of Dr. R. V. Rice)

Most interesting is the fact that chain axes are preferentially ordered normal or nearly
normal to the basal planes of the lamellae. The lamellar habit is typical of crystallites formed by homopolymers. Such structures are central to the understanding of
molecular morphology and properties. Of particular interest and importance is the
nature of the interphase between the crystalline and noncrystalline regions. This
problem will be discussed in detail in Volume 3.

When homopolymers are crystallized from very dilute solutions, either
lozenge-shaped platelets or crystals that possess a dendritic habit are formed.
Some typical electron micrographs of the crystals precipitated from dilute solution
are shown in Figs. 1.12 and 1.13. The crystal habit that is observed depends on
the molecular weight of the polymer and the crystallization conditions, such as
the temperature and the nature of the solvent. A very striking feature is that the
platelets are only about 100 to 200 Å thick. In conjunction with selected-area
electron diffraction studies, it is shown that the chains are again preferentially
oriented normal, or nearly so, to the basal plane of the platelet. Considering the

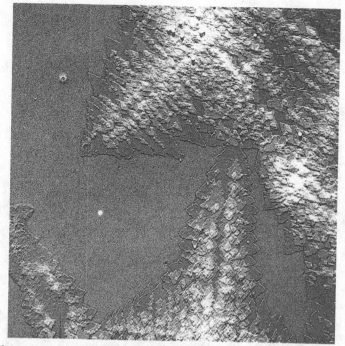

Fig. 1.13 Electron micrograph of linear polyethylene ($M_v = 50\,000$) isothermally crystallized at 60 °C from a dilute tetralin solution. (Courtesy of Dr. R. V. Rice)

high molecular weight involved, it can be concluded that a given chain must pass through these crystals many times. Hence, within the crystal, the polymer chain must assume some sort of folded structure. A very detailed discussion of the folded chain structure, and the nature of the lamellar crystals, will be given in the discussion of molecular morphology in Volume 3.

It has been recognized that there are many unique features and complications involved in delineating the detailed structure and conformation of a single, isolated long chain molecule. The organization of such molecules into a partially crystalline array poses further problems as should be apparent to the most casual observer. In subsequent chapters we endeavor to develop a systematic treatment and understanding of the nature of the crystalline state of long chain molecules. We use as a guide the schematic diagram in Fig. 1.1.

References

1. Flory, P. J., *Principles of Polymer Chemistry*, Cornell University Press (1953) pp. 3ff.
2. Mandelkern, L., *Faraday Discuss. Chem. Soc.*, **68**, 310 (1979).
3. Reference 1 pp. 399ff.
4. Flory, P. J., *Protein Structure and Function, Brookhaven Symp. Biol.*, **13**, 89 (1960).

5. Wilson, E. B., Jr., in *Advances in Chemical Physics*, vol. 11, Interscience Publishers, Inc., New York (1959).
6. Mizushima, S., *Structure of Molecules and Internal Rotation*, Academic Press, Inc., New York (1954).
7. Flory, P. J., *Statistical Mechanics of Chain Molecules*, Interscience Publishers (1969) pp. 1ff.; *ibid.*, Hansen Publishers (1988).
8. Volkenstein, M., *Configurational Statistics of Polymeric Chains* (translated from Russian, S. N. Timasheff and M. J. Timasheff eds., Interscience Publishers (1963).
9. Ising, E., *Z. Physik.*, **31**, 253 (1925).
9a. Rehahn, M., W. L. Mattice and U. W. Suter, *Adv. Polym. Sci.*, **131/132**, Springer-Verlag (1997).
10. Williams, A. D. and P. J. Flory, *J. Polym. Sci. A-2*, **5**, 417 (1967).
11. Erman, B., P. J. Flory and J. P. Hummel, *Macromolecules*, **13**, 484 (1980).
12. Flory, P. J., *Principles of Polymer Chemistry*, Cornell University Press (1953) p. 602.
13. Flory, P. J., *Faraday Discuss. Chem. Soc.*, **68**, 14 (1979).
14. Hayashi, H., P. J. Flory and D. G. Wignall, *Macromolecules*, **16**, 1328 (1983).
15. Hayashi, H. and P. J. Flory, *Physica*, **120B**, 408 (1983).
16. Lieser, G., E. W. Fischer and K. Ibel, *J. Polym. Sci.: Polym. Lett.*, **13B**, 39 (1975).
17. Fischer, E. W. and M. Dettenmaier, *J. Non-Crystalline Solids*, **31**, 181 (1978).
18. Wignall, G. D., in *Physical Properties of Polymers, Second Edition*, J. E. Mark ed., American Chemical Society (1993) pp. 313ff.
19. Graessley, W. W., *J. Polym. Sci.: Polym. Phys.*, **18**, 27 (1980).
20. Fetters, L. J., D. J. Lohse and R. H. Colby, in *Physical Properties of Polymers Handbook*, J. E. Mark ed., American Institute of Physics (1996) p. 335.
21. Fetters, L. J., D. J. Lohse and W. W. Graessley, *J. Polym. Sci.: Pt. B: Polym. Phys.*, **37**, 1023 (1999).
22. Natta, G. and P. Corradini, *Rubber Chem. Tech.*, **33**, 703 (1960).
23. Bunn, C. W. and D. R. Holmes, *Discuss. Faraday Soc.*, **25**, 95 (1958).
24. Liquori, A. M., *Acta Cryst.*, **8**, 345 (1955).
25. Bunn, C. W. and E. R. Howells, *Nature*, **174,** 549 (1954).
26. Pauling, R., R. B. Corey and H. R. Branson, *Proc. Natl. Acad. Sci. U.S.*, **37**, 205 (1951).
27. Cowan, P. M. and S. McGavin, *Nature*, **176**, 501 (1955).
28. Watson, J. D. and F. H. C. Crick, *Nature*, **171**, 737, 964 (1953); *Proc. Roy. Soc. (London), Ser. A*, **223**, 80 (1954).
29. Wilkins, M. H. F., A. R. Stockes and H. R. Wilson, *Nature*, **171**, 738 (1953).
30. Natta, G., P. Corradini and P. Ganis, *J. Polym. Sci.*, **58**, 1191 (1962).
31. Bunn, C. W., *Chemical Crystallography*, Oxford University Press, London (1946).
32. Morawetz, H., *Polymers. The Origin and Growth of Science*, John Wiley (1985) pp. 70ff.
33. Mandelkern, L., C. R. Worthington and A. S. Posner, *Science*, **127**, 1052 (1958).
34. Posner, A. S., L. Mandelkern, C. R. Worthington and A. F. Diorio, *J. Appl. Phys.*, **31**, 536 (1960); **32**, 1509 (1961).
35. Stein, R. S., in *Growth and Perfection of Crystals*, R. H. Doremus, B. W. Roberts and D. Turnbull eds., John Wiley & Sons, Inc., New York (1958) p. 549.
36. Price, F. P., *J. Polym. Sci.*, **37**, 71 (1959).
37. Eppe, R., E. W. Fischer and H. A. Stuart, *J. Polym. Sci.*, **37**, 721 (1959).
38. Voigt-Martin, I. G. and L. Mandelkern, *J. Polym. Sci.: Polym. Phys. Ed.*, **19**, 1769 (1981).

2

Fusion of homopolymers

2.1 Introduction

Characteristic changes take place in properties during the transformation of a pure homopolymer from the crystalline, or partially crystalline, state to the liquid. Major changes occur in physical and mechanical properties, in spectroscopic and scattering behavior and in the extensive thermodynamic variables. A crystalline homopolymer is typically a hard, rigid solid that possesses high strength. In contrast, in the molten state a polymer can possess the properties of a liquid of low fluidity. However, if the molecular weight is sufficiently high the liquid will exhibit rubber-like characteristics. The influence of crystallinity on mechanical properties manifests itself by a change in the modulus of elasticity by factors from about 10^3 to 10^5 upon melting. The mechanical strength of fibers can be attributed to the influence of oriented crystalline regions.

Crystalline homopolymers yield a number of wide-angle x-ray diffraction reflections that are superimposed on a diffuse halo, or haloes. These reflections disappear after melting and only broad haloes remain. Distinctive changes in infra-red and Raman spectra also occur during the transformation. Latent enthalpy and volume changes that are usually associated with a phase change of the first order are also observed. The distinct differences in thermodynamic and structural properties between these two polymeric states are very similar to those which occur during the melting of crystals of monomeric substances.

The melting–crystallization process of a system of small molecules is formally described as a first-order phase transition. Appropriate laws then follow that can be applied to a variety of problems. For a one-component system at constant pressure, the transition temperature is independent of the relative abundance of either of the two phases that are maintained in equilibrium. Melting is very sharp. The characteristic temperature of equilibrium is defined as the melting temperature. For the above conditions to be experimentally satisfied an almost perfect internal

24

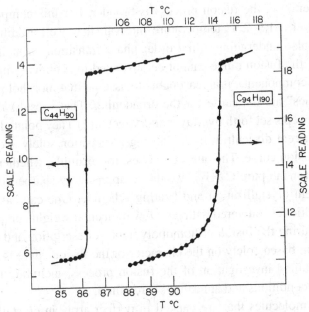

Fig. 2.1 Fusion of *n*-hydrocarbons. Plot of dilatometric scale reading against temperature for $C_{44}H_{90}$ and $C_{94}H_{190}$.(1)

arrangement of the crystalline phase is required. Moreover, crystals of large size are required to minimize any excess contribution to the free energy change caused by the surfaces or junctions between the two phases. Deviation from these idealized conditions will inevitably lead to a broadening of the melting range.

The criteria set forth above for an idealized first-order phase transition should apply equally well to the melting–crystallization of all substances. Before considering whether it is valid to apply these classical ideas to polymers it is instructive to examine the fusion of low molecular weight substances in more detail. Of particular interest in the present context is the behavior of *n*-hydrocarbons, which are pure low molecular weight chain molecules. The results of a dilatometric study of the fusion of pure $C_{44}H_{90}$ and of $C_{94}H_{190}$ are presented in Fig. 2.1.(1) Each of these pure compounds is of uniform chain length. The complete molecule participates in the crystal structure and thus molecular crystals are formed. These compounds should, therefore, behave in a classical manner. The fusion process is relatively sharp for each of the compounds. The melting temperature, representing the termination of fusion, is clearly defined and can be determined with a high degree of certainty. However, upon close scrutiny differences can be observed between the fusion of the two compounds. The $C_{44}H_{90}$ melts almost exactly according to theoretical expectations. The fusion process is relatively sharp and takes place within less than 0.25 °C. On the other hand, although the chemically pure $C_{94}H_{190}$ gives a well-defined melting temperature, the fusion process is broader. For this compound, melting takes place over a 1.5–2 °C temperature interval. It is well established that this compound is also undergoing a first-order phase transition on melting. Here the broadening of the fusion range cannot be attributed to chemical impurities (or an added second component). It is reasonable to assume that morphological or structural "impurities" are the cause for the broadening. These results illustrate quite vividly the concepts set forth by Mayer and Streeter.(2) They pointed out that there are certain inherent difficulties in classifying a transition solely according to the shape of the fusion curve. The question arises, for example, whether the transformation range for the pure $C_{94}H_{190}$ would be appreciably sharpened by adopting a more stringent crystallization and heating schedule. One can easily anticipate that the difficulties encountered with pure low molecular weight compounds would be enhanced during the fusion of homopolymers. A description and classification of the transition based solely on the character of the fusion curve is arbitrary and difficult. A detailed investigation of the fusion process, including the effects of crystallization conditions and annealing, is required.

Long chain molecules that are packed in perfect array in crystallites of sufficiently large dimensions represent a state that can be termed crystalline. The fact that a chain molecule may permeate many unit cells, in contrast to low molecular weight substances, is of no real consequence in the present context. The unique feature of chain structure, i.e. the covalent connectivity of chain atoms and repeating units, is of concern in analyzing polymer crystallization. This connectivity of hundreds to thousands of chain atoms sets the crystallization of polymers apart from other molecular systems. It is the reason for some of the differences that are observed in crystallization behavior. For example, for the flexible type chain molecule the crystallization process is rarely, if ever, complete. Depending upon molecular weight and crystallization conditions, the extent of crystallization can range from about 30 to 90% in homopolymers.(3) Because of the basic structural differences that exist one cannot tacitly assume that polymers and low molecular weight substances will display the same crystallization behavior in general and that the melting–crystallization of polymers is a first-order phase transition.

The essence of the problem is whether the ordered regions in the crystalline polymers can be treated as a separate phase. The usual thermodynamic criteria will have to be satisfied. For a pure phase of one component the chemical potential must be uniform throughout the phase and only depend on the temperature and pressure. For a poorly developed crystalline system, whether it be polymer or low molecular weight species, this condition will obviously not be fulfilled. Under these circumstances the chemical potential will also depend on the degree of order and the crystallite size. The extent to which the idealized crystalline state can be approached must ultimately be judged by the sharpness of the fusion process and the reproducibility of the melting temperature. As was pointed out above there are inherent difficulties in defining a transition by the shape of the fusion curve. A detailed investigation of the nature of the fusion and the characteristics of the transformation temperature is required. The concept that the melting of polymers is a first-order transition has important and far reaching consequences. Hence, it is important that the validity of this concept be investigated. If this postulate is not satisfied by experiment and molecular theory, then this premise will have to be discarded.

2.2 Nature of the fusion process

We examine the problem posed above by analyzing the melting of different polymers. Attention is focused on linear polyethylene as a model since the fusion of this polymer is known to be typical of other crystalline polymers. It also offers a continuity with the melting of the low molecular weight homologues that were illustrated in Fig. 2.1. In order to study the fusion process properly, procedures must be adopted that ensure conditions close to equilibrium are attained. Experience has taught us that these requirements are best fulfilled when crystallization from the melt is carried out either isothermally at temperatures as close to the melting temperature as is practical, or by protracted annealing at elevated temperatures of the already formed crystalline phase. Particular attention must also be given to the molecular constitution of the chains. In an ideal situation this involves specifying the molecular weight of narrow fractions, or the distribution for polydisperse systems.

The influence of the crystallization and melting conditions on the fusion process is illustrated in Fig. 2.2 for a very polydisperse linear polyethylene ($M_n = 1.2 \times 10^3$; $M_w = 1.5 \times 10^5$).[4] The open circles represent the results when the sample was slowly cooled from the melt to room temperature. The solid circles give the results when the same polymer was crystallized at 130 °C for 40 days and then cooled, over a 24-hr period, to room temperature, prior to fusion. When a heating rate of the order of 1 °C per day was used, represented by the open circles, the course of fusion is marked by partial melting–recrystallization. Despite this partial melting–recrystallization, the fusion process is still relatively sharp. The last trace of crystallinity disappears at a well-defined temperature, that for this sample can be taken to be 137.5 ± 0.5 °C. The isothermally crystallized sample, represented by the solid circles in Fig. 2.2, yields a higher level of crystallinity as indicated by the lower specific volume. Presumably a more perfect set of crystallites have been developed by this more rigorous crystallization procedure. On subsequent heating, the partial melting–recrystallization process that was prevalent with the nonisothermally crystallized samples is minimized. It is important to note that there is also a perceptible sharpening of the fusion curve, although the same melting temperature is obtained.

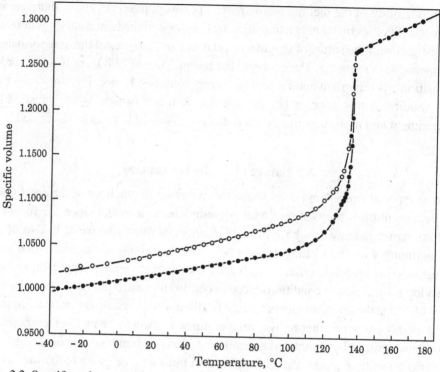

Fig. 2.2 Specific volume–temperature relations for an unfractionated linear polyethylene sample. Slowly cooled from melt ○; isothermally crystallized at 130 °C for 40 days then cooled to room temperature ●.(4)

By appropriate experiment it can be demonstrated that a well-defined temperature exists at which the last traces of crystallinity disappear. This temperature is reproducible and is independent of the crystallization conditions and the previous thermal history of the sample. Although the melting temperature of homopolymers is sharp and reproducible, the fusion process appears to violate one of the prime requirements of a first-order phase transition, namely that at constant pressure the transformation temperature should be independent of the relative abundance of the two phases. The melting range for the unfractionated polymers illustrated in Fig. 2.2 is, however, relatively narrow, being limited to a few degrees at most. Thus, for two molecularly identical systems the crystallization conditions, and presumably the resulting morphological forms, influence the course of fusion although the melting temperatures themselves are very clearly defined. There is no reason to believe that the ultimate in crystallization conditions and melting procedures has as yet been developed. Improvement in these methods will by necessity sharpen the melting range.

The use of molecular weight fractions has allowed for a major improvement in our understanding of the fusion process. A comparison between the fusion

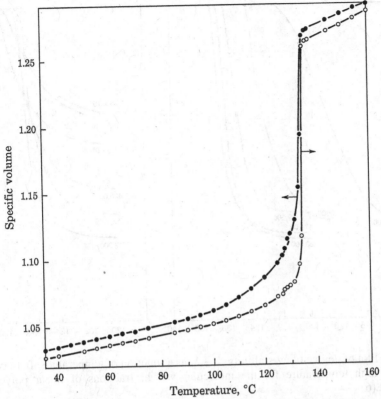

Fig. 2.3 Specific volume–temperature relation for linear polyethylene samples. Samples initially crystallized at 131.3 °C for 40 days. Unfractionated polymer, Marlex-50 ●; fraction $M_n = 32\,000$ ○. (From Chiang and Flory (5))

characteristics of the unfractionated polymer that was illustrated in Fig. 2.2 and a molecular weight fraction, $M = 3.2 \times 10^4$ crystallized under extreme isothermal conditions, is given in Fig. 2.3.(5) The melting temperatures are clearly defined and are the same for both polymers. It is evident in Fig. 2.3 that the melting of this fraction, (open circles), is appreciably sharper than the whole polymer. For the fraction, 80% of the transformation occurs over a 2 °C range. Over the same temperature interval there is only a 35–40% change in the polydisperse polymer. Molecular weight fractions appear to be able to develop a more perfectly developed crystalline state with a concomitant sharper fusion process. However, studies have shown that, depending on molecular weight, the use of fractions can also introduce complexities into the fusion process. Figure 2.4 illustrates the change in crystallinity level with temperature for fractions whose molecular weights range from 3.3×10^3 to 1.55×10^6.(6) These samples were isothermally crystallized and never cooled subsequent to the initiation of fusion. The two lowest molecular weight fractions melt very broadly. This result is easily explained by the high concentration

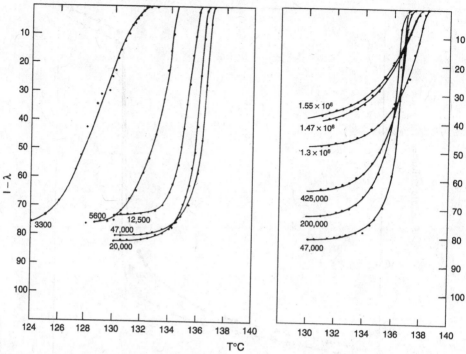

Fig. 2.4 Plot of degree of crystallinity, $1 - \lambda$, as a function of temperature after crystalliz-ation at high temperature for the molecular weight fractions of linear polyethylene indicated.(6)

of end-groups and their exclusion from the crystal lattice. There is, in effect, a significant built in impurity concentration that results in the expected broadening of the melting range. As the end-group concentration becomes insignificant an appre-ciable sharpening of the melting takes place. For molecular weights in the range 1.25×10^4 to 4.7×10^4 about 80–90% of the transformations occur over only a $2\,^{\circ}\text{C}$ interval. The curves closely resemble the one for $C_{94}H_{190}$ in Fig. 2.1. Thus behavior expected for a first-order phase transition of a pure substance is observed. However, for the highest molecular weights, including examples not illustrated, the curves broaden significantly with increasing chain length. It has also been found that if the level of crystallinity is restricted in the higher molecular weights, the melting range narrows considerably and becomes comparable to that observed for the lower molecular weight species. The factors involved in the broadening appear to be associated with the increasing level of crystallinity although for high molec-ular weights the absolute level of crystallinity that can be attained is relatively low.(3) The reasons for the broadening are probably structural and morphological in character. It could be caused by a distribution of crystallite sizes, the influence of the interfacial structure and the structure of the residual noncrystalline regions. These factors will be discussed in more detail subsequently.

The results described above give clear evidence that it is possible to develop fusion curves in homopolymers that are comparable to those obtained for low molecular weight substances. In all cases the temperature at which the last traces of crystallinity disappear is clearly defined. Molecular weight polydispersity, along with structural and morphological features, tend to broaden the fusion range. Even for monomeric substances the fusion process can be broadened by rapid cooling from the melt and the freezing in of nonequilibrium states. It is not unexpected, therefore, that these processes will be exaggerated in polymers. In order for the fusion process to be sharp in polymers the chain lengths must be highly uniform. In addition, the stringent crystallization conditions that are necessary cannot be easily employed because of kinetic restraints. The differences in the fusion process between polymers and low molecular weight substances is, therefore, one of degree rather than of kind. We can, therefore, conclude that the melting of crystalline polymers is a first-order phase transition and all of the dictates of this transition should be followed. The consequences of this conclusion are profound and have far-reaching implications.

The fusion of other polymers follows the same pattern as was found for linear polyethylene. Some representative results for the fusion of different type chain molecules are illustrated in Fig. 2.5.(7) Here the relative volume is plotted as a function of temperature. Very slow heating rates were employed subsequent to essentially uncontrolled crystallization. Characteristically, partial melting and recrystallization is again observed during fusion. Under these stringent conditions the melting process is quite sharp. The temperature at which the last traces of crystallinity disappear is well-defined in each of the examples. The abrupt termination of the fusion process is indicated. More stringent measures are needed to approach the equilibrium condition of polymers relative to low molecular weight species.

The importance of adopting slow heating rates to allow for partial melting of the unstable crystallites at a given temperature, and the subsequent recrystallization is emphasized by the Wood and Bekkedahl study on the crystallization and melting of natural rubber.(8) It was shown that if, subsequent to crystallization, fusion is carried out utilizing rapid heating rates (on the order of 0.1 °C per min) the observed melting temperature is a marked function of the crystallization temperature. The fusion curves that were obtained following isothermal crystallization at various temperatures are given in Fig. 2.6.(8) The observed melting temperatures range from about 0 to 30 °C and depend on the crystallization temperature, the melting temperature being higher for the higher crystallization temperatures. The fact that the crystallization temperature has such a decided influence on the melting temperature cannot be taken by itself as evidence of the lack of an equilibrium melting temperature in polymers. This phenomenon, that on rapid heating the observed melting temperature depends on the crystallization temperature, has now been universally observed with crystalline polymers. It has its origin in kinetic and

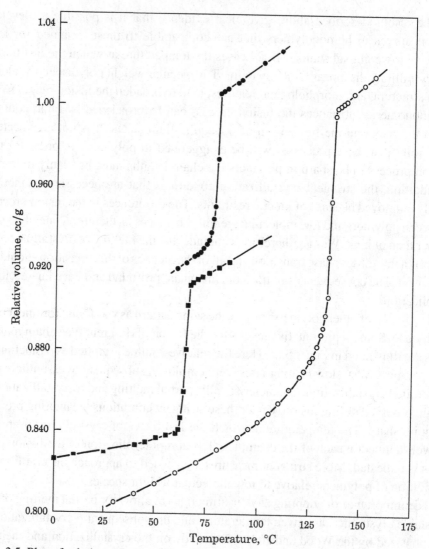

Fig. 2.5 Plot of relative volume against temperature: ○ polymethylene; ■ poly(ethylene oxide); ● poly(decamethylene adipate).(7)

morphological factors. It has been found that subsequent to the crystallization of natural rubber, as well as other polymers, if a slow heating schedule is adopted, a reproducible melting temperature that is independent of the previous thermal history of the sample is obtained.(4,7) This melting temperature is independent of crystallization conditions, including the crystallization temperature, and is invariably significantly greater than that observed with fast heating rates.

It is expected that because of the built-in disorder, such as chain ends in low molecular weight polymers and morphological and structural regions in general, the

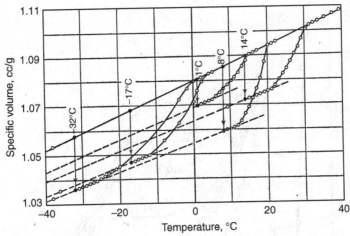

Fig. 2.6 Melting range of natural rubber as a function of the temperature of crystallization. (From Wood and Bekkedahl (8))

melting of a homopolymer must inevitably occur over a small but finite temperature range. However, the last vestiges of crystallinity should disappear at a well-defined temperature. This deduction has been amply confirmed by experiment. According to theory this temperature is defined as the melting temperature. The equilibrium melting temperature of a polymer, T_m^0, represents the melting temperature of the hypothetical macroscopic perfect crystal. Melting, in the limit of an infinite molecular weight homopolymer that forms a perfectly ordered crystalline phase, should occur sharply at a well-defined temperature.[1] We can then account for the melting characteristics in terms of a first-order phase transition. Theoretically, even a diffuse melting process can also be treated as a first-order phase transition.(4) Our discussion so far has been concerned with equilibrium and the equilibrium melting temperature. The establishment of complete equilibrium in the crystalline state with a collection of long chain molecules is a very difficult, if not impossible task. Consequently the actually measured melting temperature will differ by varying amounts from the true equilibrium value. A major task in the study of crystalline polymers is to determine or, more usually, estimate the equilibrium melting temperature. To accomplish this, one has to understand the morphological and structural features that cause deviations from the equilibrium melting temperature.

Certain results, based on general thermodynamic considerations, can be expected from a system undergoing a first-order phase transition. We consider here the consequences of equilibrium between two macroscopic phases of a one-component system. For equilibrium to be maintained between two phases at constant temperatures

---

[1] The perfectly ordered crystal is one with the lowest free energy. Since a certain amount of lattice disorder can be tolerated at equilibrium, it does not necessarily represent the crystal with perfect internal order.

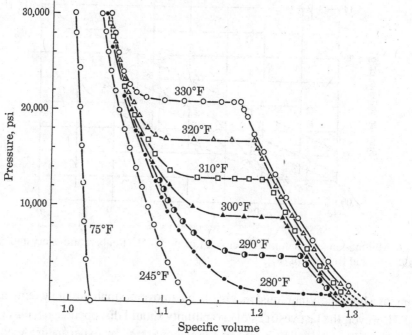

Fig. 2.7 Pressure–volume–temperature diagram for linear polyethylene (Marlex-50). (From Matsuoka (9))

the pressure must be independent of volumes, i.e. the pressure will not depend on the relative abundance of either of the phases. A pressure–volume–temperature diagram of polyethylene is given in Fig. 2.7.(9) The invariance of the pressure with the volume at the transformation temperature, a characteristic of a one-component system undergoing a first-order phase transition, is clearly evident. The inescapable conclusion is reached, without recourse to the molecular nature of the substance being studied, that the two phases must be in equilibrium. For the case being studied, the two phases are obviously the liquid and crystalline ones of polyethylene. Studies of highly oriented systems in phase equilibrium, to be discussed in detail in Chapter 8, yield the complementary result that the applied force is independent of the sample length.

The premise that the crystalline–liquid transformation in polymers possesses all the characteristics of a first-order phase transition can be subjected to further testing. Predictions can be made with respect to the influence of added species, either low molecular weight or polymeric, the incorporation of comonomers, cross-linking and chain orientation, on the equilibrium melting temperature and the crystallinity level. The analysis of such systems, following phase equilibrium theory, will be given in the following chapters. It will be found that these apparently diverse subjects can be treated from a unified point of view.

At the melting temperature equilibrium exists between the liquid and crystalline states. There is no reason in principle why equilibrium between the two phases cannot be maintained at finite levels of crystallinity as well. It is thus possible to uniquely specify the properties of the crystalline phase as long as it is recognized that the stipulation of equilibrium must be met. For crystallization under conditions removed from equilibrium this condition cannot be satisfied. The properties of the crystalline phase will depend on the mode by which crystallinity is developed. The development of higher levels of crystallinity that approach equilibrium requirements is not achieved very easily. It is primarily governed by the nature of the crystallization mechanisms. These problems are not unique to the crystallization of polymers. The same difficulties are encountered by crystallizing low molecular weight systems.

2.3 Fusion of the *n*-alkanes and other oligomers

The knowledge of the equilibrium melting temperature is fundamental to understanding the crystallization behavior of polymers. Whether the interest be in either melting, kinetics or morphology, the knowledge of the equilibrium melting temperature is crucial. As was stated, the equilibrium melting temperature T_m^0, of a crystalline polymer is the melting temperature of a perfect crystal formed by infinite molecular weight chains.(10) The melting temperature under equilibrium conditions of a chain of finite molecular weight, T_m, is also an important quantity. There are many reasons why the equilibrium melting temperature is important. One of these is that it reflects the conformational character of the chain. Another is that when the idealized equilibrium melting temperatures are compared with the melting temperature obtained for real systems important morphological and structural information can be deduced. It is also the key parameter in influencing crystallization rates because it establishes the undercooling and thus controls the very important nucleation processes that are involved. By definition T_m^0 cannot be determined by direct experiment. As will become clear in subsequent chapters, molecular weight, morphological complexities and kinetic restrictions that are placed on the crystallite size make the direct determination of the equilibrium melting temperature virtually impossible for high molecular weight chains. To experimentally determine these quantities recourse must be made to extrapolative procedures.

It is possible, however, to calculate from first principles the equilibrium melting temperatures of polymers of finite molecular weight as well as those of oligomers.(10,11) In the simplest case, that of oligomers, the chains are assumed to be of precisely the same chain length. The homologous series of *n*-alkanes represent the classical example of oligomers and the underlying requirement of uniform chain length. There is, however, a demarcation, depending on molecular weight, for

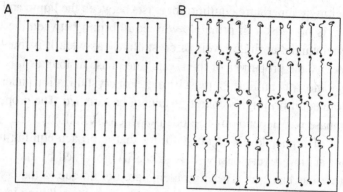

Fig. 2.8 Schematic representation of crystallite. (A) Molecular crystal where end-groups are paired. (B) Nonmolecular crystal, end sequences are disordered.(15)

crystallization to take place in either an extended or some type of chain folded form. When crystallized from the pure melt, *n*-alkanes equal to or lower than $C_{192}H_{386}$ ($M = 2688$) only form extended chain crystallites.(12,13) In contrast, $C_{216}H_{434}$ ($M = 3024$) can develop either a folded or extended chain crystallite, depending on the crystallization temperature.[2](14) When crystallized from dilute solution a folded type crystallite can be developed for $C_{150}H_{302}$ ($M = 2100$) and greater. Lower molecular weights form extended type crystallites when crystallized under these conditions.

The analysis of oligomers requires molecular crystals. Hence, it is restricted to extended chains. A schematic of the crystals is shown in Fig. 2.8A.(15) In this figure the vertical straight lines represent the ordered sequence conformation. For the *n*-alkanes it is all trans planar zigzag. For other oligimers it could very well represent some type of helical conformation. In the crystalline state depicted, the molecules are placed end-to-end so that the terminal groups are juxtaposed in successive layers of the lattice. The end-groups are paired, one to another, so that the sequence of chain units from one molecule to the next is perpetuated through successive layers of the lattice. The requirement of exactly the same chain length is crucially important if this model is to be valid. This condition cannot be satisfied by any real polymer system, no matter how well fractionated. There is an important limitation that must be clearly recognized. The results of the analysis, therefore, cannot be applied to polymers of finite chain length, unless absolute uniform chain length is achieved.(16) However, it is valid to calculate T_m^0 utilizing thermodynamic data for low molecular weight compounds which satisfy the above structural features. An extensive data set is available for the *n*-alkanes to test the analysis.

---

[2] The concept of a folded chain crystallite is used only in the context that the crystallite thickness is not comparable to the extended chain length. At this point, this phraseology carries no implications as to the interfacial structure of the basal plane of the crystallite. No assumptions are being made as to the specific nature of the folding.

It had been assumed for a long time that the melting temperatures of the n-alkanes could be explained by assuming that both the enthalpy and entropy of fusion could be represented by linear expressions comprising a term proportional to the number of carbon atoms, n, and an additive constant representing the end-group contribution.(17) Extrapolation could then be made to chains of infinite molecular weight. Flory and Vrij (11) pointed out that although the enthalpy of fusion could be reasonably taken as a linear function of n, the situation was quite different with respect to the entropy of fusion. Upon melting, the end-pairing represented in Fig. 2.8A is destroyed. This disruption leads to an additional contribution to the entropy of fusion above the usual disordering characteristics of the liquid state. The terminal segment of a chain can now be paired with any of the cn segments of another molecule, where c is a constant. An additional contribution to the entropy of fusion results, that is of the form of $R \ln cn$.

The molar free energy of a chain of n repeating units at any arbitrary temperature T can be expressed as(11)

$$n\Delta G_n = n\Delta G_u(T) + \Delta G_e(T) - RT \ln n \qquad (2.1)$$

In this equation $\Delta G_u(T)$ represents the free energy of fusion, at temperature T, of a repeating unit in the limit of infinite chain length. $\Delta G_e(T)$ is the end-group contribution which is assumed to be independent of n. The constant $R \ln c$ is incorporated into ΔG_e. This latter term plays a role analogous to that of an interfacial free energy.

The temperature dependence of ΔG_u and ΔG_e can be accounted for by performing a Taylor series expansion around the equilibrium melting temperature. By expanding ΔG_u to second order one obtains(11)

$$\Delta G_u(T) = \Delta G_u - \Delta S_u\left[T - T_m^0\right] - \left(\Delta C_p/2T_m^0\right)\left(T - T_m^0\right)^2 \qquad (2.2)$$

Here, ΔG_u and ΔS_u represent the values of these quantities at T_m^0. By defining $\Delta T \equiv T_m^0 - T$ and noting that at T_m^0, $\Delta G_u = 0$ and $\Delta S_u = \Delta H_u/T_m^0$, Eq. (2.2) reduces to[3]

$$\Delta G_u(T) = \Delta H_u \Delta T / T_m^0 - \Delta C_p(\Delta T)^2/2T_m^0 \qquad (2.3)$$

Expanding ΔG_e to first order yields

$$\Delta G_e(T) = \Delta G_e - \Delta S_e[T - T_m] \qquad (2.4)$$

with $\Delta G_e(T) = \Delta H_e - T\Delta S_e$. Both ΔG_u and ΔG_e could obviously be expanded to as high an order as desired. In this way any extremes in their respective temperature dependences can be accounted for. However, the second- and first-order

---

[3] It is necessary that the complete function ΔG_u be expanded and not the individual quantities ΔH_u and ΔS_u. Incorrect equations and serious errors result when this latter procedure is used. A detailed critique of the use of Eq. (2.2) is available.(18)

expansions, that are illustrated here, are usually adequate for most purposes. By inserting Eqs. (2.3) and (2.4) into Eq. (2.1), the free energy of fusion can be expressed as

$$n\Delta G_n = n\left[\Delta H_u \Delta T / T_m^0 - \Delta C_p (\Delta T)^2 / 2T_m^0\right] + \Delta H_e - T\Delta S_e - RT \ln n$$

(2.5)

At the melting temperatures of an n-mer, $\Delta G_n = 0$ and $T = T_m$. After rearrangement, the melting temperature of a series of homologues of length n is given to an approximation that is usually sufficient, by(11)

$$n\Delta H_u \Delta T / R - n\Delta C_p (\Delta T)^2 / 2R - T_m T_m^0 (\ln n) \simeq \left[T_m^0 / R\right](T_m \Delta S_e - \Delta H_e)$$

(2.6)

Equation (2.6) was used by Flory and Vrij to analyze the melting points of the n-alkanes that were available to them.

The melting temperatures of the n-alkanes were compiled by Broadhurst.(17)[4] The other parameters necessary to perform the analysis are also available. The value of ΔH_u has been obtained by independent methods.(19) A satisfactory estimate of ΔC_p can be made from specific heat measurements of the n-alkanes. The major objective of this analysis is to determine the value of T_m^0 from the melting point data of the n-alkanes.

Utilizing the parameters cited, the left-hand side of Eq. (2.6) can be calculated for each n-alkane for an assumed value of T_m^0. Following Flory and Vrij (11) we present in Figs. 2.9 and 2.10 plots of the left-hand side of Eq. (2.6) against $\Delta T = T_m^0 - T_m$ with T_m^0 being arbitrarily chosen as 418 K and 419 K respectively. The vertical bars in these figures illustrate the effect of shifting T_m by ± 0.5 K. These plots are very sensitive to the experimental value of T_m and to the assigned value of T_m^0 in the range of high n. The effect of changing T_m^0 by ± 1 K is also made clear by the respective plots. For $T_m^0 = 418$ K, (Fig. 2.9) the points in the upper range fall somewhat below the straight line drawn through the point representing the lower n's while for $T_m^0 = 419$ K (Fig. 2.10) they are somewhat above the line. For temperatures less than 417 K, or greater than 419 K, the divergence from linearity in the range of large n becomes quite severe. Thus, based on the premises of the theory and the thermodynamically significant melting temperatures available for the n-alkanes, a T_m^0 value of 418.5(± 1) K was deduced for linear polyethylene.(11) Sensible variations in the quantities ΔH_u and ΔC_p do not alter this conclusion. This value exceeds Broadhurst's extrapolated value of 414.3 ± 2.4 K based on the assumption that both the enthalpy and entropy of fusion are linear functions of n.(17)

---

[4] Of specific interest is the determination of T_m^0 for linear polyethylene. Hence, the melting of the orthorhombic crystalline form of the n-alkanes is pertinent here. Some of the lower members of the n-alkane series undergo a transition from the orthorhombic to the hexagonal form prior to melting.(11,17) For this case the melting temperatures of the metastable orthorhombic form can be calculated in a straightforward manner.(11) Contrary to other suggestions, the resulting melting temperatures can be used quite properly in the analysis.

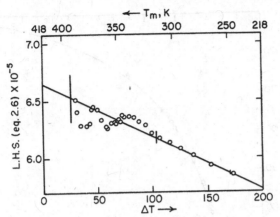

Fig. 2.9 Plot of left-hand side of Eq. (2.6) against ΔT for $T_m^0 = 418$ K. (From Flory and Vrij (11))

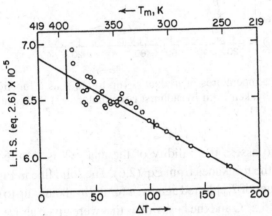

Fig. 2.10 Plot of left-hand side of Eq. (2.6) against ΔT for $T_m^0 = 419$ K. (From Flory and Vrij (11))

This difference, of about 4 K, is significant in many uses of T_m^0. The direct demonstration and observation of T_m^0 for any polymer is a very difficult matter. The highest directly observed melting temperatures of linear polyethylenes are in the range 141–146 °C.(20,21) Different extrapolative methods give a T_m^0 value of 146 °C for linear polyethylene.(22–27)

The slope and intercept of the straight line in Fig. 2.9, when analyzed according to Eq. (2.6), yields $\Delta H_e = -2200$ cal mol^{-1} and $\Delta S_e = 2.2$ cal deg mol^{-1}. The straight line in Fig. 2.10 yields similar results for these parameters. The negative value of ΔH_e signifies a decrease in the magnitude of the intermolecular crystal energy by the end-group layer. The positive value for ΔS_e has been attributed to relaxation of the precise positioning of the terminal group in the lattice.

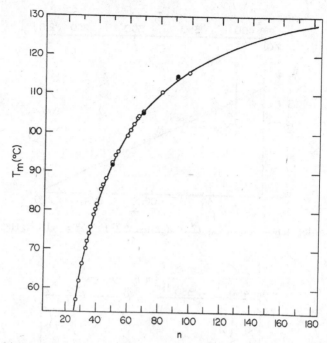

Fig. 2.11 Melting temperatures of n-alkanes (up to C_{100}) as a function of chain length. Experimental points taken from Broadhurst compilation.(17) Solid curve calculated from Flory–Vrij analysis.(1)

Another way to assess the validity of the analysis is to directly calculate the melting points of the n-alkanes from Eq. (2.6). The solid line in Fig. 2.11 represents the calculated values of T_m plotted against n for the n-alkanes up to C_{100}.(1) Here, T_m^0 was taken to be 145.5 °C and the best values that were given above were used for the other parameters. The experimental points are represented by the open and closed circles. This comparison between experiment and theory makes quite evident that the Flory–Vrij analysis gives an excellent representation of the n-alkane melting temperatures for this particular data base.

The analysis can also be assessed by comparing the observed and calculated enthalpies of fusion.(11,18) By expanding ΔH_u as a function of temperature one obtains an expression for the enthalpies of fusion of the n-alkanes. Thus

$$n\Delta H_n = n\Delta H_u - n\Delta C_p \Delta T + \Delta H_e \qquad (2.7)$$

where ΔH_n is the enthalpy of fusion of an n-mer. A comparison of the observed and calculated enthalpies of fusion is given in Table 2.1.(18) The values taken for ΔC_p and ΔH_e are the same as were given in the melting point analysis. Extremely good agreement is obtained between theory and experiment, confirming the theoretical analysis. It should be noted that there is a range of 100 °C in the melting

Table 2.1. *Enthalpies of fusion of* n-*alkanes[a]*

n	Observed	Calculated[b]	Calculated[c]
15	700	669	669
19	750	721	751
25	800	770	800
29	805	792	822
30	795	796	826
43	800	840	870
100	924	898	928

[a] From Ref.(18).
[b] Calculated assuming $\Delta H_u = 980$ cal mol^{-1}.
[c] Calculated assuming $\Delta H_u = 950$ cal mol^{-1}.

temperatures of the sample represented in Table 2.1. The agreement obtained indicates that the temperature expansion used to represent the enthalpy of fusion is more than adequate for present purposes. We conclude that the Flory–Vrij analysis quantitatively explains the enthalpy of fusion data for the *n*-alkanes.

Synthetic advances have allowed for the preparation of *n*-alkanes containing up to 390 carbon atoms ($M = 5408$).(13,14) Although the main thrust of these studies was concerned with other aspects of crystallization behavior, the melting temperatures of these compounds were also reported. These melting temperatures were determined by differential scanning calorimetry and were identified with the maximum in the endothermic peak.

A compilation of the melting temperatures that have been reported for all of the *n*-alkanes is given in Fig. 2.12.(30) The most extensive melting point data are those by Wegner and Lee, which cover the range $n = 44$ to 216.(12,14) In this data set there is a considerable overlap in carbon number with the Broadhurst compilation. Starting above $n = 160$, the measured melting temperatures are slightly lower than the Flory–Vrij theoretical values. The Wegner–Lee value for $n = 216$, the highest molecular weight *n*-hydrocarbon that formed extended chain crystallites is in good agreement with theory. The results of Ungar *et al.* (13) cover the range $n = 102$ to $n = 390$. The melting temperatures for $n = 102$ to 150 are in good agreement with theory. However, as the carbon number increases the observed melting temperatures are slightly lower than the expected values. The results of Takamizawa *et al.* (28) are in very good agreement with theory for $n = 60$ to $n = 120$. However, a deviation at $n = 160$ is again found.

The differences between theory and experiment are small when the higher molecular weight *n*-alkanes are considered. The significance of these small differences is not clear. There is concern as to the thermodynamically significant melting

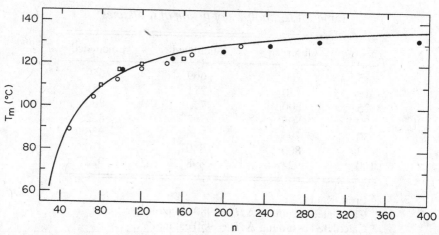

Fig. 2.12 Melting temperature of *n*-alkanes as a function of chain length. Solid curve calculated from Flory–Vrij analysis.(1) Experimental results: ● Ungar *et al.* (13); ○ Flory and Vrij (11), Lee and Wegner (14); □ Takamizawa *et al.* (28).

temperatures that are obtained by differential scanning calorimetry.(18,29) Although melting temperatures can be obtained by this technique that are comparable to those from conventional adiabatic calorimetry and dilatometry, special care and procedures need to be adopted. There is also the question of the purity of the compounds with respect to chain length, chain structure and chemical impurities. These points need to be clarified before any real shortcomings in the theory are addressed.

In the discussion of the fusion of the *n*-alkanes up to now, only the extreme states of perfect crystalline order and of a completely molten liquid have been considered. It is conceivable that, for chain molecules, the molecular crystals may undergo partial melting with disruption of the planar arrangement of the terminal CH_3 groups. This possibility is illustrated in Fig. 2.8B where *m* methylene units from the terminal sequences of each molecule are conformationally disordered, or melted. This change can be termed pre-melting and is a prelude to complete melting. Of interest here is this very specific type of pre-melting. Other types of pre-melting due to impurities, or inclusion of a second component, or the advent of the so-called "rotator phase" (31) have also been discussed in reference to the *n*-alkanes. In the present discussion of pre-melting in the *n*-alkanes, only conformational disorder, which is confined to sequences of methylene groups at the chain ends, will be of concern. Gauche conformers are introduced, causing disorder that involves trans bond orientation as well as gauche ones.

In their classical work Flory and Vrij also analyzed this pre-melting phenomenon. They let *m* CH_2 units from the terminal sequences of each molecule be melted. Thus $n - m$ consecutive units from the center of the molecule occupy a crystalline zone comprising similar sequences from neighboring chains. The fairly drastic alteration in the chain conformation associated with pre-melting can only be accomplished by the simultaneous disordering of neighbors. All of the molecules will thus be constrained to adopt approximately the same value of m. The partially melted crystal is thus envisioned as consisting of a succession of layers of crystalline and disordered (amorphous) zones, the terminal units of each chain being allocated to the disordered zone.

To quantitatively analyze the problem, both the free energy of fusion and the interfacial free energy, which are involved in the disordering process, have to be taken into account. In addition, there is also a combinatorial contribution that arises from the number of locations within the molecule that exclude the terminal disordered sequences from the interior of the crystalline zone. Consequently, the free energy change associated with partial melting will involve, in addition to a term $m \Delta G_u$ and an interfacial term, a contribution $-RT \ln(m + 1)$ that results from the $m + 1$ possible locations of the molecule such that the terminal units are excluded from the interior of the crystalline zone. Accordingly, the molecular free energy change associated with partial melting can be expressed as(11)

$$(\Delta G)_m = m \Delta G_u - RT \ln(m + 1) + 2\sigma_{eq} + \Delta G_e \qquad (2.8)$$

Here σ_{eq} is the interfacial free energy associated with each interzonal boundary and $-\Delta G_e$ is the defect free energy of the end-group layers destroyed by the pre-melting. By equating $\partial(\Delta G)_m / \partial m$ to zero and substituting $\Delta G_u = \Delta H_u \Delta T / T_m^0$, where $\Delta T = T_m^0 - T$, the optimum extent of pre-melting can be expressed as(11)

$$(m^* + 1) = RT^2 / \Delta H_u \Delta T \qquad (2.9)$$

Substituting Eq. (2.9) into Eq. (2.8) yields

$$(\Delta G)_m = RT[1 - \ln RT^2 / \Delta H_u \Delta T] + 2\sigma_{eq} + \Delta G_e \qquad (2.10)$$

By imposing the inequality $(\Delta G)_m < 0$ one finds that

$$\ln(RT^2 / \Delta H_u \Delta T) > 1 + (2\sigma_{eq} + \Delta G_e)/RT \qquad (2.11)$$

as the condition for the partial melting. The temperature for partial melting, T_P, identified with T in Eq. (2.11) depends on the quantities σ_{eq} and ΔG_e and not explicitly on the length of the n-alkane. The chain length dependence resides in either or both of the quantities σ_{eq} and ΔG_e. ΔG_e is the end-group contribution to the free energy of fusion of the molecular crystal. It can be looked on as the defect free energy for the end-group layer destroyed by partial melting. This quantity will depend on the end-group interaction and hence on the chemical nature of the terminal groups and their orientation relative to one another. Since ΔG_e depends only on the terminal group it would not be expected to be molecular weight dependent. The

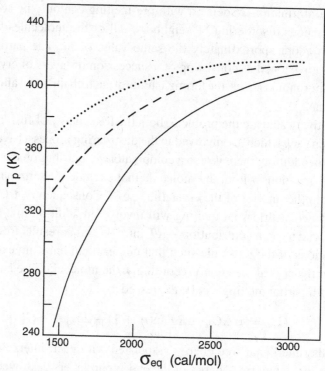

Fig. 2.13 Plot of pre-melting temperature, T_p, as a function of σ_{eq} for different values of ΔH_e
$\cdots \cdot \Delta H_e = -1500$ cal mol^{-1}; $---$ $\Delta H_e = -2150$ cal mol^{-1}; $---$ $\Delta H_e = -3000$
cal mol^{-1}.(15)

quantity σ_{eq} is the interfacial free energy associated with the boundary between the ordered and disordered region in the pre-melted chain. It is difficult to evaluate σ_{eq} theoretically. However, it can be deduced from the dependence of the equilibrium melting temperature on the chain length (see Sect. 2.4). The value of σ_{eq} is found to depend on molecular weight. It ranges from 1300 to 3500 cal mol^{-1} of sequences as the number average molecular weight increases from 570 to 5600.(15)

The consequences of Eq. (2.11) are examined in the following figures. Figure 2.13 is a plot of the calculated pre-melting temperature, T_p, as a function of σ_{eq} for different values of ΔH_e.(15) For a methyl terminated n-alkane, Flory and Vrij estimated ΔH_e to be equal to -2150 cal mol^{-1} and $\Delta S_e = 2.45$ cal mol^{-1} K^{-1}. Accordingly, ΔG_e is equal to -3200 cal mol^{-1} at $T_m^0 = 418.7$ K. The other values of ΔH_e used in Fig. 2.13 were arbitrarily selected to represent other possible types of end-groups and their interactions. The curves in Fig. 2.13 indicate that T_p is very sensitive to the value of σ_{eq}. For $\sigma_{eq} \geq 3000$ cal mol^{-1}, values that correspond to high molecular weights, T_p is predicted to be very close to T_m^0 (418.7 K). It then would be very difficult to observe. However, as σ_{eq} decreases, a precipitous drop in T_p is predicted for the n-alkanes. The pre-melting temperature can be significantly lower than the final

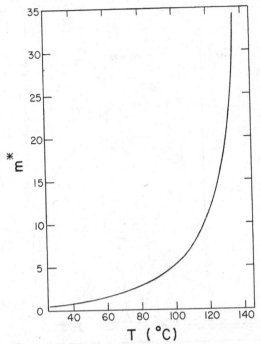

Fig. 2.14 Plot of m^*, the number of disordered units, as a function of temperature.(15)

melting temperature of an n-alkane, particularly for the higher molecular weight homologues. For example, from the curve corresponding to $\Delta H_e = -2150\,\mathrm{cal\,mol^{-1}}$, and σ_{eq} equal to $2000\,\mathrm{cal\,mol^{-1}}$, T_p is predicted to be $380\,\mathrm{K}$. This value of T_p is below the melting temperatures of n-alkanes greater than about $C_{80}H_{162}$. For σ_{eq} equal to $1500\,\mathrm{cal\,mol^{-1}}$ T_p is predicted to be $337\,\mathrm{K}$. This temperature is well below the melting temperature of $C_{25}H_{52}$ and the higher n-alkanes. There is, therefore, the expectation from theory that for n-alkanes greater than about $C_{30}H_{52}$, T_p is sufficiently below T_m that it should be observed experimentally. The other curves in Fig. 2.13 indicate that, depending on the value of ΔH_e (or ΔG_e), T_p will vary as the terminal group changes.

A plot of m^* against T, as calculated from Eq. 2.9, is given in Fig. 2.14.(15) The predicted amount of disorder is small at low temperatures. It increases rather substantially at about $95\,^{\circ}\mathrm{C}$. At this temperature m^* is about 5 units and increases to about 20 units at $130\,^{\circ}\mathrm{C}$. Thus, a significant number of units in the higher n-alkanes can be expected to be disordered prior to melting.

The theory outlined above provides a basis for pre-melting that is attributed to the conformational disorder of terminal sequences. The expectation is that this type of pre-melting should be observed in the higher n-alkane homologues that are available. The basic question to be addressed is whether the expectation of this pre-melting mechanism is actually observed experimentally.

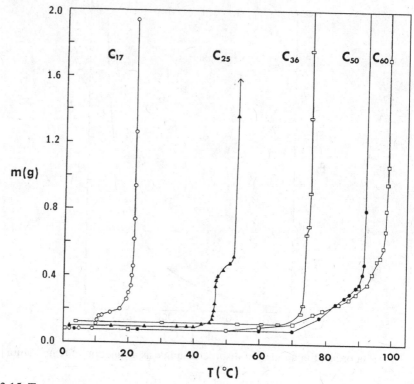

Fig. 2.15 Temperature dependence of disorder in terms of gauche bonds per chain for indicated *n*-alkanes. (From Kim, Strauss and Snyder (32))

In examining appropriate experimental results it is found that there is a large body of evidence, involving a diversity of experimental techniques, that demonstrates pre-melting of the type of interest here. The experimental techniques involve electron diffraction, small-angle x-ray scattering, nuclear magnetic resonance, lattice expansion and thermodynamic measurements. Details of the methods and the results have been reviewed.(15)

An example of pre-melting, detected by vibrational spectroscopy, taken from the work of Synder and collaborators is shown in Fig. 2.15.(32) In this figure, the chain disorder, $m(g)$, measured in terms of gauche bonds per chain, is given as a function of temperature for a set of *n*-alkanes. The onset of pre-melting is indicated by the rather abrupt increase in the gauche concentration at a well-defined temperature that is characteristic of each of the *n*-alkanes. For $C_{17}H_{36}$ and $C_{25}H_{52}$ the disorder is either just slightly below or at the temperature of the onset of the pseudo hexagonal rotator phase transition. Pre-melting in $C_{36}H_{74}$ begins around 66 °C. However, the amount of disorder although detectable remains low until the orthorhombic to hexagonal transition takes place about 8 °C higher. Disordering in $C_{50}H_{102}$ and $C_{60}H_{122}$ also take place in the vicinity of 65–70 °C. The concentration of the disorder observed at T_p is about the same as that observed with the shorter chains. The vibrational spectroscopic studies indicate that T_p increases initially at the low carbon number and then levels off.(15,33) The values of σ_{eq}, obtained by applying Eq. (2.11), are in the range 1390–1650 cal mol^{-1} for sequences of these alkanes, and are consistent with the values quoted previously. The vibrational spectra of the low carbon number alkanes and solid state ^{13}C NMR studies of *n*-hexatriacontane (33) demonstrate that pre-melting, as defined here, occurs prior to the onset of the pseudo hexagonal (rotator) phase transition. It is not clear whether the pre-melting triggers off this transition. The relaxation that occurs in the pseudo hexagonal phase cannot be attributed solely to the rotation of a rigid molecule. Contributions from the motion of the disordered end sequences must be taken into account.

Table 2.2 is a compilation of the pre-melting temperature, T_p, of the *n*-alkanes as determined by the different methods.(15) The values of T_p, as determined by the different methods, are in excellent agreement with one another. A composite plot of T_p against the carbon number is given in Fig. 2.16. The pre-melting temperature increases rapidly with carbon number and then effectively levels off for *n* approximately greater than 80. The agreement that is obtained between the different methods and the many investigators is impressive. Except for a few minor exceptions there is virtually complete agreement. This extensive compilation gives strong support to the pre-melting phenomenon of interest for the large range of *n*-alkanes that are available for study. It is a universal characteristic of the *n*-alkanes irrespective of chain length. The dependence of T_p on chain length is a reflection of the variation in the quantity σ_{eq}. These results give strong support to the Flory–Vrij analysis of pre-melting in chain molecules and the basic understanding for this phenomenon. Arguments have been presented against this type of pre-melting.(34) However, they cannot be substantiated in view of the overwhelming amount of experimental evidence that is available from a diversity of sources.

The *n*-alkanes as a class have presented a large body of experimental data suitable for analysis. A similar pre-melting phenomenon should be observed for methylene chains that are terminated by end-groups other than methyl. The pre-melting temperature will depend on the chemical nature of the end-group(s) and their influence on the parameters that appear in Eq. (2.11). Oligomers of other type chains should also display a similar melting behavior.

Figure 2.17 gives plots of the dependence of the melting temperature on the number of repeating units for different oligomers. The results for the different oligomers are qualitatively similar to one another. Initially, there is a rapid rise in T_m with the first few repeating units, followed by only a very slow increase in T_m and a leveling off with chain length. Oligomers of poly(dimethyl siloxane) (not illustrated) behave in a similar manner.(43) For molecular weights greater than about 2500 there is essentially no change in the melting temperature. However,

Table 2.2.[a] *Compilation of pre-melting temperatures (K) as determined by different methods*

Carbon number	Vibrational spectroscopy	NMR	Calorimetry	Small-angle x-ray scattering	Lattice expansion
17	283				
18			288		
19	295	296–298	293		
		288–298			
23			308		
24		318–321			
25	315				
26			323		
28			328		
32		333–343			323, 330
33				338	
36	339	344–346	341	335	343
					340–343
					338–343
37					
38					343
40					
44		333–343			343–345
45				339	
48				338	
50	338–343				348
60	333–343	~347		355	348
61				351	
62				≥343	
65				361	
69				363	
70				343–356	
82					353
94				≥353	353
100			365		
168		<360			
		>300			
192			<363		353

[a] Data from Ref. (15).

there is a precipitous drop in this temperature as the chain length is decreased. The melting temperatures of symmetrical ketone oligomers show a similar behavior.(44)

    Theoretically, it is possible and very tempting to extrapolate these data to their respective values of T_m^0. Prudence should be exercised, however, since, except for ethylene oxide, the data available are limited to just a few repeating units. Despite this concern, the analysis of the thermodynamic data of the set of normal

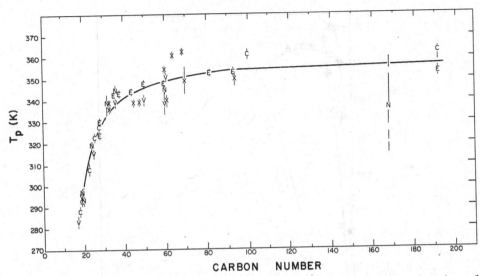

Fig. 2.16 Composite plot of pre-melting temperatures, T_p against carbon number of *n*-alkanes. T_p determined by vibrational spectroscopy V; by NMR N; by calorimetry C; by small-angle x-ray scattering X; by expansion E.(15)

perfluoroalkanes, from $m = 6$ to 24, by the Flory–Vrij equation, led to a T_m^0 value in close accord with estimates from experimental studies.(36) The necessity of including the $R \ln n$ term in the total entropy of fusion was demonstrated. Failure to include this term lowers the expected T_m^0 by about $40\,°C$. The measured enthalpies of fusion of the low molecular weight homologues were found to be related to that of the repeating unit of the infinite chain by Eq. (2.7).

Empirical and semi-empirical methods have also been used to extrapolate the *n*-alkane oligomer data to T_m^0.(18,45–49) In these analyses T_m^0 values in the range 141–$142\,°C$ were obtained for linear polyethylenes, which are virtually identical to that given by Broadhurst.(17) In other examples, extrapolation of the oligomer data leads to a T_m^0 value of $69.3\,°C$ for poly(ethylene oxide),(48) while the more acceptable value is in the range 76–$80\,°C$. However, another extrapolation method gave $75\,°C$ for T_m^0 for this polymer.(49)

The analyses of the fusion of the *n*-alkanes, as well as the other oligomers, are of interest by themselves. However, they also play an important role in that they serve as a connection to the equilibrium aspects of polymer crystallization. This problem is addressed in the next section.

2.4 Polymer equilibrium

It is apparent that when dealing with chains of high molecular weight the thickness of a crystallite, ζ, defined in terms of the number of repeating units will be less

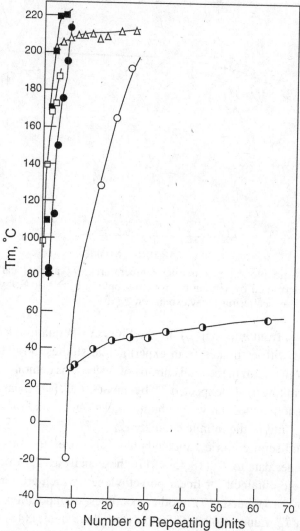

Fig. 2.17 Plot of the melting temperature against the number of repeating units for different oligomers. △ ε-amino caproic acid (From Rothe and Dunkel (35)); ○ perfluoro alkanes (From Starkweather (36)); ◑ ethylene oxide (From Yeates, Teo, Mobbs and Booth (37)); ● phenylene sulfide (From Bourgeois and Fonassi (38); Montando, Bruno, Maravigna, Finocchiaro and Centineo (39); Koch and Heitz (40)); ■ ethylene terephthalate-hydroxy terminated (From Zahn and Krzikalla (41)); □ tetramethylene terephthalate (From Hasslin, Dröscher and Wegner (42))

than x, the number of repeating units in the chain. It also must be recognized that molecular weight fractions of polymers, no matter how well fractionated, are not monodisperse. In contrast to the n-alkanes, and other oligomers, the chain lengths in a given fraction are not uniform. Hence, the model of end-pairing, as illustrated in

Fig. 2.8A cannot be applied. The actual situation is similar to that shown in Fig. 2.8B, with the proviso of the nonuniformity of chain length. Thus, when calculating or discussing the melting temperatures of polymers, the Flory–Vrij molecular crystal model should not be used, except in the limit of infinite molecular weight.(16) This is an important stricture that needs to be adhered to. The inherent polydispersity needs to be taken into account in a specific manner. Despite these restraints, an equilibrium theory of the fusion of polymers has been developed and will be discussed in the following. Fractions, as well as polydisperse systems with defined distributions, are treated. In this model it is recognized that the chain ends, as well as the contiguous noncrystalline sequences, are excluded from the crystal lattice. Thus, for polymers of finite molecular weight there is in essence a "built in" set of impurities. Taking cognizance of these facts, Flory (10) developed a quantitative description of the semi-crystalline state of unoriented polymers. Although the direct application of the equilibrium conditions to polymer crystallization is limited, the theory sets forth what can be achieved, and establishes a set of important reference points. Moreover, it can be used to describe situations where the crystallites are not of equilibrium size.(50) It can also be adapted to develop nucleation theory appropriate to long chain molecules.(51,52)

Based on a lattice model, Flory (10) treated the general case of N homopolymer molecules, each having exactly x repeating units, admixed with n_1 molecules of a low molecular weight species.[5] The composition of the mixture is characterized by the volume fraction of polymer v_2. Since in general the diluent will be structurally different from the polymer repeating unit it is assumed to be excluded from the crystal lattice.[6] The objective of the calculation is to calculate the free energy of fusion under equilibrium conditions.

The entropy of the unoriented semi-crystalline polymer is assumed to arise solely from the number of configurations that are available to the polymer. Contributions from the random orientation of the crystallites, or their further subdivision into small crystallites, are neglected. Since the size and flexibility of the repeating unit is important in lattice type calculations it is necessary to distinguish between the configurational segment and the structural repeating unit. If x' represents the number of configurational segments per molecule and z represents the number of segments per structural unit, then x' equals zx. Thus, if z_s is equal to the ratio of the molar volume of the solvent to the volume of segment then n_1', which is equal to $z_s n$, is the number of lattice cells occupied by solvent molecules. The configurational properties of the semi-crystalline polymer can be conveniently described by using a lattice with coordination number Z. The size of the lattice cell is chosen to

---

[5] Although this chapter is only concerned with pure polymers the general derivation is given here for future use.
[6] An assumption of this type is always necessary in multicomponent-multiphase systems. For polymer–diluent mixtures it has been found that the diluent is excluded from the lattice for the vast majority of systems studied.

accommodate one segment. Definite regions in the lattice must be reserved for occupancy by the crystallite. There are assumed to be ν crystallites, each with an average length of ζ repeating units, and a cross-section of σ chains. The total number of crystalline sequences, m, is thus equal to $\nu\sigma$. The total number of chain units which participate in the crystallization is then $m\zeta$. The major problem to be solved is the calculation of the configurational entropy of the semi-crystalline polymer. When the entropy of the completely disordered mixture is subtracted from this quantity the entropy of fusion, ΔS_f, results.

For calculation purposes it is convenient to randomly join the polymer and diluent molecules to form a single linear chain. This corresponds to an entropy change of

$$S_1 = k\{-n_1 \ln[n_1/(n_1 + N)] - N \ln[N/(n_1 + N)]\} \tag{2.12}$$

A single chain is then introduced into the lattice while observing the conditions stipulated by the reserved regions. For each segment whose location relative to its predecessor is not restricted there will be contribution to the entropy of $k \ln[(Z - 1)/e]$. All segments except those beyond the first in a crystalline sequence are unrestricted in this respect. The configurational entropy of the chain on the lattice is thus

$$S_2 = k[n_1' + x'N - (\zeta - 1)m] \ln[(Z - 1)/e] \tag{2.13}$$

In this arrangement, chain ends and diluent have been allowed to enter the lattice cells reserved for crystallites. However, a given arrangement is acceptable only if these cells are occupied by polymer segments. The probability of fulfilling the latter condition has been calculated for low degrees of crystallinity and leads to an entropy contribution of

$$S_3 = km\{\ln[x'N/(n_1' + x'N)z + \ln[(x - \zeta + 1)/x]\} \tag{2.14}$$

The results of the calculation are thus limited to low levels of crystallinity. In the last step, the severing of linkages between molecules leads to an entropy contribution

$$S_4 = -k(n_1 + N) \ln[(Z - 1)/e] + (n_1 + N) \ln[(n_1 + N)/(n' + x'N - \zeta'm)] \tag{2.15}$$

The configurational entropy, S_c, of the semi-crystalline polymer–diluent mixture is the sum of the above four entropy contributions. By subtracting this sum from the entropy S_1 of the completely disordered mixture, ($m = 0$), one obtains the entropy of fusion ΔS_f. In terms of molar quantities ΔS_f can be expressed as

$$\Delta S_f/xN = (1 - \lambda)\Delta S_u - R[(V_u/V_1)(1 - v_2)/v_2 + 1/x] \ln[1 - v_2(1 - \lambda)]$$
$$- R[(1 - \lambda)/\zeta]\{\ln v_2 D + \ln[(x - \zeta + 1)/x]\} \tag{2.16}$$

In Eq. (2.16) λ is the fraction of polymer that is noncrystalline (amorphous) and is equal to $(xN - \zeta m)/xN$. The entropy of fusion per repeating unit, ΔS_u, is formally defined as $kz \ln(Z-1)/e$. The parameter D is defined as $(Z-1)/ze$. V_u and V_1 are the molar volumes of the repeating unit and diluent respectively. In calculating ΔS_f it was assumed that the boundary between the crystalline and amorphous regions was sharp. It was recognized, however, that this is a physically untenable situation since some degree of order must persist for some distance beyond the crystalline boundary. The effect of the diffuseness of this boundary on the configurational entropy can be formally accounted for by redefining the parameter D. Equation (2.16) represents a melting and dilution process. However, even in the absence of diluent ΔS_f will depend on the degree of crystallinity, $1-\lambda$, and the crystallite thickness ζ. It is, therefore, not an inherent property of the polymer chain. On the other hand, the entropy of fusion per structural unit, ΔS_u, is a characteristic of a given polymer, irrespective of the actual characteristics of the crystallite.

The enthalpy of fusion consists of a contribution from the melting of the crystallites and the mixing of these previously crystalline segments with the amorphous disordered mixture. The former contribution can be expressed as $\zeta m \Delta H_u$, where ΔH_u is the enthalpy of fusion per repeating unit. The effect of the lower energy that would be expected at the crystallite boundary can also be incorporated in the parameter D. The heat of mixing can be expressed in the van Laar form as is customary in polymer solution theory.(53) The free energy change that occurs on fusion, ΔG_f, can then be expressed as

$$\begin{aligned} \Delta G_f/xN = (1-\lambda)(\Delta H_u - T\Delta S_u) + RT[(V_u/V_1)(1-v_2)/v_2 + 1/x] \\ \times \ln[1 - v_2(1-\lambda)] + [(1-\lambda)/\zeta]\ln v_2 D + \ln(x - \zeta + 1)/x \\ + \chi_1^*(1-v_2)^2(1-\lambda)/(1-v_2+v_2\lambda) \end{aligned} \qquad (2.17)$$

Here χ_1^* is related to the conventional thermodynamic interaction parameter χ_1 by the relation

$$\chi_1^* = \chi_1(V_u/V_1) \qquad (2.18)$$

The quantities that comprise the parameter D play a role analogous to that of an interfacial free energy. It can, therefore, be redefined as

$$-\ln D = \frac{2\sigma_{eq}}{RT} \qquad (2.19)$$

so as to correspond to conventional notation. In this equation, σ_{eq} is the interfacial free energy characteristic of the equilibrium crystallite.[7]

---

[7] Several physically different interfacial free energies are involved in polymer crystallization. These different quantities must be clearly distinguished from one another since they are not *a priori* identical. In the present context we are concerned with the interfacial free energy associated with an equilibrium crystallite. There is also the interfacial free energy associated with the nonequilibrium crystallite σ_{ec}, as well as the one involved in nucleation σ_{en}. None of these quantities can be assumed to be equal to one another.

The equilibrium, or most stable crystalline state, characterized by $\lambda = \lambda_e$ and $\zeta = \zeta_e$, is obtained by maximizing ΔG_f. Accordingly ζ_e is given by

$$-\ln v_2 D = \zeta_e/(x - \zeta_e + 1) + \ln[(x - \zeta_e + 1)/x] \qquad (2.20)$$

and λ_e by

$$1/T - 1/T_m^0 = R/\Delta H_u[(V_u/V_1)(1 - v_2) + v_2/x]/[1 - v_2(1 - \lambda_e)]$$
$$+ 1/(x - \zeta_e + 1) - \chi_1^*\{(1 - v_2)/[1 - v_2(1 - \lambda_e)]\}^2 \qquad (2.21)$$

In the absence of diluent Eqs. (2.20) and (2.21) reduce to

$$-\ln D = \frac{2\sigma_{eq}}{RT} = \zeta_e/(x - \zeta_e + 1) + \ln[(x - \zeta_e + 1)/x] \qquad (2.22)$$

and

$$1/T - 1/T_m^0 = R/\Delta H_u[1/x\lambda_e + 1/(x - \zeta_e + 1)] \qquad (2.23)$$

The equilibrium melting temperature of the pure polymer of infinite chain length, T_m^0, is identified with the ratio $\Delta H_u/\Delta S_u$. The fact that ζ_e is independent of λ_e is a consequence of approximations made in calculating ΔS_f. Physically significant values of ζ_e occur only when $v_2 D$ is less than unity.

Equation (2.23) leads to the expectation that even for chains of exactly the same length, fusion will occur over a finite temperature range. This expectation contrasts with the melting of pure monomeric species. Experimental observations support this conclusion. The breadth of melting for the two lowest molecular weight fractions, as illustrated in Fig. 2.4, can be attributed to the built in impurities of the end-groups. However, as predicted by theory, the fusion interval for linear polyethylene decreases with increasing molecular weight. A very perceptible sharpening of fusion in linear polyethylene occurs up to molecular weights of about 5×10^4. However, the experimental evidence in Fig. 2.4 indicates that at the high molecular weight $\geq 5 \times 10^5$ the melting range broadens appreciably with increasing chain length. This disparity is probably due to the extreme difficulty in approaching the equilibrium conditions at the high molecular weights.

Despite the fact that the melting of a semi-crystalline polymer occurs over a finite temperature range, even under equilibrium conditions, as λ_e approaches unity $d\lambda_e/dT \neq 0$. Therefore, the last traces of crystallinity will appear at a well-defined temperature. Above this temperature $d\lambda_e/dT = 0$. Thus the thermodynamic melting temperature T_m can be defined. At $\lambda_e = 1$, $T \equiv T_{m,e}$ so that

$$\frac{1}{T_{m,e}} - \frac{1}{T_m^0} = \frac{R}{\Delta H_u}\left(\frac{1}{x} + \frac{1}{x - \zeta_e + 1}\right) \qquad (2.24)$$

with ζ_e being defined as

$$2\sigma_{eq} = RT_{m,e}\left\{\frac{\zeta_e}{x - \zeta_e + 1} + \ln\left(\frac{x - \zeta_e + 1}{x}\right)\right\} \tag{2.25}$$

Equations (2.24) and (2.25) represent the dependence of the equilibrium melting temperature of a molecular weight fraction on chain length for a pure polymer system that resembles the crystallite model given in Fig. 2.8B. Here σ_{eq} represents the interfacial free energy associated with the basal plane of the equilibrium crystallite. $T_{m,e}$ is the equilibrium melting temperature of a crystallite of size ζ_e, formed by chains that are x units long. It cannot, and should not, be identified with the melting temperature of a molecular crystal formed by chains of exactly the same size, and whose end-groups are paired. Equation (2.24) can be rewritten in more compact form as

$$1/T_{m,e} - 1/T_m^0 = (R/\Delta H_u)(1 + b)/x \tag{2.26}$$

where $b \equiv [1 - (\zeta_e - 1)/x]^{-1}$. If σ_{eq} is independent of chain length, b will be approximately constant, unless x is very small. Under these very special circumstances

$$\frac{1}{T_{m,e}} - \frac{1}{T_m^0} = \frac{R}{\Delta H_u}\frac{c}{x} \tag{2.27}$$

This inverse relationship, for a constant interfacial free energy, between melting temperature and chain length contrasts markedly to the results obtained for molecular crystals.

It is tempting to apply Eq. (2.27) to experimental data in order to extrapolate to the melting temperature of the infinite chain. However, the requirements that lead to Eq. (2.27) need to be satisfied. The experimental quantities that are usually determined are T_m and x for a given fraction. Equations (2.24) and (2.25) make clear that this information is not sufficient to determine T_m^0 without arbitrary assumptions being made with respect to either ζ_e or σ_{eq}. The relationship expressed by Eq. (2.27) also requires that σ_{eq} be independent of x. It needs to be established whether this assumption is valid before attempting the extrapolation. In addition, in order for the crystallites to attain their equilibrium size it is required that their thicknesses be comparable to the extended chain length. Moreover, they must also be of uniform thickness. These exacting conditions are extremely difficult, if not impossible to fulfill over the complete molecular weight range. The molecular weight range over which extended chain crystals can be formed is limited. Suitable data for analysis are thus restricted to very low molecular weights. Appropriate data for fractions of polyethylene (55) and poly(ethylene oxide) (18) are available for analysis.[8]

---

[8] Crystallization under conditions of high pressure and high temperature usually produces thick crystallites for high molecular weight polymers. The thicknesses of these crystallites are not, however, comparable to the extended chain lengths, as had been supposed.(54)

Table 2.3. *Parameters governing fusion of linear polyethylene^a(55)*

M_n	x	$T_{m,e}$ (°C)	ζ_e	ζ_e/x	σ_{eq}(cal mol^{-1})	$x - \zeta_e$
1586	113	124.5	95 ± 1	0.84 ± 0.01	1298 ± 200	18
2221	159	126.0	140 ± 2	0.88 ± 0.01	2024 ± 200	19
3769	269	132.0	242 ± 3	0.90 ± 0.01	2551 ± 300	27
5600	400	134.2	368 ± 4	0.92 ± 0.01	3485 ± 500	32

^{a} Uncertainties calculated by assuming $T_{m,e} = \pm 1\,°C$.

Table 2.4. *Parameters governing fusion of poly(ethylene oxide)^a*

			$T_m^0 = 80\,°C$				$T_m^0 = 76\,°C$			
M_n	x	T_m (°C)	ζ_e	ζ_e/x	σ_{eq} (cal mol^{-1})	$x - \zeta_e$	ζ_e	ζ_e/x	σ_{eq} (cal mol^{-1})	$x - \zeta_e$
1110	25	43.3	23	0.90	1413	2	22	0.88	1186	3
1350	31	46.0	28	0.91	1734	3	28	0.90	1447	3
1890	43	52.7	39	0.91	1995	4	38	0.89	1588	5
2780	63	57.6	58	0.93	2567	5	57	0.90	1954	6
3900	89	60.4	84	0.94	3410	5	82	0.92	2523	7
5970	136	63.3	129	0.95	4776	7	127	0.93	3389	9
7760	176	64.3	169	0.96	6080	7	166	0.94	4261	10

^{a} Melting temperature data from Ref. (18).

The results of the analyses for linear polyethylene and poly(ethylene oxide) are summarized in Tables 2.3 and 2.4 respectively. In order to perform the calculation it is necessary to assume a value for T_m^0. For polyethylene $145 \pm 1\,°C$ was taken for T_m^0, while for poly(ethylene oxide) either $76\,°C$ or $80\,°C$ was assumed. Although the precise values of the parameters deduced will depend on the value of T_m^0 the trends with molecular weight are unaffected by the choice. Similar results are obtained for both polymers. Over the molecular weight range for which appropriate data are available σ_{eq} varies three- to four-fold. Put another way, the parameter b in Eq. (2.26) varies by about a factor of two over the molecular weight range appropriate for analysis. Booth and coworkers performed a similar analysis with the low molecular weight poly(ethylene oxides).(56) The σ_{eq} values were of the same magnitude as reported here, and they also increased with chain length. It is evident that Eq. (2.27) cannot be used to extrapolate the melting temperatures of chains of finite length to T_m^0.

The increase in σ_{eq} with chain length is caused by the first term on the right in Eq. (2.25). It stems from the number of ways the equilibrium sequence of crystalline units ζ_e can be chosen from among the x units, if the chain ends are excluded. It is thus a unique and important property of chain molecules. It can, therefore, be expected that σ_{eq} will reach a limiting value at some higher molecular weight. The ratio ζ_e/x increases with x and appears to reach a limiting value which is of the order of 0.92–0.95 for both polymers. Until extended chain crystals of high molecular weights are available for analysis these limiting values can only be anticipated. The différence between x and ζ_e, given in the tables, increases with molecular weight. This trend would be expected to continue with increasing molecular weights.

It is of interest to compare the theoretical melting temperature–chain length relation for molecular crystals, obtained with the Flory–Vrij relation, with the unpaired, disordered interface model just discussed. In order to make this comparison values have to be taken for the parameters ΔG_e and σ_{eq}. A comparison of the melting temperatures between the two models is given by the curves in Fig. 2.18.(1) Here the ΔG_e value has been taken from the Flory–Vrij analysis. For the unpaired model σ_{eq} values of 1200 and 4600 cal mol^{-1} were considered. Curve A in Fig. 2.18 is a repeat and continuation of the solid curves from Figs. 2.11 and 2.12 and represents the melting of molecular crystals. Curves B and C are for the disordered chain end model. The curves in Fig. 2.18 clearly indicate that the stability of the particular model will depend on the relative values of the two parameters. The value of ΔG_e should be essentially independent of chain length, while it has been found that σ_{eq} increases with chain length. As indicated in the figure, the unpaired model will be stable at all chain lengths for the low value of σ_{eq}. In contrast, for the high value of σ_{eq} the end-paired molecular weight model will be stable for all molecular weights. It

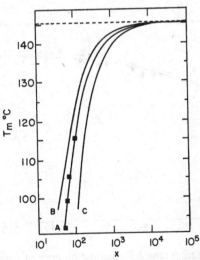

Fig. 2.18 Plot of melting temperature as a function of the number of carbon atoms in chain. Curve A: Flory–Vrij analysis. Curves B and C: theoretical calculations for disordered end sequences with $\sigma_e = 1200$ and 4600 respectively. ■ values for *n*-paraffins.(1)

is, therefore, theoretically possible that for low values of x the melting temperatures would be comparable to or somewhat greater than those of the corresponding *n*-alkanes. At the higher chain lengths, with the corresponding increase in σ_{eq}, the molecular crystal model would become more stable. Although the theoretical expectations appear reasonable, the boundaries for stability of the two crystallite types need to be established by experiment.

An appropriate way to examine this problem is to experimentally compare the equilibrium melting temperatures of *n*-alkanes with linear polyethylene fractions of comparable molecular weights. In comparing melting temperatures, it must be ensured that only extended chain crystallites are being considered. This requirement has been established for the *n*-alkanes up to and including $C_{390}H_{782}$.(12,13,14,18) Extended chain crystals have also been formed with linear polyethylene fractions up to and including $M_n = 5600$, by appropriate choice of crystallization temperatures.(55) Extended chain crystallites could not be formed in a fraction $M_n = 8000$. There is, therefore, a molecular weight range where the necessary comparison can be made. The manner in which the melting points are determined is also important in making the comparison. The same methods have not always been used. In particular, when differential scanning calorimetry is used the observed melting temperatures need to be extrapolated to zero heating rates, a procedure that has not always been adopted.(13,30) Despite these possible shortcomings the comparison between the two types of extended chain crystals is illuminating.

Figure 2.19 is a compilation of T_m values plotted against the carbon number 29 to 390 for the *n*-alkanes and a similar range for linear polyethylene fractions.(30)

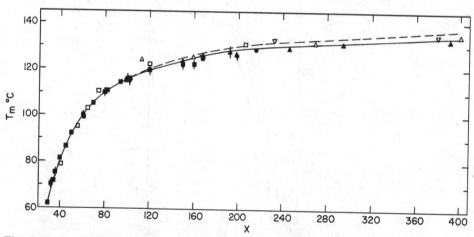

Fig. 2.19 Plot of melting temperature T_m against number of carbon atoms x for *n*-alkanes (solid symbols) and low molecular weight polyethylene fractions (open symbols).(30) *n*-Alkanes: ● Lee and Wegner (14); ▲ Ungar *et al.* (13); ▲(57); ■ Flory and Vrij (11); ▉(29); Takamizawa *et al.* (58); ◆(59). Polyethylene: □ (60); △ (55); ▽ (6).

The data in this figure give a very interesting set of results. Within 1 °C, all of the melting temperatures, for both the n-alkanes and the polymer fractions, fall on the common curve described by the solid line. Consideration of the different sources of the data, the different experimental techniques used, and some common molecular weights between the two species, makes the agreement noteworthy and real. The question then is what is the reason for these results?

It has already been deduced that with polymer fractions, molecular crystals with end pairing cannot be formed. End sequences of the polymers are disordered as indicated by ^{13}C NMR(33,61,62) and Raman studies.(32) Abundant evidence for pre-melting and end-group disorder has been presented for the n-alkanes (Section 2.3). The fact that the n-alkanes and polymers of the same molecular weight have the same melting temperatures should not be surprising since they have the same crystallite structure prior to melting.

The dashed curve in Fig. 2.19 represents the Flory–Vrij calculation for the end-pairing model, i.e. where pre-melting does not take place prior to melting. Up to a carbon number of about 160 the melting temperatures of the pre-melted alkanes and polymers and those for end pairing are indistinguishable from one another. The theoretical melting temperatures for molecular crystals (end pairing) are only slightly higher than those observed experimentally for carbon numbers 160 to 390. The differences are not beyond experimental error. The calculated melting temperatures for higher carbon number chains will depend on the relation between σ_{eq} and ΔG_e. As was pointed out, the latter quantity is expected to be independent of chain length, while in the range accessible to measurements σ_{eq} depends on molecular weight. A plot of σ_{eq} against molecular weight, as is shown in Fig. 2.20, is quite illuminating.(30) In the low molecular weight range, $M_n \leq 2000$, the σ_{eq} value is fairly constant at about 1700 cal mol^{-1}. However, for $M_n \geq 2000$, σ_{eq} increases monotonically and reaches a value of 3000–3500 cal mol^{-1} for $M_n = 5600$. Although σ_{eq} can be expected to reach an asymptotic value with molecular weight, it is clear that it has not done so for the highest molecular weight linear polyethylene where extended crystallites, required for equilibrium, are formed.

As was pointed out previously the variation in σ_{eq} with molecular weight makes it difficult to use the observed melting temperatures of the extended chain crystallites of the n-alkanes and fractions to extrapolate to the equilibrium melting temperature of polyethylene, by means of Eq. (2.27). A similar problem would be expected to be encountered with other type repeating units. The problem can be seen in Eq. (2.26). If b is assumed to be constant, independent of molecular weight, then a plot of $1/T_{m,e}$ against $1/x$ should be linear and extrapolated to $1/T_m^0$ as $x \to \infty$. Although recognizing that b is not constant it is instructive to analyze the available data in terms of the above assumption. Consequently, the appropriate plot is made in Fig. 2.21.(30) Careful scrutiny of this plot indicates that the data points are not sufficiently linear

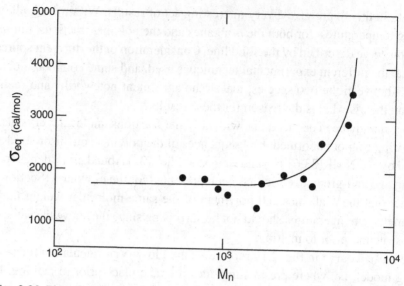

Fig. 2.20 Plot of σ_{eq} (cal mol^{-1}) against M_n for linear polyethylene fractions.(30)

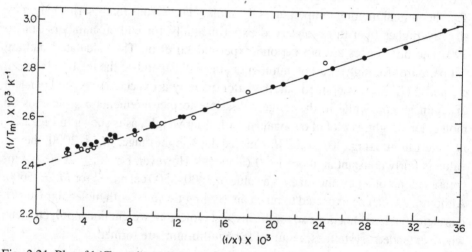

Fig. 2.21 Plot of $1/T_m$ against $1/x$ for n-alkanes (●) and low molecular weight polyethylene fractions (○).(30)

for a reliable extrapolation to T_m^0. Melting temperatures of higher molecular weight extended chain crystals are necessary in order to carry out the extrapolation with any reliability. The straight line that is drawn gives $T_m^0 = 144.3\,°C$ with a large uncertainty. Other suggested extrapolation methods have similar problems.

The analysis that has been given for polymer fractions can be extended to polydisperse systems. Modification in the theory only needs to be made in the expression for S_3, Eq. (2.14), which reflects the probability that the lattice cells reserved for crystallites are properly occupied. However, the molecular weight distribution that describes the polydisperse system needs to be explicitly specified. The behavior of two such distributions has been analyzed. In principle, the methods used can be extended to any defined distribution.

One distribution that has been analyzed is the "most probable" one.[9](10) This distribution is defined by

$$w_x = x(1-p)^2 p^{x-1} \tag{2.28}$$

Here, w_x is the weight fraction of the species comprised of x repeating units and p is a parameter which represents the probability of the continuation of the chain from one unit to the next. The entropy of fusion for this polydisperse system is (10)

$$\Delta S_f/xN = (1-\lambda)\Delta S_u - R[(z/z_s)(1-v_2)/v_2 + 1/\bar{x}_n]\ln[1 - v_2(1-\lambda)]$$
$$- R[(1-\lambda)/\zeta][\ln(v_2 D/p) + \zeta \ln p] \tag{2.29}$$

The correspondences between this equation and the corresponding one for fractions is made clear by substituting the expression

$$p = 1 - 1/\bar{x}_n \tag{2.30}$$

The equilibrium conditions are obtained from the free energy of fusion. Therefore, in the absence of diluent

$$1/T - 1/T_m^0 = (R/\Delta H_u[1/\bar{x}_n\lambda - (1/\zeta)\ln D + (1 - 1/\zeta)/\bar{x}_n] \tag{2.31}$$

with $\lambda = \lambda_e$ and $\zeta = \zeta_e$. At the melting temperature $T = T_m$ and $\lambda_e = 1$ so that

$$1/T_m - 1/T_m^0 = R/\Delta H_u[2/\bar{x}_n - (\ln D)/\zeta] \tag{2.32}$$

At equilibrium the term in $1/\zeta$ should be vanishingly small, so that Eq. (2.32) reduces to

$$1/T_m - 1/T_m^0 = 2R/\bar{x}_n \Delta H_u \tag{2.33}$$

The quantity $2/\bar{x}_n$ represents the mole fraction of the noncrystallizing terminal units. Since these units are distributed at random in the melt, Eq. (2.33) can also be derived from the condition for phase equilibrium with the impurities being restricted to the liquid phase.(63) The melting temperature expressed by Eq. (2.33) is characteristic of very long crystalline sequences. Such sequences will be formed from the larger molecular weight species, even for a very low number average molecular weight,

---

[9] Strictly speaking Eqs. (2.28) and (2.29) only apply to the case where a chain supplies only one sequence to a crystallite. These equations will thus be valid for the lower molecular weights where the last two terms are important. For higher molecular weights Eq. (2.29) can be shown to be an excellent approximation to the situation where many sequences from the same chain participated in a given crystallite. This also includes the hypothetical model where the chains are regularly folded within the crystallite. In this case ζ is identified with the length of each sequence and an additional interfacial free energy needs to be added to account for the folds.

and will be rare. Therefore, their equilibrium melting temperature will be difficult to determine experimentally. It should be emphasized at this point that Eq. (2.33), the relation between $1/T_m$ and $1/x_n$, only applies to systems that possess a "most probable" molecular weight distribution. It cannot be applied indiscriminately to any polydisperse polymers, or to molecular weight fractions.

Booth and coworkers have adapted the Flory treatment to a polydisperse system that has an exponential molecular weight distribution.(56) For this distribution

$$w(x) = \frac{b^{(a+1)}}{a!} x^a e^{-bx} \tag{2.34}$$

Here $b = a/\bar{x}_n$, $a = \bar{x}_n/(\bar{x}_w - \bar{x}_n)$ and $\bar{x}_n$ and $\bar{x}_w$ are the number average chain lengths respectively. Following the procedures outlined above, the equilibrium melting temperature for this distribution is (56)

$$T_m = T_m^0 \left(1 - \frac{2\sigma_{eq}}{\Delta H_u \zeta_e}\right) \bigg/ \left[1 + \frac{RT_m^0}{\Delta H_u}\left(\frac{1}{\bar{x}_n} - \frac{\ln I}{\zeta_e}\right)\right] \tag{2.35}$$

where

$$I = \frac{b^{(a+1)}}{a!} \int_{\zeta_e}^{\infty} x^{(a-1)} e^{-bx}(x - \zeta_e + 1)\, dx \tag{2.36}$$

This analysis is based on the assumption that cocrystallization of all species occurs, i.e. there is no fractionation or segregation.

Evans, Mighton and Flory (64) have studied the melting temperatures of a series of poly(decamethylene adipate) polymers that were prepared in such manner as to have "most probable" molecular weight distributions. The polymers were terminated in the conventional manner with hydroxyl and carboxyl end-groups as well as with bulky end-groups such as benzoate, α-naphthoate and cyclohexyl moieties. The melting temperatures of these polyesters were independent of the chemical nature of the end-groups for all molecular weights studied. It would not be expected that such bulky end-groups would participate in the crystal lattice. These results provide the underlying basis for the analysis, namely the exclusion of the terminal groups from the crystal lattice. The melting temperatures for the hydroxyl–carboxyl terminated samples are plotted according to Eq. (2.33) in Fig. 2.22.(64) It is clear from the figure that the functional form of this equation is obeyed over the complete molecular weight range studied. Moreover, the value for ΔH_u obtained from the slope of the straight line in Fig. 2.22 is in very good agreement with the value obtained by other methods.

Calculations based on Eq. (2.35) for the exponential molecular weight distribution indicate that the equilibrium melting temperatures of the poly(ethylene oxides), and presumably other polymers as well, are sensitive to the width of the distribution. Significant changes in the equilibrium melting temperature can occur. For example,

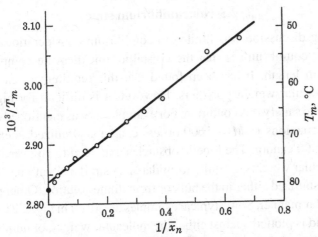

Fig. 2.22 $1/T_m$ against $1/\bar{x}_n$ for poly(decamethylene adipates) terminated with hydroxyl and carboxyl groups. (From Evans, Mighton and Flory (64))

for σ_{eq} equal to $1500\,\text{cal mol}^{-1}$ and $M_n = 1000$, T_m^0 increases by about $10\,^{\circ}\text{C}$ as M_w/M_n increases from one to two. When M_n is increased to 6000, T_m^0 increases by about three degrees for the same change in the distribution.(56) It can be expected that T_m^0 will be affected in a similar manner for other type distributions. Experimental data are needed that assess the influence of different types of molecular weight distributions on the equilibrium melting temperature.

Although the equilibrium melting temperatures for the poly(decamethylene adipates) with "most probable molecular weight" distribution are independent of the chemical nature of the end-group, it does not follow that this is a general principle applicable to all polymers. Even when excluded from the crystal lattice, specific interactions between end-groups in the interfacial layer could alter the value of σ_{eq}. Thus, even under equilibrium conditions the melting temperature could be affected. This problem was addressed by Booth and coworkers (65–68) who determined the equilibrium melting temperatures of low molecular weight poly(ethylene oxides), $M = 1000$–3000, with different end-groups, that were crystallized in extended form. No significant difference in T_m^0 was found among the polymers terminated by hydroxy, methoxy or ethoxy groups. This trend continued with alkyl groups that contained up to seven carbons. In contrast, chains terminated with chloro, phenoxy, acetoxy and trimethyl siloxy end-groups had T_m^0 values that were lower by about 5–$7\,^{\circ}\text{C}$. Density measurements indicate that the hydroxyl groups, as well as the other end-groups, were excluded from the crystal lattice. Thus, it is possible that although the terminal groups are excluded from the lattice the equilibrium melting temperature could be altered. These results demonstrate that the disordered, interfacial layer can be influenced by the size of interactions between the end-groups. The value of σ_{eq} is increased accordingly.

2.5 Nonequilibrium states

The foregoing discussion was limited to equilibrium considerations. A basic requirement for equilibrium is that the crystallite thickness be comparable to the extended chain length. It has been found that this requirement can be satisfied with low molecular weight polymers. However, it is difficult for high molecular weight chains to satisfy this condition. For example, as was mentioned earlier, linear polyethylene fractions of $M = 5600$ or less can be crystallized in extended form while $M = 8000$ cannot. The kinetic obstacles that need to be overcome for even modest molecular weights are quite formidable. A similar situation is found for the n-alkanes crystallized either in the bulk or from dilute solution. Other polymers behave in a similar manner. Some typical examples are given in Fig. 2.23a and b. Here, the long-period is plotted against either the molecular weight, or number of repeating units for poly(ϵ-caprolactone)(69) and a diol urethane.(70) In both examples there is initially a linear increase in long-periods with chain length that corresponds to the formation of extended chain crystallites. However, above a certain molecular weight, $M = 1300$ for poly(ϵ-caprolactone) and 1200 for the poly(urethane), the long-periods become constant or only increase with chain length very slightly indicating that extended chain crystallites are no longer formed. Rather, some type of folded chain crystallites develop. This change in crystallite structures also affects the relation between the observed melting temperature and molecular weight.(69,70) Although in general it is quite important, at this point, the detailed structure of the folded chains is not of concern. This important matter will be discussed in detail in subsequent volumes. The important matter here is the fact that above a certain low molecular weight a nonequilibrium or metastable state develops.

Although the primary discussion has been directed to equilibrium states, some of the analyses that have been developed can be adapted to nonequilibrium metastable states. In particular, one can examine the thermodynamic properties of a crystallite whose thickness is very much smaller than the extended chain. The finite crystallite thickness has been found to be the major contributor to the reduced melting temperature that is observed. It is, therefore, opportune to discuss the subject at this time.

The free energy of fusion of a nonequilibrium crystallite can be obtained from Eq. (2.17) prior to maximizing the free energy function. The free energy of fusion for a pure system, $v_2 = 1$, can be written as

$$\frac{\Delta G_f}{xN} = \frac{\zeta\rho}{xN}\Delta G_u + RT\left\{\ln\left(1 - \frac{\zeta\rho}{xN}\right) + \frac{\zeta\rho}{xN}\frac{1}{\zeta}\left[\ln D + \ln\frac{(x - \zeta + 1)}{x}\right]\right\}$$

(2.37)

for a crystallite of ζ repeating units thick that has ρ sequences in cross-section. It is assumed here that the mature crystallite is sufficiently large in the directions normal to the chain axis so that the influences of the lateral surface free energies

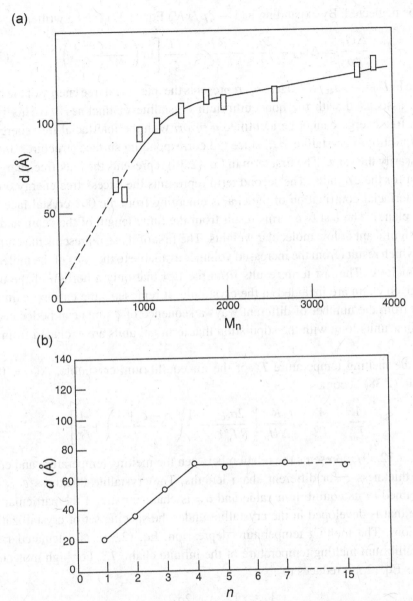

Fig. 2.23 Representative plots of long-period against chain length. (a) Long-period against molecular weight for poly(ϵ-caprolactone) (From Perret and Skoulios (69)). (b) Plot of long-period against the number of repeating units, n, for a diol urethane polymer. (From Kern, Davidovits, Rauterkus and Schmidt (70))

can be neglected. By expanding $\ln(1 - \zeta\rho/xN)$ Eq. (2.37) can be written as

$$\frac{\Delta G_f}{\zeta\rho} = \Delta G_u - \frac{2\sigma_{ec}}{\zeta} - \frac{RT}{x} + \frac{RT}{\zeta}\left[\ln\left(\frac{x - \zeta + 1}{x}\right)\right] \qquad (2.38)$$

where $\ln D \equiv -2\sigma_{ec}/RT$. Here σ_{ec} represents the interfacial free energy of the basal plane associated with the nonequilibrium crystallite of thickness ζ. This interfacial free energy cannot be identified *a priori* with the interfacial free energy of the equilibrium crystallite, σ_{eq}, since the corresponding surface structures are not necessarily the same. The first term in Eq. (2.38) represents the bulk free energy of fusion for the $\zeta\rho$ units. The second term represents the excess free energy due to the interfacial contribution of the chains emerging from the 001 crystal face (the basal plane). The last two terms result from the finite length of the chain and are only significant at low molecular weights. The first of these represents the entropy gain which results from the increased volumes available to the ends of the molecule after melting. The last term results from the fact that only a portion of the units of a given chain are included in the crystallite. It represents the entropy gain that arises from the number of different ways a sequence of ζ units can be located in a chain x units long with the stipulation that terminal units are excluded from the lattice.

At the melting temperature T_m^* of the nonequilibrium crystallite, $\Delta G_f = 0$ so that Eq. (2.38) becomes

$$\frac{1}{T_m^*} - \frac{1}{T_m^0} = \frac{R}{\Delta H_u}\left[\frac{2\sigma_{ec}}{RT_m^*\zeta} - \frac{1}{\zeta}\left(\frac{x - \zeta + 1}{x}\right) + \frac{1}{x}\right] \qquad (2.39)$$

Equation (2.39) represents the relation between the melting temperature and crystallite thickness ζ for different chain lengths. The crystallite thickness ζ is not constrained to its equilibrium value and σ_{ec} is characteristic of the particular interface that is developed in the crystallite under the specific set of crystallization conditions. The melting temperature depression, Eq. (2.39), is calculated from the equilibrium melting temperature of the infinite chain, T_m^0. For high molecular weights Eq. (2.39) reduces to

$$\frac{1}{T_m^*} - \frac{1}{T_m^0} = \frac{2\sigma_{ec}}{\Delta H_u T_m^* \zeta} \qquad (2.40)$$

or

$$T_m^* = T_m^0[1 - 2\sigma_{ec}/\Delta H_u\zeta] \qquad (2.41)$$

Equation (2.40) is identical to the classical Gibbs–Thomson expression for the melting of crystals of finite size. Thus, following the Flory theory (10) nonequilibrium crystallites of high molecular weight chains obey the same melting point relation as do low molecular weight substances. However, corrections need to be made for lower molecular weight chain molecules.

Equation (2.40) suggests a method by which T_m^0 can be determined. If T_m^* is measured as a function of ζ, then a plot of T_m^* against $1/\zeta$ should be linear and extrapolate to T_m^0 as $\zeta \to \infty$. The use of Eq. (2.40) requires that σ_{ec} be the same for each sample, i.e. for the samples of varying thickness the interfacial structure must be the same. However, this condition may be difficult to fulfill. Moreover, T_m^0 represents an extended chain equilibrium crystallite, while for high molecular weights T_m^* represents some type of folded chain crystallite. The use of the Gibbs–Thomson relation to obtain T_m^0, as well as other extrapolative methods that have been suggested will be examined in detail when the morphology and structure of the crystalline state is discussed. At this point it can be stated that the reliable and accurate determination of the equilibrium melting temperature of high molecular weight polymers is a formidable and difficult task. Analysis of the melting temperature of high molecular weight oligomers appears to be a promising path.

In summary, the major conclusion to be made from the discussion of the fusion of homopolymers is that a first-order phase transition governs the melting–crystallization process. Despite the difficulties in establishing equilibrium and in determining the equilibrium melting temperature the underlying guiding principle is phase equilibrium. The consequences of phase equilibrium are invoked in the chapters that follow. The fusion of polymer–low molecular weight-diluent mixtures, polymer–polymer mixtures, copolymers, and the influence of deformation will be discussed from this point of view.

References

1. Mandelkern, L., in *Comprehensive Polymer Science, Volume 2 Polymer Properties*, C. Booth and C. Price eds., Pergamon Press (1989) p. 363.
2. Mayer, J. E. and S. F. Streeter, *J. Chem. Phys.*, **7**, 1019 (1939).
3. Ergoz, E., J. G. Fatou and L. Mandelkern, *Macromolecules*, **5**, 147 (1972).
4. Mandelkern, L., *Rubber Chem. Technol.*, **32**, 1392 (1959).
5. Chiang, R. and P. J. Flory, *J. Am. Chem. Soc.*, **83**, 2857 (1961).
6. Fatou, J. G. and L. Mandelkern, *J. Phys. Chem.*, **69**, 417 (1965).
7. Mandelkern, L., *Chem. Rev.*, **56**, 903 (1956).
8. Wood, L. A. and N. Bekkedahl, *J. Appl. Phys.*, **17**, 362 (1946).
9. Matsuoka, S., *J. Polym. Sci.*, **42**, 511 (1960).
10. Flory, P. J., *J. Chem. Phys.*, **17**, 223 (1949).
11. Flory, P. J. and A. Vrij, *J. Am. Chem. Soc.*, **85**, 3548 (1963).
12. Lee, K. S., Ph.D. Thesis, Freiburg, FRG 1984.
13. Ungar, G., J. Stejny, A. Keller, I. Bidd and M. C. Whiting, *Science*, **229**, 386 (1985).
14. Lee, K. S. and G. Wegner, *Makromol. Chem., Rapid Commun.*, **6**, 203 (1985).
15. Mandelkern, L., R. G. Alamo and D. L. Dorset, *Acta Chimica Hungarica – Models in Chemistry*, A. P. Schubert ed. (1993) p. 415.

16. Hoffman, J. D., L. J. Frolen, G. S. Ross and J. I. Lauritzen, *J. Res. Nat. Bur. Stand.*, **79A**, 671 (1975).
17. Broadhurst, M. G., *J. Res. Nat. Bur. Stand.*, **66A**, 241 (1962).
18. Mandelkern, L. and G. M. Stack, *Macromolecules*, **17**, 871 (1984).
19. Quinn, F. A., Jr. and L. Mandelkern, *J. Am. Chem. Soc.*, **80**, 3178 (1958).
20. Rijke, A. M. and L. Mandelkern, *J. Polym. Sci.*, *A-2*, **8**, 225 (1970).
21. Alamo, R. G., B. D. Viers and L. Mandelkern, *Macromolecules*, **28**, 3205 (1995).
22. Gopalan, M. and L. Mandelkern, *J. Phys. Chem.*, **71**, 3833 (1967).
23. Chivers, R. A., P. J. Barham, I. Martinez-Salazar and A. Keller, *J. Polym. Sci.: Polym. Phys. Ed.*, **20**, 1717 (1982).
24. Brown, R. J. and R. K. Eby, *J. Appl. Phys.*, **35**, 1156 (1964).
25. Huseby, T. W. and H. E. Bair, *J. Appl. Phys.*, **39**, 4969 (1968).
26. Hoffman, J. D., G. T. Davis and J. I. Lauritzen, Jr., *Treatise in Solid State Chemistry*, vol. 3, N. B. Hannay ed., Plenum Press (1976) p. 497.
27. Fujiwarra, Y. and T. Yoshida, *J. Polym. Sci., Polym. Lett.* **1B**, 675 (1963).
28. Takamizawa, K., Y. Ogawa and T. Uyama, *Polym. J.*, **14**, 441 (1982).
29. Mandelkern, L., G. M. Stack and P. J. M. Mathieu, *Anal. Calorim.*, **5**, 223 (1984).
30. Mandelkern, L., A. Prasad, R. G. Alamo and G. M. Stack, *Macromolecules*, **23**, 3696 (1990).
31. A. Müller, *Proc. R. Soc. London*, **A 127**, 417 (1930); **A158**, 403 (1937); **A174**, 137 (1940).
32. Kim, Y., H. L. Strauss and R. G. Snyder, *J. Phys. Chem.*, **93**, 7520 (1989).
33. Stewart, M. J., W. L. Jarrett, L. J. Mathias, R. G. Alamo and L. Mandelkern, *Macromolecules*, **29**, 4963 (1996).
34. Hoffman, J. D., *Kolloid Z. Z. Polym.*, **231**, 449 (1969).
35. Rothe, M. and W. Dunkel, *J. Polym. Sci.: Polym. Lett.*, **5B**, 589 (1967).
36. Starkweather, H. W., Jr., *Macromolecules*, **19**, 1131 (1986).
37. Yeates, S. G., H. H. Teo, R. H. Mobbs and C. Booth, *Makromol. Chem.*, **185**, 1559 (1984).
38. Bourgeois, E. and A. Fonassi, *Bull. Soc. Chim. France*, **9**, 941 (1911).
39. Montando, G., G. Bruno, P. Maravigna, P. Finocchiaro and G. Centineo, *J. Polym. Sci.: Polym. Chem. Ed.*, **11**, 65 (1973).
40. Koch, W. and W. Heitz, *Makromol. Chem.*, **184**, 779 (1983).
41. Zahn, H. and R. Krzikalla, *Makromol. Chem.*, **23**, 31 (1957).
42. Hasslin, H. W., M. Dröscher and G. Wegner, *Makromol. Chem.*, **179**, 1373 (1978).
43. Clarson, S. J., K. Dodgson and I. A. Semlyen, *Polymer*, **26**, 933 (1985).
44. Nakasone, K., Y. Urabe and K. Takamizawa, *Thermochem. Acta*, **286**, 161 (1996).
45. Wunderlich, B. and A. Czornyj, *Macromolecules*, **10**, 960 (1977).
46. Atkinson, C. M. L. and M. J. Richardson, *Trans. Faraday Soc.*, **65**, 949 (1969).
47. Hay, J. N., *J. Polym. Sci.: Polym. Chem. Ed.*, **14**, 2845 (1976).
48. Carlies, V., J. Devaux, R. Legras and D. J. Blundell, *J. Polym. Sci.: Pt. B: Polym. Phys.*, **36**, 2563 (1998).
49. Hay, J. N., *Makromol. Chem.*, **177**, 2559 (1976).
50. Mandelkern, L., *J. Polym. Sci.*, **47**, 494 (1960).
51. Mandelkern, L., J. G. Fatou and C. Howard, *J. Phys. Chem.*, **68**, 3386 (1964).
52. Mandelkern, L., J. G. Fatou and C. Howard, *J. Phys. Chem.*, **69**, 956 (1965).
53. Flory, P. J., *Principles of Polymer Chemistry*, Cornell University Press (1953) pp. 495ff.
54. Mandelkern, L., M. R. Gopalan and J. F. Jackson, *J. Polym. Sci.*, **B5**, 1 (1967).
55. Stack, G. M., L. Mandelkern and I. G. Voigt-Martin, *Macromolecules*, **17**, 321 (1984).

56. Beech, D. R., C. Booth, D. V. Dodgson, R. R. Sharpe and J. R. S. Waring, *Polymer*, **13**, 73 (1972).
57. Stack, G. M., L. Mandelkern, C. Kröhnke and G. Wegner, *Macromolecules*, **22**, 4351 (1989).
58. Takamizawa, K., Y. Sasaki, K. Kono and Y. Urabe, *Rep. Prog. Polym. Phys. Jpn.*, **19**, 285 (1976).
59. Mandelkern, L. and G. M. Stack, unpublished observations.
60. Prasad, A. and L. Mandelkern, *Macromolecules*, **22**, 914 (1989).
61. Jarret, W. L., L. J. Mathias, R. G. Alamo, L. Mandelkern and D. Dorset, *Macromolecules*, **25**, 3468 (1992).
62. Möller, M., H. J. Cantow, H. Drotloff, D. Emeis, K. S. Lee and G. Wegner, *Makromol. Chem.*, **187**, 1237 (1986).
63. Flory, P. J., *Principles of Polymer Chemistry*, Cornell University Press (1953) p. 570.
64. Evans, R. D., H. R. Mighton and P. J. Flory, *J. Am. Chem. Soc.*, **72**, 2018 (1950).
65. Booth, C., J. M. Bruce and M. Buggy, *Polymer*, **13**, 475 (1972).
66. Ashman, P. C. and C. Booth, *Polymer*, **14**, 300 (1973).
67. M. J. Fraser, D. R. Cooper and C. Booth, *Polymer*, **18**, 852 (1977).
68. Cooper, D. R., Y. K. Leung, F. Heatley and C. Booth, *Polymer*, **19**, 309 (1978).
69. Perret, R. and A. Skoulios, *Makromol. Chem.*, **156**, 157 (1972).
70. Kern, W., J. Davidovits, K. J. Rauterkus and G. F. Schmidt, *Makromol. Chem.*, **106**, 43 (1961).

3

Polymer–diluent mixtures

3.1 Introduction

The fusion of polymers with low molecular weight diluents will be discussed in this chapter. When considering phase equilibria of multicomponent systems several *a priori* assumptions have to be made. These assumptions are universal and are applicable to all types of molecular systems and are not unique to polymers. It is necessary to specify whether the disordered, or liquid, state is homogeneous or heterogeneous, i.e. does liquid–liquid phase separation occur. The composition of the crystalline phase and in particular whether it remains pure, also needs to be specified. If the crystalline phase is not pure it is necessary to know whether the diluent enters the crystal lattice as a result of equilibrium considerations and if compound formation occurs. Also to be considered is whether the diluent enters the lattice as a defect. All of these possibilities need to be considered separately. The appropriate expression for the free energy of mixing that applies in each specific case has to be known. All of these factors will be considered in the following.

3.2 Melting temperature: concentrated and moderately dilute mixtures

The concentrated and moderately dilute concentration range is one where the Flory–Huggins free energy of mixing is applicable. This implies that there is a uniform distribution of diluent and polymer segments in the melt.(1) The most general considerations lead to the expectation that the addition of a lower molecular weight diluent to a homogeneous melt will result in a lowering of the melting temperature and a broadening of the melting range of all molecular species, including polymers. With but a few exceptions, in most of the polymer diluent mixtures that have been studied the crystalline phase remains pure. We shall also have occasion to refer specifically to melting into a heterogeneous, two-phase melt. Limiting ourselves at present to the concentration range where the Flory–Huggins theory is valid we return to Eq. (2.17). By maximizing the expression for ΔG_f, and setting $\lambda = \lambda_e$

70

and $\zeta = \zeta_e$, the equilibrium melting temperature of a polymer–diluent mixture can be expressed as(2)

$$1/T_m - 1/T_m^0 = R/\Delta H_u\{(V_u/V_1)(1 - v_2)$$
$$+ (1/x)[v_2 + x/(x - \zeta_e + 1)] - \chi_1(1 - v_2)^2\} \qquad (3.1)$$

for a homogeneous melt with a diluent being excluded from the crystalline lattice. The values of ζ_e is defined by Eq. (2.20). For large values of x Eq. (3.1) reduces to

$$1/T_m - 1/T_m^0 = (R/\Delta H_u)(V_u/V_1)[(1 - v_2) - \chi_1(1 - v_2)^2] \qquad (3.2)$$

Equation (3.2) is very similar to the classical expression for the depression of the melting temperature of low molecular weight binary systems, i.e. the freezing point depression equation. The only difference results from the expression for the activity of the crystallizing polymer component in the molten phase. Consequently, Eq. (3.1) can also be derived by the application of phase equilibrium requirements.(3)

From the Flory–Huggins mixing expression, the chemical potential μ_1 of the diluent in the melt, relative to that of the pure component μ_1^0, can then be written as (3)

$$\mu_1 - \mu_1^0 = RT\left[\ln(1 - v_2) + \left(1 - \frac{1}{x}\right)v_2 + \chi_1 x v_2^2\right] \qquad (3.3)$$

The chemical potential of a polymer molecule, taking the pure liquid polymer as the reference state, can be expressed as

$$\mu_2 - \mu_2^0 = RT[\ln v_2 - (x - 1)(1 - v_2) + \chi_1 x(1 - v_2)^2] \qquad (3.4)$$

In Eqs. (3.3) and (3.4), x is the number of segments per molecule. The chemical potential per mole of polymer structural units is obtained by dividing Eq. (3.4) by $x V_1/V_u$, the number of units per molecule. Thus,

$$\mu_u^1 - \mu_u^0 = RT\frac{V_u}{V_1}\left[\frac{\ln v_2}{x} - \left(1 - \frac{1}{x}\right)(1 - v_2) + \chi_1(1 - v_2)^2\right] \qquad (3.5)$$

The chemical potentials of other components in the crystalline phase cannot be derived with such generality. These potentials will depend on the components present, and the mixing law that is involved. An important distinction can be made, however, as to whether the diluent is either present or absent in the crystalline phase. When present, the exact role of the diluent must be enunciated in order to specify the chemical potential of the components. The analysis is greatly simplified when the diluent is excluded from the crystalline lattice. This is an example of a binary liquid mixture of which only one component crystallizes over the whole composition range. With this restraint for equilibrium the chemical potential of the crystallizing component in the two phases must be equal. At the melting point of a polymer–diluent mixture, with the crystalline phase being pure, it is required that

$$\mu_u^c - \mu_u^0 = \mu_u^l - \mu_u^0 \tag{3.6}$$

where the superscripts refer to the crystalline and liquid phases, respectively, and the pure molten polymer is taken as the reference state. The difference in the chemical potential between a polymer unit in the pure crystalline and liquid states can be written as

$$\mu_u^c - \mu_u^0 = -\Delta G_u = -(\Delta H_u - T\,\Delta S_u) \tag{3.7}$$

By defining the ratio $\Delta H_u/\Delta S_u$ as T_m^0, Eq. (3.7) can be written as

$$\mu_u^c - \mu_u^0 = -\Delta H_u \left(1 - \frac{T}{T_m^0}\right) \tag{3.8}$$

The tacit assumption has been made that the ratio of ΔH_u to ΔS_u does not vary with temperature. By utilizing Eq. (3.3) for $\mu_u^l - \mu_u^0$, one obtains

$$\frac{1}{T_m} - \frac{1}{T_m^0} = \left(\frac{R}{\Delta H_u}\right)\left(\frac{V_u}{V_1}\right)\left[-\frac{\ln v_2}{x} + \left(1 - \frac{1}{x}\right)(1 - v_2) - \chi_1(1 - v_2)^2\right] \tag{3.9}$$

For large x, Eq. (3.2) is regenerated.

There is the expectation from Eq. (3.2) that the melting temperature should be well-defined and be systematically depressed with the addition of low molecular weight diluent. This expectation has been observed for all polymer–diluent mixtures that have been studied, even though equilibrium melting temperatures are not always used. A set of examples is given in Fig. 3.1 for the fusion of poly(decamethylene adipate) and its mixtures with diluents.[4] Here, the specific volume is plotted against the temperature for different diluents and concentrations. In each example the melting temperature is clearly defined, and is depressed by the addition of diluent. The melting range of the pure polymer is relatively sharp. However, the melting interval progressively broadens as the diluent concentration increases. These results are a natural consequence, for both polymers and low molecular weight substances, of the type of binary systems being analyzed. The statistical mechanical analysis indicates that the melting range will depend on the value of the product $v_2 D$.[2] It is also possible to extend the thermodynamic analysis to ternary systems, consisting of polymer and two different diluents.[5]

According to Eq. (3.2) the depression of the melting temperature depends on the volume fraction of diluent in the mixture, its molar volume and the thermodynamic interaction between polymer and diluent. The melting point depression is a colligative property. Therefore, as Eq. (3.2) indicates the depression will be larger with diluents of smaller molar volume. It is also expected that the larger the value

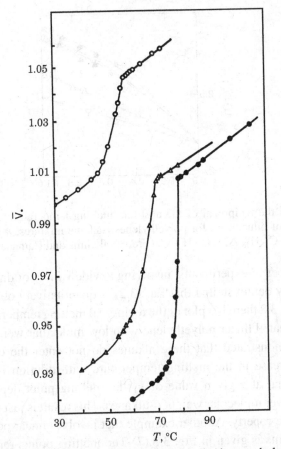

Fig. 3.1 Specific volume–temperature curves for pure poly(decamethylene adipate), ●; for its mixtures with dimethyl formamide ($v_1 = 0.60$), ○; and for its mixtures with diphenyl ether ($v_1 = 0.18$), △.(4)

of ΔH_u the smaller the melting point depression. ΔH_u represents the enthalpy of fusion per mole of repeating unit. It is an inherent and characteristic property of the repeating unit of crystalline polymers and does not depend on the level of crystallinity or any other morphological feature. This quantity should not be identified with the enthalpy of fusion, ΔH_u^*, obtained by direct calorimetric measurements. These two quantities are quite different. The latter quantity depends on the level of crystallinity and crystallite structure. The former depends only on the nature of the chain repeating unit. From it one can deduce important thermodynamic characteristics of the chain. The quantity ΔH_u can be obtained from experiment by means of Eq. (3.2). Thus, given ΔH_u and T_m^0 one can obtain ΔS_u. With other quantities being equal a larger depression of the melting temperature should be observed with good solvents (smaller values of χ_1) than with poor ones. It should be noted that the quantity χ_1 depends on both the temperature and composition.

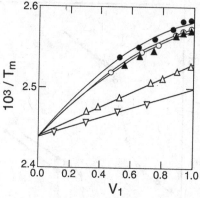

Fig. 3.2 Plots of the reciprocal of the absolute melting temperatures, $1/T_m$ against the volume fraction of diluent, v_1, for polyethylene–n-alkane mixtures. n-C_6H_{14} ●; n-C_7H_{16} ○; n-C_8H_{18} ▲; n-$C_{18}H_{38}$ △; n-$C_{32}H_{66}$ ▽. (From Nakajima and Hamada (8))

A large number of experiments, involving a wide variety of different polymers, has conclusively demonstrated that Eq. (3.2) is quantitatively obeyed.(6,7) As an example, in Fig. 3.2 there is a plot of the reciprocal melting temperature of mixtures of an unfractionated linear polyethylene with low molecular weight n-alkanes.(8) It has been demonstrated that these alkanes do not enter the crystal lattice. A continuous decrease in the melting temperature with dilution is observed. It is also apparent that at a given value of v_1 the melting point depression is much larger for the lower molecular weight n-alkanes. This result is just what is expected for a colligative property. Another example that involves linear polyethylene with other type diluents is given in Fig. 3.3.(7) The melting points represented by the lower curve are for tetralin and α-chloronaphthalene as diluents. A continuous decrease of the melting temperature is observed as increasing amounts of diluent are added. Such behavior is expected for these relatively good solvents. Coincidentally, these two diluents behave in an almost identical manner. The melting temperature–composition relation for the upper two curves, which represent n-butyl phthalate and o-nitrotoluene as diluents, behave quite differently. With the initial addition of diluent there is only a very small decrease in the melting point. However, when a critical diluent concentration is reached, the melting temperature remains invariant with further dilution. The reason for the invariance in the melting temperature for these and other polymer–diluent mixtures will be discussed shortly.

For mixtures that display a continuous depression of the melting temperature with decreasing polymer concentration, a direct comparison with Eq. (3.2) can be made. Establishing the validity of Eq. (3.2) is important since it is a potential method for determining ΔH_u. According to this equation, the initial slope of the plot of $1/T_m$ against the diluent concentration, $(1 - v_2)$, should be inversely proportional to ΔH_u. A quantitative analysis of experimental data by means of Eq. (3.2) requires

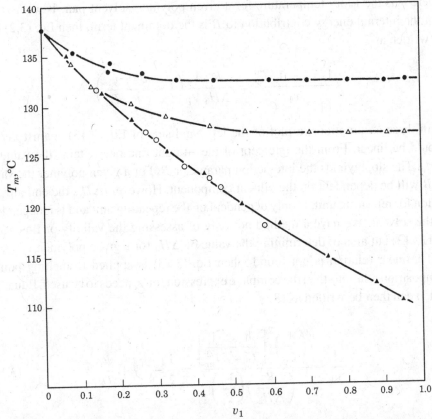

Fig. 3.3 Plot of melting temperature–composition relations of linear polyethylene for different diluents. ● *n*-butyl phthalate; △ *o*-nitrotoluene; ▲ α-chloronaphthalene; ○ tetralin. v_1 is volume fraction of the diluent present in the mixture.(7)

the decomposition of χ_1 into its enthalpic and entropic contributions. According to the standard analysis of binary polymer mixtures χ_1 can be expressed as (1)

$$\chi_1 = \kappa_1 - \psi_1 + 1/2 \qquad (3.10)$$

where κ_1 and ψ_1 are heat and entropy parameters such that the partial molar enthalpy $\Delta H_1 = RT\,\kappa_1\,v_2^2$ and the partial molar entropy $\Delta S_1 = RT\,\psi_1\,v_2^2$. The enthalpic term can also be represented as

$$\kappa_1 = \frac{BV_1}{RT} \qquad (3.11)$$

where B is the interaction energy density character of the polymer–diluent pair. It can also be represented as (1)

$$\kappa_1 = \frac{\psi_1 \theta}{T} \qquad (3.12)$$

where θ is the Flory temperature for a given polymer–solvent pair. If we assume that the internal energy contribution to B is the dominant term, then Eq. (3.2) can be written as

$$\frac{1/T_\mathrm{m} - 1/T_\mathrm{m}^0}{v_1} = \frac{R}{\Delta H_\mathrm{u}} \frac{V_\mathrm{u}}{V_1}\left(1 - \frac{BV_1}{R}\frac{v_1}{T_\mathrm{m}}\right) \qquad (3.13)$$

With these assumptions, a plot of the left-hand side of Eq. (3.13) against v_1/T_m should be linear. From the intercept of the straight line one obtains the value of ΔH_u. The slope yields the interaction parameter B. For a given polymer the value of B will be dependent on the diluent component. However, ΔH_u, the enthalpy of fusion for repeating unit, is only dependent on the repeating unit and is independent of the solvent. We have a method not only of assessing the validity of Eqs. (3.2) and (3.13) but also of determining the value of ΔH_u for a given polymer.

If a linear relation is not found when Eq. (3.13) is applied to melting point–composition relations then the complete expression for χ_1 needs to be used. Equation (3.13) can then be written as (8)

$$\frac{1}{v_1} - \frac{\left(\dfrac{1}{v_1}\right)^2\left[\dfrac{1}{T_\mathrm{m}} - \dfrac{1}{T_\mathrm{m}^0}\right]}{\left(\dfrac{R}{\Delta H_\mathrm{u}}\right)\left(\dfrac{V_\mathrm{u}}{V_1}\right)} = \frac{1}{2} - \psi_1 + \frac{\psi_1}{T_\mathrm{m}} \qquad (3.14)$$

Under these circumstances ΔH_u cannot be obtained in any simple manner, when data is analyzed over the complete composition range. However, it can be determined if the limiting slope is established. The neglect of the entropic term in Eq. (3.13) can usually be justified by the small temperature range that is encompassed by experiment. The value of ΔH_u should be scarcely affected by the approximation introduced. However, the value of B may be. Hence, it should not be unexpected that it often does not agree with the values obtained by other methods. Melting point depression studies are not always accurate in determining thermodynamic interaction parameters between polymer and diluent.

Representative plots of experimental data, treated in accordance with Eq. (3.13), are given in Figs. 3.4, 3.5 and 3.6 for three different polymers, linear polyethylene (8), natural rubber (poly 1-4 cis isoprene) (9), and poly(decamethylene terephthalate) (10), respectively. The data used to prepare Fig. 3.4 are the same as given in Fig. 3.2 and thus cover wide composition range. A set of linear relations result when these data were treated according to Eq. (3.13). The ΔH_u values obtained for a given polymer agree among the different diluents used (see below). A more detailed analysis, following Eq. (3.13), indicates that for all these mixtures κ_1 is

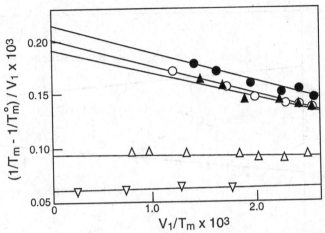

Fig. 3.4 Plot of $(1/T_m - 1/T_m^0)/v_1$ against v_1/T_m for polyethylene–n-alkane mixtures. n-C$_6$H$_{14}$ ●; n-C$_7$H$_{16}$ ○; n-C$_8$H$_{18}$ ▲; n-C$_{18}$H$_{38}$ △; and n-C$_{32}$H$_{66}$ ▽. (From Nakajima and Hamada (8))

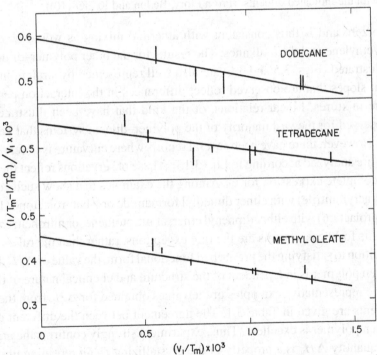

Fig. 3.5 Plot of quantity $(1/T_m - 1/T_m^0)/v_1$ against v_1/T_m for natural rubber mixed with the indicated diluents.(9)

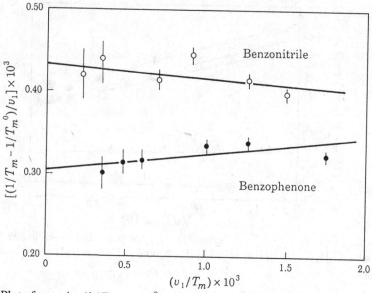

Fig. 3.6 Plot of quantity $(1/T_m - 1/T_m^0)/v_1$ against v_1/T_m for polydecamethylene tereph-thalate with the indicated diluents. (From Flory, Bedon and Keefer (10))

close to zero and is thus consistent with athermal mixing as would be expected for polyethylene and the *n*-alkanes. The results for the other polymer–diluent mixtures illustrated (Figs. 3.5 and 3.6) are also well represented by straight lines. The different slopes that are observed reflect differences in the interaction parameters for these mixtures. Linear relations, of the kind that have been illustrated, have been observed for the vast majority of the polymer–diluent systems that have been studied. However, there have been several reports where curvature is observed when the data are analyzed according to Eq. (3.13). These observations reflect the need to use the complete expression for κ_1. Among the examples that show such curvature are poly(acrylonitrile) with either dimethyl formamide or γ-butyrolactone (10a) and poly(caprolactam) with either diphenyl ether, or nitrotoluene, or nitrobenzene.(10b) The results for these systems are the rare exceptions, rather than the rule.

In addition to satisfying the prescribed functional form, the value of ΔH_u deduced for a given polymer is independent of the structure and chemical nature of the diluent. Some representative examples of the values obtained for ΔH_u from these type experiments are given in Table 3.1. The agreement between the different diluents for a given polymer is excellent. Thus, experiment strongly confirms the argument that the quantity ΔH_u is a property of the crystallizing chain repeating unit and is independent of the nature of the diluent used. Since the necessary requirements are fulfilled, it can be concluded from this straightforward analysis that, at the melting temperature, equilibrium between the pure crystalline polymeric phase and the two-component homogeneous liquid phase is established. We thus have a method of

Table 3.1. *Typical results for ΔH_u as determined from Eq. (3.13)*

Polymer	Diluent	ΔH_u, cal mol^{-1} of repeating unit
Polyethylene[a]	Ethyl benzoate	930
	o-Nitrotoluene	935
	Tetralin	990
	α-Chloronaphthalene	970
Polyethylene[b]	*n*-Hexane	1 085
	n-Heptane	1 050
	n-Octane	1 000
	n-Octadecane	1 015
	n-Ditricontane	980
Natural rubber[c]	Tetradecane	1 040
	Methyl oleate	980
	Dodecane	1 100
Poly(decamethylene terephthalate)[d]	Benzonitrile	11 600
	Benzophenone	10 400
Poly(chloro trifluoroethylene)[e]	Toluene	1 220
	Mesitylene	1 100
	o-Chlorobenzotrifluoride	1 260
	Cyclohexane[†]	1 330
Poly(oxymethylene)[f]	*p*-Chlorophenol	1 570
	Tetralin	1 465
	Phenol	1 760
	m-Cresol	1 775
	α-Chloronapthalene	1 400

[†] Since the data for T_m scatter, the value obtained for ΔH_u is only approximate.

[a] F. A. Quinn, Jr. and L. Mandelkern, *J. Am. Chem. Soc.*, **80**, 3178 (1958); *ibid.* **81**, 6533 (1959).

[b] A. Nakajima and F. Hamada, *Kolloid Z. Z. Polym.*, **205**, 55 (1965).

[c] D. E. Roberts and L. Mandelkern, *J. Am. Chem. Soc.*, **77**, 781 (1955).

[d] P. J. Flory, H. D. Bedon and E. H. Keefer, *J. Polym. Sci.*, **28**, 1511 (1958).

[e] A. M. Bueche, *J. Am. Chem. Soc.*, **74**, 65 (1952).

[f] T. Korenga, F. Hamada and A. Nakajima, *Polym. J.*, **3**, 21 (1972).

relating the chain structure to a thermodynamic parameter governing crystallization behavior. This relation will be discussed in detail for many polymers in Chapter 6.

When analyzing melting temperature–composition relations according to Eq. (3.13) the implicit assumption is made that the crystallite structure and size do not vary over the composition range studied. It is also assumed that the interfacial free energy associated with the crystallites remains constant. Since the crystallization of the polymer was conducted from the mixture, there could be concern that these factors vary with composition. However, there are no problems when

Polymer–diluent mixtures

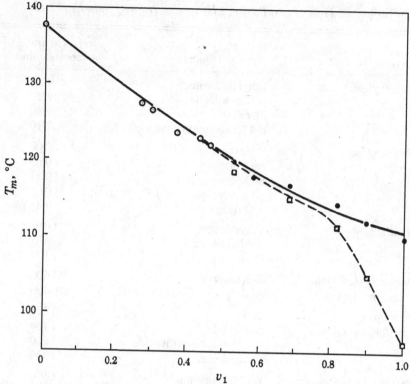

Fig. 3.7 A plot of the solubility temperature against volume fraction diluent v_1 for polyethylene in tetralin. $\bigcirc$, melting points determined dilatometrically; $\bullet$, solubility point of melt crystallized samples; $\square$, solubility point of solution crystallized sample. (From Jackson, Flory and Chiang (11))

the polymer is crystallized external to the diluent and the melting temperature of the mixture is determined at a given concentration. For example, the melting temperature–concentration relations for the dissolution in tetralin of finely divided samples of polyethylene, originally crystallized in the bulk at high temperatures, are given in Fig. 3.7.(11) These observations are indicated by the filled circles in the plot. These data points fall on the same smooth curve as the melting temperatures obtained dilatometrically at the higher polymer concentrations when crystallization takes place from the binary mixture. The melting points for this latter procedure are indicated by the open circles. The melting temperature–composition relation, from pure polymer to dilute solution, can be represented by a continuous function. However, when the crystallization takes place within the polymer–diluent mixtures, prior to determining the melting temperature, complications can develop. When limited to concentrated solutions, i.e. $v_2 \geq 0.30$, no serious difficulty presents itself. At lower polymer concentrations, however, different melting temperatures (or solubility points) are found depending on the procedure. In this range polymers crystallized from solution invariably display a lower melting temperature. Melting temperatures determined by the latter method for polyethylene–tetralin mixtures are given by the open squares of Fig. 3.7. These differences become progressively more pronounced with dilution. For the very dilute solutions, about a 12 °C difference in the melting temperature is observed. A similar difference in melting temperature is observed when extended and folded chain crystallites of linear polymers are compared in dilute solution.(12) Since the liquid state is the same for the two cases, irrespective of the mode of crystallization, the disparity in the melting temperatures must reside in differences in the nature of the crystalline phases. The lower melting temperatures observed after crystallization from solution indicate that a metastable crystalline form is obtained. This metastability could in principle arise for a variety of reasons. However, thin plate-like crystals are the usual morphological form observed after crystallization from dilute solution. As the polymer concentration decreases the crystallite structure and size change and eventually reach those typical of crystallites formed in dilute solution. The thickness of crystallites formed from dilute solutions is much smaller than when crystallized from the pure melt. A detailed discussion of the structure and morphology of these and related structures will be presented in Volume 3.

For present purposes it suffices to take cognizance of these observations so that caution is exercised in analyzing the experimentally observed melting temperature–composition relations. The melting temperature of the most stable species is required at all concentrations. For purposes of determining ΔH_u the dilute range should be avoided unless bulk crystallized polymers are utilized.

The analysis that has been given is based on the Flory–Huggins expression for the free energy of mixing long chain molecules with low molecular weight species in the disordered melt. Besides the concentration restraints, there are other formal ways of expressing this mixing free energy. Among them is the principle of corresponding states.(13–19) The application of corresponding state theory in the present context is in the evaluation of the thermodynamic interaction parameters between polymer and diluent.(20) There are many examples where the Flory–Huggins interaction parameter χ_1 is not constant, but depends on concentration and temperature. The use of corresponding state theory alleviates many of these problems.(15) Since the determination of ΔH_u is dependent only on the initial slope of Eq. (3.13), or related expressions, its value is not affected.

In analyzing polyethylene–diluent mixtures it was noted that for certain diluents and concentrations the melting temperature remained invariant with composition. This is not an isolated observation. Besides linear polyethylene, this phenomenon has also been observed in long chain branched polyethylene,(21) poly(chlorotrifluoroethylene),(22) poly(N,N′-sebacoyl piperazene),(23) isotactic poly(propylene),(24) and poly(acrylonitrile)(25) when the polymers are admixed

with appropriate diluents. The invariance of the melting temperature in these diluent mixtures can be given a simple explanation. When this phenomenon occurs,
the molten state always consists of two immiscible liquid phases rather than a
homogeneous one. Therefore, three phases coexist in equilibrium at the melting
temperature. It then follows, as a consequence of the Phase Rule, that the melting
temperature must be invariant with composition since the system has no degrees of
freedom. At the melting temperature, therefore, crystalline polymers obey one of
the fundamental tenets of phase equilibria.

Typical phase diagrams that illustrate these principles are presented in Fig. 3.8
for isotactic poly(propylene)–alky phenol mixtures (24) and in Fig. 3.9 for poly
(N,N′-sebacoyl piperazine) (23) with various diluents. At the higher isotactic poly
(propylene) compositions only liquid–solid curves are observed that result in
typical melting temperature depression. Although at lower polymer concentrations

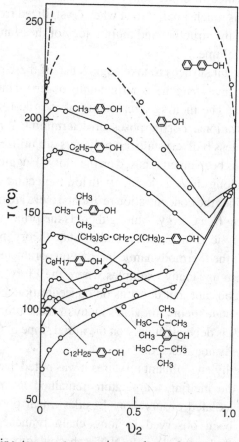

Fig. 3.8 Plot of melting temperature against volume fraction of polymer for isotactic
poly(propylene)–alkyl phenol mixtures. (From Nakajima and Fujiwara (24))

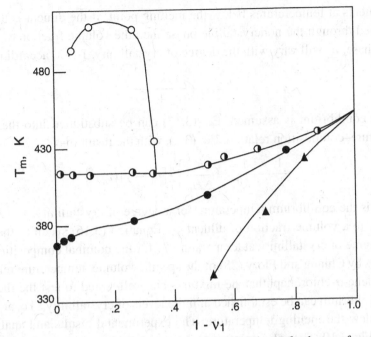

Fig. 3.9 Plot of melting temperatures against volume fractions of polymers for mixtures of poly(N,N′ sebacoyl piperazine) with different diluents. Diphenyl ether ◐, ○; *o*-nitrotoluene ●; *m*-cresol ▲.(23)

the liquid–solid curve is monotonic with dilution, many of the mixtures undergo liquid–liquid phase separation, as is indicated by the binodials. The expected invariance in melting temperatures is then observed. The phase diagrams for the poly(amide), shown in Fig. 3.9 illustrate the specific role of the diluent. As the polymer–diluent interaction become less favorable eventually liquid–liquid phase separation occurs and, as is illustrated, at the same time the melting temperature becomes constant. These types of phase diagrams can also be calculated from first principles.(26) What at first glance appears to be a surprising and puzzling observation can, however, receive a straightforward explanation based on the consequences of phase equilibrium.

An interesting situation exists when the melting temperature–composition curve lies above the binodial for liquid–liquid phase separation. On an equilibrium basis the phase boundaries do not intersect. However, depending on kinetic factors it is often possible to supercool the homogeneous melt into the two-phase region without crystallization intervening. Under these circumstances crystallization can occur in the heterogeneous melt. In this nonequilibrium situation the melting temperature is not invariant with composition.(27)

The equilibrium conditions that have been discussed involved the relation of the melting temperature to composition. It is also possible to test the equilibrium requirements at temperatures below the melting point. If the diluent is uniformly distributed through the noncrystalline phase then the volume fraction of polymer in this phase, v_2' will vary with the degree of crystallinity, $1 - \lambda$, according to

$$v_2' = \frac{v_2\lambda}{1 - v_2 + v_2\lambda} \tag{3.15}$$

If phase equilibrium is assumed, Eq. (3.15) can be substituted into the melting temperature–composition relation, Eq. (3.2), with the result that

$$\frac{1}{T_\lambda} - \frac{1}{T_m^0} = \frac{R}{\Delta H_u} \frac{V_u}{V_1}[v_1' - \chi_1(v_1')^2] \tag{3.16}$$

Here T_λ is the equilibrium temperature for a degree of crystallinity, $1 - \lambda$, corresponding to a volume fraction of diluent v_1'. Equation (3.16) specifies the equilibrium degree of crystallinity at temperature T_λ for the nominal composition v_2.

Studies by Chiang and Flory (28) of the specific volume–temperature relation of polyethylene-α-chloronaphthalene mixtures are well-suited to test the thesis that phase equilibrium can be established at finite levels of crystallinity,[1] i.e. at temperatures below the melting temperature. The experimental results and analysis are given in Fig. 3.10. The solid points in this figure represent the experimental observations, while the dashed lines are calculated from Eq. (3.16) using values for ΔH_u and χ_1 appropriate to linear polyethylene and the diluent. The agreement between the theoretical expectation and the experimental observation is excellent over the composition range studied, zero to about 50% crystallinity. Slight deviations between theory and experiment were observed for mixtures that contained smaller amounts of diluent. These differences can be attributed to the enhanced difficulties of establishing equilibrium. This formal thermodynamic analysis demonstrates conclusively that equilibrium can be established between the two distinct phases even when appreciable levels of crystallinity are developed. One of these phases has the thermodynamic properties of the liquid mixture (at the appropriate composition) while the other has that of the pure crystalline polymer.

The analysis of the melting temperature–composition relations so far has been limited to the most common case where the crystalline phase remains pure. The situation, similar to that found in low molecular weight systems, where the second component enters the lattice, is also observed with polymers. As an example, both structural and thermodynamic evidence indicates that diluents can enter the lattice of poly(2,6-dimethyl-1,4-phenylene oxide).(29,30) Although the occurrence of a mixed crystalline phase is rare for synthetic polymers it is quite common among macromolecular systems of biological interest.

---

[1] These experiments were conducted under carefully controlled crystallization conditions. For this system the specific volume is easily converted to the degree of crystallinity.

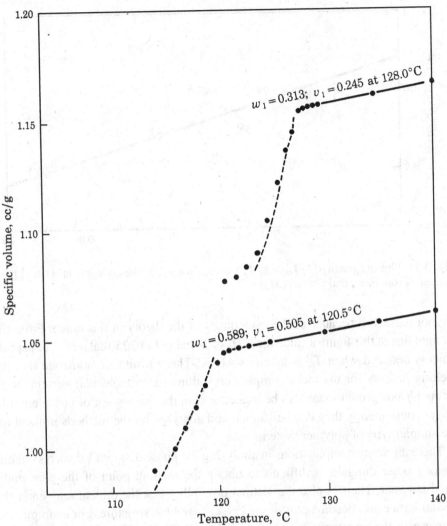

Fig. 3.10 Relationship of specific volume to temperature for mixtures of polyethylene ($M = 50\,000$) with α-chloronaphthalene for the indicated weight (w_1) and volume (v_1) fractions. Dashed lines represent calculations according to Eq. (3.16), assuming equilibrium between the crystalline and liquid phases. (From Chiang and Flory (28))

Melting point depressions by diluent have been successfully employed by Flory and Garrett (31) in studying the thermodynamics of the crystal–liquid transformation of the fibrous protein collagen. By means of sensitive dilatometric techniques, the melting temperatures of collagen (from rat-tail tendon and beef Achilles tendon)–anhydrous ethylene glycol mixtures were determined over a wide composition range. As illustrated in Fig. 3.11, the melting temperature–concentration relations are apparently in accord with Eq. (3.13). Data points in the dilute range

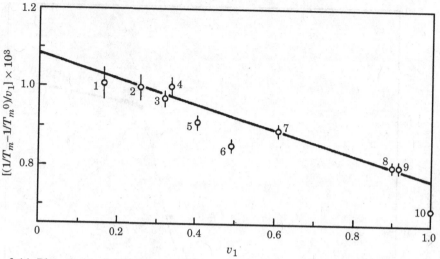

Fig. 3.11 Plot of quantity $(1/T_m - 1/T_m^0)/v_1$ against v_1 for the collagen–ethylene glycol system. (From Flory and Garrett (31))

are not preserved because of the inadequacy of the theory in this range. From the straight line of this figure a value for ΔH_u of $24\,\mathrm{cal\,g^{-1}}$ or $2250\,\mathrm{cal\,mol^{-1}}$ of peptide units is deduced when T_m^0 is taken as $418\,\mathrm{K}$. These results demonstrate that the melting process for the more complex crystalline macromolecular systems such as the fibrous proteins can also be treated within the framework of phase equilibrium. Consequently they can be studied and analyzed by the methods utilized for the simpler type of polymer systems.

There are some complications in analyzing the phase diagram for such systems. Among other things it is difficult to obtain the melting point of the pure undiluted polymer. There is also the distinct possibility that the diluent may enter the crystal lattice and become part of the crystallographic structure. For example, the increase in the equatorial x-ray spacing is indicative that the diluent (water) is entering the ordered phase of collagen and also the nucleic acids.(32,33) Under these circumstances, the conditions for equilibrium stipulated by Eq. (3.6) are no longer sufficient, and an additional condition must be fulfilled. Specifically, it is now required that

$$\mu_1^1 = \mu_1^c \tag{3.17}$$

Moreover, if there are any interactions or mixing of polymer units and diluent molecules in the crystalline phase Eq. (3.17) may no longer be satisfactory. The simplest example of this type would be the formation of a solid solution in the crystalline state. Hence, when dealing with the two-phase equilibrium of a two-component system, where both components are present in each phase, the activity as a function of the composition of each component in each of the phases must be specified in order to arrive at the melting temperature–composition relation. For these situations, analysis of the experimental observations is more complex than is indicated by Eq. (3.2) and is not easily generalized.

Certain simplifying assumptions were made in analyzing the results for collagen.(31) It was assumed that a fixed amount of diluent (independent of total composition) is firmly bound to the protein while the remainder is loosely held. Therefore, at the melting temperature the latter can be relegated to the amorphous region. With this assumption, the chemical potential of the polymer unit in the crystalline phase will be constant, independent of the total composition. Under these conditions Eq. (3.2) will again stipulate the requirements for equilibrium. However, T_m^0 will not represent the melting point of the pure undiluted polymer but that of the polymer–diluent complex. Hence it is not independent of the nature of the added second component. Utilization of this approximate procedure must necessarily lead to a greater uncertainty in the deduced value of ΔH_u than is usually expected when the aforementioned complications do not exist.

In the binary mixtures of interest, it is possible that besides the polymer the diluent component can also crystallize in the temperature range of interest. If this happens a classical eutectic type phase diagram results. Theoretical analysis, and possible phase diagrams involving polymers that contain eutectics have been presented.(26,34) An experimental example is given in Fig. 3.12 for mixtures of linear polyethylene and 1,2,4,5-tetrachlorobenzene.(35,36) This is a textbook type phase diagram. The eutectic composition is 55% w/w of polyethylene and the temperature is 120 °C. Similar phase diagrams have been reported for other polyethylene mixtures,(37) isotactic poly(propylene),(38) poly(ethylene oxide)(39,40) and poly(ε-caprolactone).(41) In all cases the diagrams are classical ones, but have interesting morphological implications.

3.3 Crystallization from dilute solution: flexible chains

The melting, or dissolution, of long chain molecules at high dilution is a natural consequence of phase equilibrium. The dissolution process results in the separation of the solute molecules and is usually accompanied by a change in the molecular conformation of the chain from an ordered structure to a statistical coil. However, it is also possible for the individual polymer molecules to maintain the conformation in solution that is typical of the crystalline state. This is particularly true if the steric requirements that favor the perpetuation of a preferred bond orientation or the ordered crystalline structure can be maintained by intramolecular bonding, such as hydrogen bonds. Further alterations in the thermodynamic environment can cause a structural transformation in the individual molecules. Each molecule is then

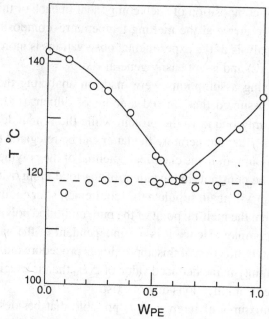

Fig. 3.12 Phase diagram of the binary mixture of polyethylene with 1,2,4,5-tetrachloroben-zene. (From Smith and Pennings (36))

converted to a conformation typical of the disordered state. This intramolecular transformation has been popularly termed the helix–coil transition because of the two different chain conformations that are involved in dilute solution. There are, therefore, two distinctly different situations that need to be considered in treating dilute solutions. We discuss first the case where the dissolution involves the direct change in the chain conformation from that in the ordered crystalline state to that in the disordered, or liquid, state.

Melting in dilute solution is not a simple or obvious extension of the analysis of the more concentrated system. The reason is that a homogeneous dilute solution is characterized by a nonuniform polymer segment distribution throughout the medium. In the analysis that has been given heretofore, the tacit assumption has been made that the polymer segments are uniformly distributed through the solution as is characteristic of concentrated systems. Under these circumstances, the use of the Flory–Huggins expression for the free energy of mixing, and the derived chemical potentials, is appropriate. However, since a dilute solution of flexible chain molecules is characterized by a nonuniform polymer segment distribution through the medium, the use of the Flory–Huggins free energy function is no longer appropriate.(1) θ solvents provide an exception, since under these circumstances the molecules can interpenetrate one another freely.

In order to develop a theory for crystallization from dilute solution it is necessary to express the chemical potential of the polymer species in the disordered state. In the general theory of dilute solutions the chemical potential of the solvent species can be expressed in virial form. Consequently, the chemical potential of the solvent species can be given quite generally as(1)

$$\mu_1 - \mu_1^0 = -\frac{RTv_1}{M}\left[\frac{v_2}{\bar{v}} + \frac{\Gamma_2 v_2^2}{\bar{v}^2} + \frac{g\Gamma_2^2 v_2^3}{\bar{v}^3}\right] \tag{3.18}$$

Here $\bar{v}$ is the partial specific volume of the polymer. In the development that follows, we utilize the theory of Flory and Krigbaum.(42) The second virial coefficient can then be expressed as

$$\Gamma_2 = x\bar{v}(1/2 - \chi_1)\,F(X) \equiv MA_2 \tag{3.19}$$

The complicated function $F(X)$ has been explicitly formulated,(42) and x again represents the number of segments per molecule. The final result does not depend on the specific form that is used for the second virial coefficient. We use the Flory–Krigbaum formulation here for convenience. This formulation explicitly accounts for the nonuniform nature of the solution. The chemical potential of the polymer obtained by the application of the Gibbs–Duhem relation also maintains the nonuniform segment distribution, statements to the contrary not withstanding.(42a) By applying the Gibbs–Duhem equation it is found that

$$\frac{\mu_2 - \mu_2^0}{RT} = (\ln v_2 - v_2 + 1) + x(\chi_1 - 1) + 2x(1/2 - \chi_1)\,F(X)v_2$$
$$+ \left[\frac{3}{2}gx\chi_1^2(1/2 - \chi_1)^2\,F^2(X) - x(1/2 - \chi_1)\,F(X)\right]v_2^2 \tag{3.20}$$

The value of the integration constant is obtained from the lattice theory as $v_2 \to 0$. After dividing by the number of units per molecule, xV_1/V_u, Eq. (3.20) can be recast as

$$\frac{\mu_u - \mu_u^0}{RT} = -\frac{V_u}{V_1}\left\{\frac{-\ln v_2}{x} + (1 - 1/x)(1 - v_2) - \chi_1(1 - v_2)^2\right.$$
$$- 2(\chi_1 - 1/2)[1 - F(X)]v_2 \tag{3.21}$$
$$\left. - \left[\frac{3g}{2}(\chi_1 - 1/2)^2 x\,F(X) - (\chi_1 - 1/2)F(X) - \chi_1\right]v_2^2\right\}$$

Here g is a constant (less than unity) that relates the third virial coefficient to the second. By invoking the equilibrium requirement between the liquid and crystalline states, the expression for the melting temperature of flexible chains in dilute solution is given by

$$\frac{1}{T_m} - \frac{1}{T_m^0} = \frac{RV_u}{\Delta H_u V_1}(1 - 1/x)(1 - v_2) - \frac{\ln v_2}{x}$$
$$- \chi_1(1 - v_2)^2 - 2(\chi_1 - 1/2)(1 - F(X)v_2) \tag{3.22}$$
$$- \left[\left(\frac{3g}{2}\right)(\chi_1 - 1/2)^2 F(X)^2 - \chi_1 - 1/2F(X) - \chi_1\right]v_2^2$$

For most cases of interest, the available experimental data indicate that only the first three terms on the right-hand side of Eq. (3.22) are important. Equation (3.22) thus reduces to

$$\frac{1}{T_m} - \frac{1}{T_m^0} = \frac{R}{\Delta H_u}\frac{V_u}{V_1}\left[-\frac{\ln v_2}{x} + (1 - 1/x)v_1 - \chi_1 v_1^2\right] \tag{3.23}$$

Surprisingly Eq. (3.23) is the same expression as given earlier, Eq. (3.9), for more concentrated systems. Thus, for practical purposes the second and higher virial coefficients should have a negligible influence on the melting point depression.

We conclude, therefore, that although the Flory–Huggins lattice treatment is clearly not adequate in dilute solution, the change in chemical potential of the polymer species with dilution is too small to have any appreciable effect on the melting temperature. Hence, the same equation can be used over the complete concentration range. Therefore, the melting point–composition relation, or solubility relations, can be expressed as a continuous function that encompasses the complete composition range. As $x \to \infty$ Eq. (3.23) reduces to Eq. (3.2). Deviation from the limiting form of Eq. (3.2) would only be expected at extremely high dilutions and low molecular weights.

A similar approach to the problem has been given by Beech and Booth (43) who did not, however, have occasion to actually calculate the melting point–composition relation. Pennings,(44) following a similar procedure, did not specifically arrive at Eqs. (3.22) and (3.23) because the higher virial coefficients were expressed in a different manner. It is not necessary to adapt empirical methods to this problem,(45) since the melting temperature–composition relation can be obtained analytically.

To examine the influence of molecular weight in the dilute region it is convenient to recast Eq. (3.23) into the form

$$\frac{1}{T_m} - \frac{1}{T_m^0} = \frac{R}{\Delta H_u}\frac{V_u}{V_1}[v_1 - \chi_1 v_1^2] - \frac{R}{\Delta H_u}\left[\frac{\ln v_2}{y} + \frac{v_1}{y}\right] \tag{3.24}$$

Here, y is the number of repeating units per molecule as opposed to x, the number of segments. The term within the first bracket represents the limiting form and is

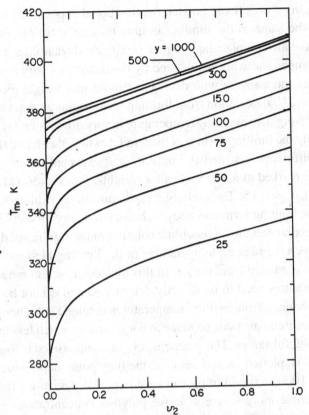

Fig. 3.13 Plot of theoretically expected melting temperature calculated from Eq. (3.24) for indicated chain lengths.(46)

independent of molecular weight. The second bracketed term represents the influence of chain length. It is only important at low molecular weights at high dilution. For lower molecular weights we identify T_m^0 with the melting temperature of the pure species of finite molecular weight. Since Eq. (3.24) represents equilibrium it is only applicable to extended chain crystals. Within the present crystallization capabilities, the analysis of experimental data is, therefore, restricted to low molecular weights.

A graphical representation of Eq. (3.24) is given in Fig. 3.13 for a model system in the low molecular weight range.(46) The parameters used are applicable to polyethylene. For convenience in these illustrations χ_1 was taken to be zero and the melting temperatures of the extended chains of pure polymers were determined experimentally.(46) The vertical displacement of the melting temperatures along the $v_2 = 1$ axis represents their molecular weight dependence. The melting temperature differences are maintained over the complete composition range. However, the shapes of the curves in the dilute range are different depending on the chain length.

For chain lengths $y = 500$ and greater, the functional dependence on composition is essentially the same as the limiting infinite molecular weight form, Eq. (3.2). However, for the lower molecular weights, significant deviations are to be expected in the dilute range. These are manifested by the downward curvature of the plots, which becomes more severe with decreasing molecular weight. For example for $y = 150$ ($M = 2100$) deviation from limiting form begins at about $v_2 = 0.07$. For this molecular weight, the melting temperature in very dilute solution is predicted to be 363 K if only the limiting form is considered. On the other hand, theory predicts 357 K. This difference between the limiting and expected melting temperatures becomes more marked at lower molecular weights. For $y = 50$ ($M = 700$) deviations begin at $v_2 = 0.15$. The melting temperature in very dilute solution would be 333 K. If the limiting form was obeyed, however, theory predicts 310 K. Hence, for lower molecular weights in the dilute solution range significant differences are expected between the melting temperatures of the limiting high molecular weight form and those predicted from theory. In this molecular weight range equilibrium melting temperatures need to be directly determined and cannot be extrapolated, in any simple manner, from melting temperatures obtained at higher polymer concentration. Measurements must be made at compositions much less than $v_2 = 0.05$ to obtain meaningful values. These conclusions are emphasized in Fig. 3.14, where the same data is plotted as the relative melting point depression. Taking the curve for $y = 1000$ as the limiting form we note that deviations from this reference curve become more severe at lower polymer concentrations as the molecular weight decreases. Simple linear extrapolations cannot be made to infinite (low concentration) from more concentrated systems.

The theoretical expectations from Eq. (3.24) can be tested experimentally by the study of melting point–composition relations of low molecular weight species that form extended chain crystals. Appropriate data are available for polyethylene–*p*-xylene mixtures for molecular weight fractions in the range 574 to 2900.(46) Extended chain crystallites are formed in this molecular weight range so that the experimental data are suited to test the theory. The results are summarized in Fig. 3.15. In this figure the melting temperatures of the pure extended chain species were determined experimentally. The solid lines represent the theory with χ_1 being taken as 0.2. This figure makes clear that except for the two highest molecular weights excellent agreement is obtained between theory and experiment over the complete composition range. The expectation that deviations of the experimental data from the limiting case should become more severe as the molecular weight and polymer concentration decrease is confirmed. These deviations can be quite substantial for the lower molecular weights. The two highest molecular weight fractions, $M = 1674$ and $M = 2900$ show the expected behavior in that the limiting law is being approached. Only minor deviations begin to appear at very low polymer

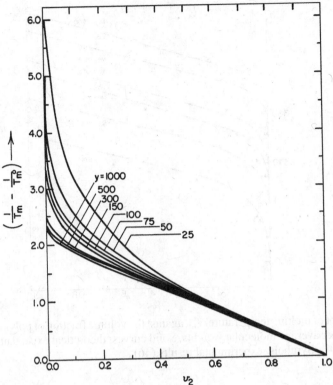

Fig. 3.14 Plot of $1/T_m - 1/T_m^0$, calculated from Eq. (3.24), as a function of polymer volume fraction, v_2, for indicated chain lengths.(46)

concentrations. Quantitatively, although the theoretical and observed melting temperatures agree quite well in the concentrated range, $v_2 \geq 0.2$, the observed values are always slightly higher than calculated in the more dilute region. These small discrepancies could be attributed to molecular weight uncertainties (which would not generally affect the concentrated region) and to variations of χ_1 with temperature and composition. Despite these small differences, the major conclusions remain that the theory as embodied in Eq. (3.24) quantitatively explains the equilibrium data for low molecular weight polymers that are available for analysis.

Qualitatively similar results have been obtained in a study of the dissolution temperatures of a set of high molecular weight n-alkanes crystallized in extended form.(42a) The dissolution temperatures of the higher molecular weight alkanes studied, $C_{198}H_{398}$ ($M = 3170$), are very similar to those of the polymer ($M = 2900$) shown in Fig. 3.15. The precipitous drop in the dilute range is not observed in either case. In contrast, the melting temperatures of the lowest alkanes studied, $C_{102}H_{206}$ ($M = 1634$) show the characteristic decrease, similar to that shown in Fig. 3.15. The dependence of the melting temperature on concentration is similar

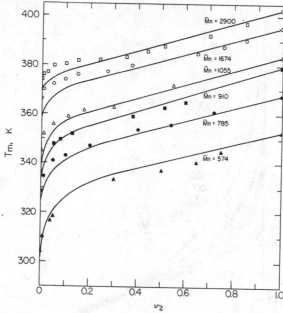

Fig. 3.15 Plot of melting temperature, T_m, against the volume fraction of polymer, v_2, for indicated number average molecular weights. Solid curves: theoretical expectants, calculated from Eq. (3.24). Symbols: experimental results.(46)

for both low molecular weight polyethylene fractions and the high molecular weight *n*-alkanes. A more quantitative comparison requires a specification of the chemical potential of the latter in solution.

The melting point–composition relation, Eq. (3.13), is in effect an expression of the temperature limit of the solubility of a crystalline polymer in a given solvent. Theoretically and experimentally, at a given concentration, the solubility temperature is not very sensitive to molecular weight, except for the very low molecular weight species. Therefore the crystallization of a polymer from a dilute solution cannot provide a very effective method of molecular weight fractionation even if equilibrium solubility conditions are achieved. It is more likely, however, that as the liquid phase is cooled, the crystallization of the polymer will be governed primarily by kinetic factors. In fact, it turns out that by taking advantage of the difference in crystallization rates from dilute solution an effective separation of molecular weight species can be achieved.

When the complete composition range, from pure polymer to very dilute solution, is studied with a good solvent as the added component, the range in melting temperatures can be quite large. The tacit assumption that the ratio of ΔS_u to ΔH_u is independent of temperature can then be seriously questioned. The entropy contribution to χ_1 must now also be taken into account. Although Eq. (3.6) still formally stipulates the equilibrium requirement, the appropriate thermodynamic quantities may no longer remain constant. An analysis more appropriate to this specific situation needs to be given.

The free energy change accompanying the process of dissolving n_2 moles of crystalline polymer in n_1 moles of solvent can be expressed as(46a)

$$\Delta G = RT\left(n_2 x \frac{V_1}{V_u} g - \ln Q_m\right) \tag{3.25}$$

Here RTg represents the free energy change per repeating unit that is associated with the disruption of the crystalline structure. Q_m is the partition function of the disordered mixture and is given by

$$Q_m = \frac{q_1^{n_2} q_2^{n_2}\left[\epsilon(\epsilon - 1)^{n_2}(n_1 + xn_2)!(n_1 + xn_2)^{-n_2(x-1)} Z^{n_2}\right]}{n_1! n_2!} \exp\left\{\frac{-\chi_{ix} n_1 n_2}{n_1 + xn_2}\right\} \tag{3.26}$$

In this expression ϵ is the lattice coordination number, q_1 and q_2 are the internal partition functions and Z is the chain configurational partition function. For high molecular weights, Z can be expressed quite generally as(47)

$$Z = z^{m-1} \tag{3.27}$$

where z is the bond rotational partition function and m is the number of rotatable bonds per chain.[2]

At equilibrium the difference in chemical potential between the dissolved and crystalline polymer must be zero. By differentiating Eq. (3.25) to obtain this difference in chemical potential, and setting the resultant equal to zero, the requirement of phase equilibrium leads, in the limit of infinite molecular weight, to

$$\left(\frac{V_u}{V_1}\right)[v_2 + \chi_1(1 - v_2)^2] + g = \beta \ln z \tag{3.28}$$

Here β is defined as the ratio m/x', where x' is the number of repeating units per molecule. Equation (3.28) represents the condition for phase equilibrium. In the limit of infinite dilution

$$\left[\frac{(V_u/V_1)\psi_1\theta + b}{T}\right] = \beta \ln z - \left[\left(\frac{V_u}{V_1}\right)(1/2 - \psi_1) - a\right] \tag{3.29}$$

where g has been resolved into its entropic and enthalpic components so that

$$g = -a + b/t \tag{3.30}$$

[2] For present purposes only the form of Eq. (3.27) is required. Detailed formulation of Z can be found in Ref. (47) For chains with independent rotational potentials Z is equal to the bond rotational partition function. For high molecular weight, with interdependent rotational potential, Z is the largest eigenvalue of the statistical weight matrix describing this interdependence.

and

$$\frac{1}{2} - \chi_1 = \psi_1 - \kappa_1 = \psi_1(1 - \theta/T) \tag{3.31}$$

Here κ_1 and ψ_1 are the conventional entropy and enthalpy parameters and θ is the Flory temperature for the polymer–diluent mixtures. We next examine Eq. (3.29) for physically meaningful solutions of T ($T > 0$), under the assumption that the quantities ψ, θ, a and b are independent of temperature.

Two cases can be distinguished, depending upon whether the quantity $(V_u/V_1)\psi_1\theta + b$ is positive or negative. The quantity z monotonically decreases with $1/T$, varying from a maximum value at $1/T = 0$ to an asymptotic limit of unity as $1/T \to \infty$. Thus, when $(V_u/V_1)\psi_1\theta + b$ is positive, a single solution for T is possible only if $[(V_u/V_1)(1/2 - \psi_1) - a] \leq \beta \ln z_{max}$. This case corresponds to conventional dissolution or melting. For the other situation of interest, where $(V_u/V_1)\psi_1\theta + b$ is negative, two possibilities exist. For this condition a single solution is obtained if $[(V_u/V_1)(1/2 - \psi_1) - a] > \beta \ln z_{max}$, which again corresponds to conventional solubility. However, if $(V_u/V_1)\psi_1\theta + b$ is still negative but if $[(V_u/V_1)(1/2 - \psi_1) - a] \leq \beta \ln z_{max}$ there is the possibility for two real solutions of T. If these two solutions exist, the lowest temperature will represent one of inverted solubility. Although not common, cases of inverted solubility, or melting, have been reported. The first condition for inverted solubility

$$(V_u/V_1)\psi_1\theta + b < 0 \tag{3.32}$$

will invariably involve a negative value for ψ_1. Although the value of the parameter b is not generally known, it would be expected to be positive and small since it represents the intermolecular contribution to the enthalpy of fusion. The other requirement that needs to be satisfied for inverted solubility is

$$\left(\frac{V_u}{V_1}\right)\left(\frac{1}{2} - \psi_1\right) \leq \beta \ln z_{max} + a \tag{3.33}$$

and focuses attention on the role of the chain conformation. To quantitatively analyze inverted solubility a large set of parameters, ψ_1, θ, a, b and z_{max} need to be independently determined. Despite these formidable obstacles many of the observed cases of inverted solubility involving crystalline polymers, such as cellulose nitrate in ethanol(48), poly(L-proline) in water(49–51), and poly(ethylene oxide) in water(52,52a), to cite but a few examples, follow the principles outlined.

3.4 Helix–coil transition

Although dispersed polymer chains usually adopt the random coil configuration in dilute solution, there are some important exceptions. These exceptions occur primarily, but not solely, among macromolecules of biological interest. Under certain circumstances an ordered structure is maintained by the isolated molecule so that a highly asymmetrical geometric shape is found in solution. The length of the molecule can be several orders of magnitude greater than its breadth. The structure is, therefore, quite different from that of the random chain. Physical-chemical measurements have established that many synthetic polypeptides are capable of existing as independent alpha-helices at high dilution in appropriate solvent media.(53,54,55) Similarly, the ordered structures of the polynucleotides, among which are compound helices comprised of two or three interwoven polymer chains, can also be maintained in dilute solution.(56–59) The solubilization of the naturally occurring nucleic acids, as well as many fibrous proteins, with the preservation of the molecular organization has been demonstrated. For example, the dissolution of the fibrous protein collagen can be accomplished, with the characteristic ordered structure of the collagen protofibril being preserved.(60,61) In these examples, the preservation and stability of the ordered structure can be attributed to the action of specific secondary bonding. For the alpha-helical structures intramolecular hydrogen bonds between peptide groups along the main chain are also involved. Interchain hydrogen bonding is also involved in the compound helical structures. The ordered helical structure of the two-stranded polynucleotides derives its stability from interchain base pairing and from the base stacking along the chain.

Specific inter- and intramolecular bonding are not necessary for ordered structures to persist in dilute solution. Ordered structures, that lead to highly asymmetric molecules, can be perpetuated by severe steric repulsions of substituents or an inherent restraint to rotations about single bonds. Such structures are known, even among synthetic macromolecules, and they form liquid-crystal systems. Some examples are polymeric aramides, poly(N-alkyl isocyanates) and some cellulose derivatives.

When individual, isolated molecules exist in helical, or other ordered forms, environmental changes, either in the temperature or solvent composition, can disrupt the ordered structure and transform the chain to a statistical coil. This conformational change takes place within a small range of an intensive thermodynamic variable and is indicative of a highly cooperative process. This reversible intramolecular order–disorder transformation is popularly called the helix–coil transition. It is an elementary, one-dimensional, manifestation of polymer melting and crystallization.

Many examples of this type of transformation are available in the literature, particularly among polymers of biological interest. One example is the polypeptide poly-L-glutamic acid that exists as a coiled molecule in dilute neutral or alkaline solutions. However, when the pH is lowered below about 5.0 the ordered alpha-helical form is generated. This molecular transformation results in large changes in

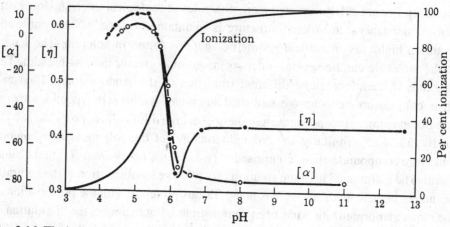

Fig. 3.16 The helix–coil transition in poly-L-glutamic acid as it is affected by the variation in pH. (From Doty (55))

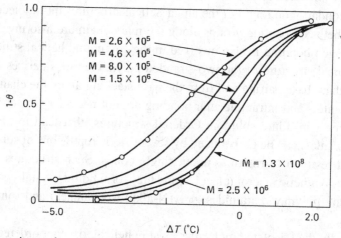

Fig. 3.17 Experimental transition curves for T2 DNA after varying amounts of shear degradation, showing the fraction of bases unbonded as a function of temperature. The molecular weight M is indicated for each curve. (From Crothers, Kallenbach and Zimm (62))

many physical-chemical properties as is illustrated in Fig. 3.16.(55) As the charge on the molecule is decreased a large increase in the magnitude of the optical rotation and in the intrinsic viscosity takes place. These changes occur within a very small pH range and are indicative of the cooperative nature of the transformation.

An example of a similar type of cooperative transformation is shown in Fig. 3.17 for different molecular weight nucleic acids (obtained by shear degradation) from T2 bacteriophage.(62) This temperature induced transformation is quite clear in the figure. At low temperature all the appropriate bases are bonded, one to another; at high temperature, in the disordered state, the bases are no longer bonded. The transition is, however, relatively broad. The termination is not sharp, or as well-defined, as in the melting of pure polymer systems where the transition is three-dimensional. The collapse of the ordered structure of multi-chain molecules in dilute solution results concomitantly in the separation of the individual chains, each in coiled form. Under careful experimental conditions regeneration of the native ordered structure can be accomplished even for such complex systems as collagen(63) and deoxyribonucleic acid.(64)

This one-dimensional intramolecular structural transition, the helix–coil transition, has received extensive theoretical treatment by many investigators.(65–75) Although a variety of models and mathematical techniques have been brought to bear on this problem the basic conclusions have been essentially the same. The methods involved, and the results, have been eloquently summarized in the treatise by Poland and Scheraga.(76) As an example, we will outline the theoretical basis for the transformation in dilute solution of an isolated polypeptide chain from the alpha-helical to the coil form.

The existence of the ordered alpha-helical structure can be attributed to the stability given to the molecule by the intramolecular hydrogen bonds between neighboring units. Specifically, according to Pauling and Corey,(77) the hydrogen atom of each main chain amide group forms a hydrogen bond with the oxygen atom of the third preceding group. Hence the bond orientation of successive units is dependent upon one another and there is a tendency for this conformation to be sustained along the chain. Schellman (65) has pointed out that the stereochemistry of an alpha-helix requires that three successive hydrogen bonds, involving three peptide units, be severed in order that one repeating unit of the ordered structure be disrupted. The necessary fulfillment of this condition, in order for a repeating unit to gain the conformational freedom of the random coil state, is the basis for the cooperative nature of the transformation. Once the enthalpy has been expended for the realization of the greater entropy of the random coil state, the latter conformation is favored. The transformation from the helix to the coil should thus be relatively abrupt, with changes in such intensive variables as either temperature, pressure, or composition.

The simplest, but very illuminating quantitative formulation of this problem is due to Schellman.(65) It is assumed that the individual molecules exist completely in either the helical (H) or the coil (C) form. The juxtaposition of both conformations in the same chain is not allowed in this model. For the equilibrium process

$$H \rightleftharpoons C \qquad (3.34)$$

$$\frac{(C)}{(H)} = K = \exp \frac{-\Delta G_t}{RT} \qquad (3.35)$$

Here ΔG_t is the difference in free energy per molecule between the ordered and random structures. For a sufficiently high molecular weight, so that the influence of the terminal residues can be neglected, $\Delta G_t = x\Delta G_u$, where ΔG_u is the change in free energy per repeating unit, there being x repeating units per molecule. When ΔG_u is zero, the concentration of molecules in each of the forms is identical. Since the number of units per molecule x is assumed to be large, a small change in ΔG_u, in the vicinity of $\Delta G_u = 0$, can cause the ratio of (C)/(H) to change dramatically. The development of one structure at the expense of the other, with the alteration of an independent variable, could be sufficiently sharp as to resemble a phase transition. Since x is large, the enthalpy change for the molecule as a whole will be large. Consequently the equilibrium constant K must change very rapidly with temperature. A measure of the breadth of the transition is the rate of change with temperature of the fraction of the molecules in randomly coiled form. At the transition temperature, $T = T_t$, $\Delta G_t = 0$, and this rate of change can be expressed as(65)

$$\left(\frac{d\{(C)/[(C) + (H)]\}}{dT}\right)_{T=T_t} = \frac{x\Delta H_u}{4RT_t^2} \tag{3.36}$$

Since ΔH_u is estimated to be of the order of several kilocalories per mole, the transition is relatively sharp and appears to possess the characteristics of a first-order phase transition. However, only in the limit of pure polypeptide chains of infinite molecular weight is the transition infinitely sharp. If the molecular weight is not large, the range of the transition will be considerably broadened.

The formulation of the problem, as presented above, gives a good insight into the problem, and is based on the assumption that the individual molecules exist in either one or the other of the two possible conformations. For molecules of high molecular weight this is not a satisfactory hypothesis. Although the helical form clearly represents the state of lowest enthalpy, whereas the random coil represents the one with the greatest conformational entropy, intermediate chain structures comprised of alternating random coil and helical regions could represent the thermodynamically most stable configurational state, the one of minimum free energy.[3] Although the disruption of one conformational sequence and the initiation of the other is not strongly favored, neither is it completely suppressed. Therefore, the more general situation, where sequences of helical and coil structures are allowed to co-exist within the same chain needs to be analyzed. In doing so, we follow the method proposed by Flory.(75)

---

[3] The problem posed here differs fundamentally from that discussed in the previous chapter. In the present case a one-dimensional system is being treated. Previously, the problem involved a three-dimensional crystallite which required all crystalline sequences to terminate in the same place.

If within a very long chain there are ν helical sequences, the molecular partition function is given by

$$Z_{x_C, x_H, \nu} = z_C^{x_C} z_H^{x_H} \alpha^{\nu} \tag{3.37}$$

Here z represents the partition function of the residue in each form and x_C and x_H are the number of residues in each form respectively. The quantity α is the factor, much less than unity, by which the partition function is diminished for each helical sequence. This term must be included because it is more difficult (in terms of free energy) to initiate an alpha-helix than it is to perpetuate it. The term $-RT \ln \alpha$ is analogous to an interfacial free energy, so α is often referred to as the nucleation parameter. If the coil state is taken as the reference then

$$Z_{x_H, \nu} = S^{x_H} \alpha^{\nu} \tag{3.38}$$

where $S = \exp(\Delta G^0_{H \to C}/RT)$ with $\Delta G^0_{H \to C}$ being standard state free energy change per residue for the conversion of helix to coil. The complete partition function for the system is the sum of Eq. (3.38) over all possible combinations of x_H and ν. It can be evaluated by standard methods.[75] From the partition function the fraction of units helical, p_H, can be obtained as a function of S. The results, in the limit of an infinite chain, are shown graphically in Fig. 3.18 for different values of the parameter α.[75]

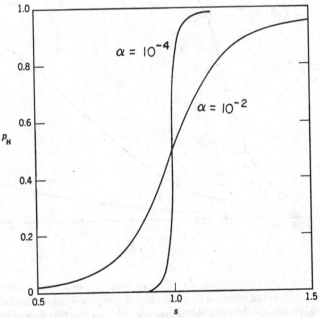

Fig. 3.18 Fraction p_H of helical units in the limit $x \to \infty$ calculated according to theory for two different values of the parameter α. (From Flory (75))

The point $S = 1$ denotes the mid-point of the transition, there being an equal number of repeating units in each of the two conformations. The transition is relatively broad for the larger value of α. However, it becomes sharper as α decreases. Only in the limit of $\alpha = 0$ does the transition actually become discontinuous. However, since α must exceed zero for any real chain the helix–coil transition is a continuous process. In this respect it differs from a true first-order phase transition. This conclusion is in accord with the general axiom enunciated by Landau and Lifshitz (78) that a one-dimensional transition must be continuous. The two phases must mix with one another to some extent. This characteristic of a one-dimensional system causes the transition to be diffuse and permits the co-existence of the two phases over a finite temperature range. Strictly interpreted, therefore, helix–coil transitions do not qualify as true phase transitions.

These theoretical expectations have been satisfied by experimental observations when proper theoretical account is taken of finite chain length. Another example of the coil→helix transition is given in Fig. 3.19 for poly-γ-benzyl glutamate in dichloroacetic acid–ethylene dichloride mixtures.(79) The transition in this case is

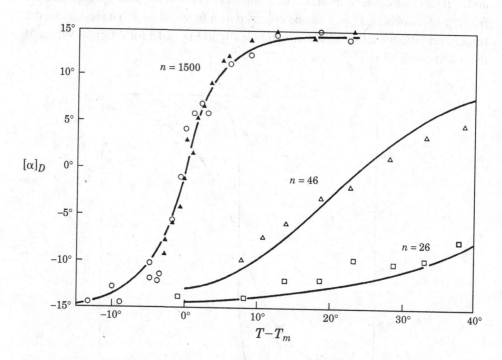

Fig. 3.19 Theoretical and experimental comparison of the helix–coil transition of poly-γ-benzyl-L-glutamate in dilute solution of an ethylene dichloride–dichloroacetic acid mixture. The experimental points are the optical rotation $[\alpha_D]$ plotted as a function of the temperature T minus the transition temperature T_m. The solid curves represent the best fit of theory for samples of various degrees of polymerization n. (From Zimm, Doty and Iso (79))

induced by varying the temperature. In the vicinity of the transition temperature, T_m, the quantity S can be approximated by

$$\Delta T = T - T_m = -(RT^2/\Delta H_u^0)(S - 1) \qquad (3.39)$$

The solid curves in the figure represent the best fit between experiment and theory for the three different molecular weight samples. Good agreement is obtained and the increased sharpness of the transition with increasing molecular weight is apparent. The best fit is obtained with $\alpha = 2 \times 10^{-4}$ and $\Delta H_u^0 = 900$ cal mol^{-1}.

The theoretical treatment of this transition for compound helices made up of more than one chain, such as are found in polynucleic acids, follows the principles that were outlined above. The details of the problem are more complicated since the helices derive their stability from interchain hydrogen bonding (base pairing) as well as the other interactions along the same chain. In addition, partial melting representing states of intermediate order produces loops of randomly coiled units which introduces mathematical complexities. The problem can, however, be treated adequately within the framework of the methods outlined above.

The helix–coil transition is unique in that the coordinated action of many molecules is not required. It is by necessity restricted to the very dilute portion of the phase diagram. As the concentration of polymer molecules in the helical conformation increases, intermolecular interactions begin to manifest themselves. The cooperative character of the transition will be further enhanced. The dimensional interdependence will increase from one in the case of a dilute solution to three in the more concentrated system. The transition will then become formally identical to the melting of the dense crystalline phase that has been discussed previously.

3.5 Transformations without change in molecular conformation

When a polymer molecule possesses an ordered structure, it is by necessity restricted to a unique conformation. A highly asymmetric, rodlike molecule results, characterized by a length many times greater than its breadth. Such a collection of molecules, wherein the individual species are uncorrelated and randomly arranged relative to one another, can exist as independent entities in a sufficiently dilute solution. However, such rodlike molecules of high axial ratio cannot be randomly arranged at high density because of space requirements; i.e. as the density of polymer is increased, sufficient volume is no longer available to allow for the maintenance of a disordered array. This qualitative concept leads to the conclusion that at high concentration a completely disordered or isotropic solution of asymmetrically shaped macromolecules is not possible. Hence either a change in molecular conformation must occur or the arrangement becomes more ordered as the polymer concentration is increased.

Asymmetry of molecular shape is a feature that is common to all substances that exhibit liquid crystallinity. The study of liquid crystals involving macromolecules is a major subject in itself. It has been discussed in several reviews (80–83) and books.(84–88) It is not the purpose here to discuss liquid crystals involving polymers in any detail. Rather, efforts will be directed to place the behavior of such highly asymmetric molecules, and the transitions that they undergo without any conformational change, in perspective in terms of polymer crystallization. Thus, the effort will be in outlining the theoretical basis for the behavior and highlighting the unique features that result.

There are several different theoretical approaches to the problem. The Landau molecular field theory was applied by de Gennes to liquid-crystal phase transitions.(89) The Maier–Saupe theory focuses attention on the role of intermolecular attractive forces.(90) Onsager's classical theory is based on the analysis of the second virial coefficient of very long rodlike particles.(91) This theory was the first to show that a solution of rigid, asymmetric molecules should separate into two phases above a critical concentration that depends on the axial ratio of the solute. One of these phases is isotropic, the other anisotropic. The phase separation is, according to this theory, solely a consequence of shape asymmetry. There is no need to involve the intervention of intermolecular attractive forces. Lattice methods are also well suited for treating solutions, and phase behavior, of asymmetric shaped molecules.(80,92,93)

The lattice method is used to enumerate the number of configurations available to n_2 rigid, rodlike polymer molecules, with an asymmetry x (the ratio of molecular length to its breadth) and partial orientation about an axis, and n_1 monomeric solvent molecules. When the usual Van Laar heat of mixing term is employed, the free energy of mixing can be expressed as (80,92)

$$\frac{\Delta G_m}{kT} = n_1 \ln v_1 + n_2 \ln v_2 - (n_1 + y n_2) \ln\left[1 - v_2\left(1 - \frac{y}{x}\right)\right]$$
$$- n_2[\ln(xy^2) - y + 1] + \chi_1^{x n_2 v_1} \tag{3.40}$$

where y is a parameter that is a measure, or index, of the disorientation of the molecules. This parameter can vary from unity, characteristic of a perfectly ordered array, to x typifying a state of complete disorder. When $y = 1$, Eq. (3.40) reduces to the free energy of mixing for a regular solution. When $y = x$ the result is essentially identical to that for the mixing of rigid polymer chains.(92) Thus for a fixed molecular asymmetry x, Eq. (3.40) is an expression for the free energy of mixing as a function of the composition and the disorientation index y.

When the composition and molecular asymmetry are kept constant, Eq. (3.40) goes through a minimum, and then a maximum, as the disorientation parameter y increases. Since there are no external restraints on the disorientation index, y

assumes the value that minimizes ΔG_m. By appropriate differentiation, it is found that for a given v_2 and x the value of y which fulfills this condition is the lesser of the two solutions to the equation

$$v_2 = \frac{x}{x-y}\left[1 - \exp\left(-\frac{2}{y}\right)\right] \tag{3.41}$$

If v_2^* is defined as the minimum concentration that allows for a solution of this equation, a necessary condition for the existence of an isotropic phase (a state of complete molecular disorder) is $v_2 < v_2^*$. It can be shown that(92)

$$v_2^* \simeq \frac{8}{x}\left(1 - \frac{2}{x}\right) \tag{3.42}$$

represents the maximum concentration allowable for the stable existence of an isotropic phase or the minimum concentration required for stable anisotropy (a state of partial equilibrium order of the molecules). Thus, for large x the maximum concentration at which the molecules can exist in random arrangement relative to one another is inversely related to the axial ratio. This conclusion depends only on the asymmetry of the molecules. It is reached without invoking the action of any intermolecular forces. In the absence of diluent ($v_2 = 1$), it is calculated that a length–diameter ratio of about $2e$ would be sufficient to cause spontaneous ordering of the phase.

From the free energy function given by Eq. (3.40), together with the equilibrium stipulation of Eq. (3.41), the chemical potentials of each of the components in the two phases can be calculated. The two phases are the isotropic one with $y = x$ and the phase where the molecules are in a state of equilibrium disorder (not completely disordered). The conditions for equilibrium between the two phases at constant temperature and pressure can then be established by equating the chemical potentials of each of the components in each phase. The expected phase diagram, calculated on the basis of the lattice theory, is shown in Fig. 3.20, for rodlike molecules that have axial ratio $x = 100$.(80,92) Here χ_1 is plotted as ordinate against the volume fractions of the co-existing phases. The ordinate can also be regarded as an inverse measure of the temperature. Some general features of this unusual phase diagram should be noted. At the low v_p values, all the mixtures are isotropic. For relatively small positive values, and all negative ones for χ_1 there is a narrow biphasic region that is often referred to as a biphasic chimney. This biphasic region encompasses only a relatively narrow composition range. There is only a small difference in composition between the two phases that are in equilibrium. Depending on the value of χ_1, as the polymer concentration increases, either a single anisotropic phase or a broad two-phase anisotropic region develops.

Examining this phase diagram in more detail we consider athermal mixing, i.e. where χ_1 is equal to zero. In this case there is no net interaction between the

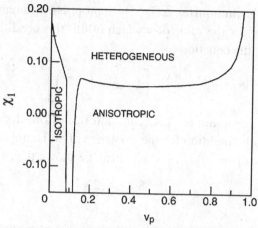

Fig. 3.20 Volume fraction of co-existing phases, for rodlike molecules of axial ratio $x = 100$ subject to interactions denoted by the parameter χ_1. The binodial for isotropic phases is on the left; that for anisotropic phases is on the right. The minimum of the shallow concave branch of the latter binodial is a critical point marking the emergence of two additional anisotropic phases. The cusp marks a triple point where three phases co-exist. Calculations carried out according to Ref. (92). (From Flory (80))

polymer and solvent. Under these circumstances, the separation into two phases, one isotropic and one ordered, must occur at relatively high dilution. For $x = 100$ the compositions of the two phases in equilibrium are $v_2 = 0.0806$ for the dilute phase and $v_2' = 0.1248$ for the slightly more concentrated one. The more dilute phase is isotropic, the orientation of the particles being uncorrelated with those of their neighbors. The more concentrated phase is highly anisotropic. Particles in a given region are fairly well aligned relative to a common axis. This anisotropic phase is commonly termed a nematic one. Phase separation in this instance occurs solely as a consequence of particle asymmetry, unassisted and unabetted by any favorable intermolecular interactions. As the molecular asymmetry is increased, the polymer concentrations in both phases diminish; however, the concentration of the ordered phase is never much greater than that of its isotropic conjugate. The polymer concentration ratio of the two phases appears to approach a limit of 1.56 as x increases.

The narrow biphasic gap in the diagram is essentially unaffected by interactions for negative values of χ. On the other hand, if the interaction between solute segments is attractive then the biphasic region is abruptly broadened when χ exceeds a small positive value. A critical point emerges at $\chi_1 = 0.055$. For χ_1 values immediately above this critical value, the shallow concave curve delineates the loci of co-existing anisotropic phases, in addition to the isotropic and nematic phases at lower concentration within the narrow biphasic gap. At $\chi_1 = 0.070$ these phases co-exist at this triple point.

The general features of this unusual phase diagram have been confirmed by experiment. Figure 3.21 is the experimentally determined phase diagram for the

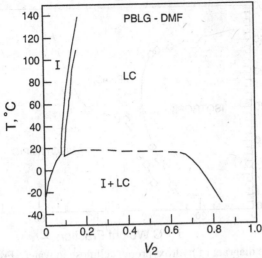

Fig. 3.21 Temperature–composition phase diagram for poly(benzyl-L-glutamate) in dimethyl formamide. Dashed line indicates areas of insufficient data. (From Miller *et al.* (94))

binary system poly(benzyl-L-glutamate), in alpha-helical form, and dimethyl formamide.(94) The weight average molecular weight of the polymer is 310 000 with rigid rod ratio of about 135. The partial phase diagram of the alpha-helical polypeptide poly(carbobenzoxy lysine), axial ratio of about 190, in dimethyl formamide shows similar features.(94) The general features of the lattice theory, the narrow biphasic region and the broad anisotropic region have also been confirmed by other studies with polypeptides.(95,96)

Another example of the phase behavior of asymmetric molecules is given in Fig. 3.22 for aqueous solutions of hydroxypropyl cellulose.(97) The phase diagram for this system shows all of the major features expected from the Flory theory for an asymmetric polymer solute. The slight tilting of the narrow biphasic region could possibly be attributed to some molecular flexibility as well as anisotropic interaction.(98) The phase diagram for the ternary system, polymer and two solvents, for poly(*p*-phenylene terephthalamide) also shows the major features expected from theory.(99)

Another test of the theory is to compare the experimentally determined dependence of v_p, the volume fraction at which the nematic phase separates, on the axial ratio.[4] The agreement between theory and experiment is particularly good with the alpha-helical polypeptide poly(γ-benzyl-L-glutamate).(94–96) Studies of solutions of the polymeric aramides, such as poly(*p*-benzamide) and poly(*p*-phenylene terephthalamide) indicate a qualitative accord between theory and experiment. Studies

---

[4] The quantity of interest, v_p, can be identified with v_2^* of Eq. (3.42) with only minor error.(80)

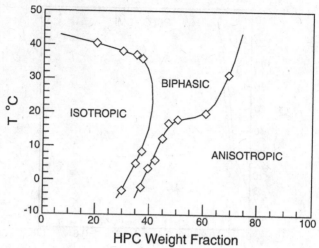

Fig. 3.22 Phase diagram of hydroxypropyl cellulose in water. (From Gido (95))

with poly(N-alkyl isocyanates) in toluene show that when the alkyl group is *n*-hexyl or *n*-octyl the threshold volume v_p is about twice those calculated from the respective axial ratios.(100,101) The discrepancy can be attributed to the fact that there is sufficient flexibility in these chains so that the effective axial ratio is lowered relative to the calculated value. Shear degraded DNA gives rodlike particles whose lengths are such that liquid crystallinity can be observed. The observed and calculated values of v_p for this system are also in good agreement.(100) Two of the basic expectations from the lattice theory, the character of the unique phase diagram and the volume fraction at which the nematic phase separates are fulfilled by experiment. This agreement, and the demonstration of the unusually shaped phase diagram is quite remarkable when it is recognized that the only information required is the axial ratio of the polymer and a reasonable value of the interaction parameter, χ_1. The development of a well-ordered anisotropic phase can now be understood in the dilute region, with $\chi_1 = 0$, based solely on the molecular asymmetry, and in the more concentrated region when χ exceeds a small positive value. The concentrated anisotropic phase can be regarded as the prototype of the crystalline state with only uniaxial order. The development of three-dimensional orders characteristic of the true crystalline state involves the introduction of specific interactions.

In analogy to the melting temperature–composition relation for an isotropic melt a similar relation for melting into an anisotropic or nematic melt can be calculated. Based on the Flory (92) and Flory and Ronca (93) theories, Krigbaum and Ciferri showed that (103)

$$\frac{1}{T_m} - \frac{1}{T_m^{0'}} = \frac{Rx}{\Delta H_f'}\left[-\frac{1}{x}\left(\ln \frac{v_2}{x} + (y-1)v_2 - \ln y^2 \right) - \chi_1 v_1^2 \right] \qquad (3.43)$$

Here $T_m^{0'}$ and $\Delta H_f'$ are the equilibrium melting temperature, and enthalpy of fusion between the crystalline polymer and its anisotropic melt. Equation (3.43) is analogous to Eq. (3.44) for an isotropic melt.

$$\frac{1}{T_m} - \frac{1}{T_m^0} = \frac{Rx}{\Delta H_f}\left[\left(1 - \frac{1}{x}\right)v_1 - \frac{1}{x}\ln v_2 - \chi_1 v_1^2\right] \qquad (3.44)$$

Although experimental data to directly test Eq. (3.43) is not available it is still of interest to compare the two melting temperature–composition relations.(104) The two expressions have the term $-\chi_1 v_1^2$ in common. However, all the remaining terms within the square brackets of Eq. (3.43) for the anisotropic melt will be small, due to the common factor $1/x$. However, for the isotropic case, the term v_2 survives even at large x, and makes a significant contribution to the melting point depression. This difference arises from the smaller disorientation entropy for the transition to the isotropic melt. As a result of this difference the melting temperature depression expected from an anisotropic melt will be minimal, unless there is a very strong interaction with solvent ($\chi_1 < 0$).

The different phase equilibria and transformations in polymer–diluent mixtures that have been discussed can be illustrated by the schematic diagram given in Fig. 3.23.(105) Process [1] represents the usual melting or crystallization of polymers with a conformational change occurring during the transformation. A diluent may or may not be present in the amorphous, or liquid, state, III, while state I represents the pure crystalline phase. Transformations in this category were discussed

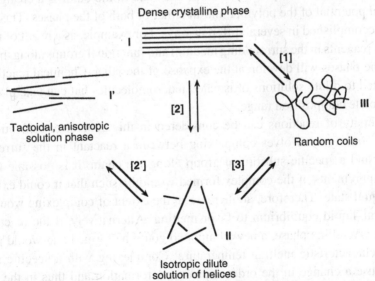

Fig. 3.23 Schematic representation of transitions and phase equilibria involving polymer chains in ordered configurations. (From Flory (105))

in Sects. 3.2 and 3.3. The formation of an isotropic dilute solution, II, wherein the molecules maintain the conformation characteristic of state I, is designated by process [2]. This process can be thought of as dissolution. However, in distinction to process [1], the molecular conformation is maintained. The inverse process represents the formation of a pure ordered phase from a dilute solution of anisotropic molecules. The helix–coil transition is then represented by process [3]. The dilute tactoidal anisotropic, nematic phase I' is formed from the dilute isotropic phase by [2'] with a slight increase in the polymer concentration. This schematic diagram points out certain similarities between the various processes and the importance of considering the complete composition range to describe the behavior adequately. For example, the helix–coil transition is seen as a manifestation of process [1] in dilute solution. The continuity between the helix–coil transition and the usual melting has been established for collagen. For certain polymer systems a point in the phase diagram may exist at low polymer concentrations where the three phases representing the pure ordered phase, the randomly coiled state, and the state of the individual asymmetric molecules co-exist. This bears an analogy to the triple point for the co-existence of solid, liquid, and vapor of monomeric substances.

3.6 Chemical reactions: melting and compound formation

Melting and crystallization can also be governed by appropriate chemical reactions and interactions between the polymer and low molecular weight species. All that is required to shift the equilibrium from one state to the other is a change in the chemical potential of the polymer unit in either or both of the phases. This change can be accomplished in several different ways. For example, as a result of reaction between reagents in the surrounding medium and functional groups along the chain, one of the phases will develop at the expense of the other. Chemical reactions are not limited to dilute solutions of isolated macromolecules but can take place over the complete composition range.

A diversity of reactions can be considered in this general classification. For example, one type involves complexing between a reactant in the surrounding medium and a specific substituent group along the chain. It is possible that the steric requirements of the complex formed would be such that it could exist only in the liquid state. Therefore, an increase in the extent of complexing would shift the crystal–liquid equilibrium to favor melting. Alternatively, if the reactant entered the crystalline phase, a new compound could be formed that would possess its own characteristic melting temperature. Complexing with a specific solvent could cause a change in the ordered chain conformation and thus in the crystal structure. Another possibility is that only some of the chain units are structurally altered. From a crystallization point of view a homopolymer would be converted to a copolymer.

For crystallizable polyelectrolytes, electrostatic effects will affect the crystal–liquid equilibrium. It is unlikely that a charged substituent and its associated counterion could be accommodated in the usual crystal lattice. Experiment indicates that melting can indeed be induced in such polymers by changes in the ionic nature of the surrounding medium. The dilute solution helix–coil transition of poly(L-glutamic acid) and poly(L-lysine) is influenced by alterations in the pH of the medium. Poly(L-lysine) has an amino group in the side chain that is positively charged at pH values below about 9.5 and is neutral above about pH 10.5. It is observed that the helical form is stable only in the uncharged state. Thus, as the pH is lowered isothermal transformation to the random coil state occurs. It has already been noted that the alpha-helical form of poly(L-glutamic acid) is stable below pH 5, where the carboxyl side groups are largely unionized. Transformation of the random coil form occurs as the pH is raised. The stability of the ordered structure of other polyelectrolytes is affected in a similar manner. The melting temperature of DNA from calf thymus is lowered from 86 °C to about 25 °C by a reduction in the ionic strength or pH. In the absence of an added electrolyte, T_m falls below room temperature. Similar behavior has also been noted in the synthetic poly(ribonucleotides). We can conclude, therefore, that the accommodation of a charged substituent in the ordered state is thermodynamically less favored than in the amorphous state. A shift in the equilibrium between the two states can thus be accomplished by control of the pH of the medium.

Chemical reactions that cause either the formation or severance of intermolecular crosslinks will also affect the stability of ordered chain structures. The role of crosslinks will be discussed in detail in Chapter 7. For polypeptides and proteins this is of importance in view of the relative chemical ease with which intermolecular disulfide bonds can be controlled.

The quantitative formulation of the coupling of the crystal–liquid transformation with a chemical reaction involves specifying the phases in which the reaction occurs and the modifications induced in the chemical potential of the repeating unit. The reaction can be treated by the usual methods of chemical equilibrium, the results of which are then imposed on the conditions for phase equilibrium. The different possibilities must be individually treated following this procedure.

As an example we consider a simple type of complexing reaction that is restricted to the liquid phase. Consider a polymer molecule P containing n substituents each capable of complexing in the amorphous phase with reactant C according to the scheme

$$P + rC \rightleftharpoons P \cdot C_r \qquad (3.45)$$

where the total concentration of polymer species, P, and P·C_r remain unchanged. The equilibrium constant for the reaction, K_r can be written as

$$K_r = \frac{n!}{(n-r)!r!} K^r \tag{3.46}$$

where K is the equilibrium constant for each of the individual complexing reactions. All of the reaction sites are assumed to be independent of one another. The combinatorial factor represents the number of ways in which the reactant can be distributed among the n possible substituents. The extent of the reaction r can be expressed as

$$r = \frac{n a_c K}{1 + a_c K} \tag{3.47}$$

where a_c is the activity of species C. The free energy change due to the reaction is

$$\Delta G_{react} = -nRT \ln(1 + K a_c) \tag{3.48}$$

The change in chemical potential per repeating unit is

$$\mu_\mu^* - \mu_\mu = -N_A RT \ln(1 + K a_c) \tag{3.49}$$

where μ_μ^* represents the chemical potential of the complexed unit and N_A is the mole fraction of chain units bearing the reactive substituent. If more than one reactive site per chain unit exists, the appropriate numerical factor must be appended to Eq. (3.49). By applying the usual conditions for phase equilibria, the melting point equation becomes

$$\frac{1}{T_m} - \frac{1}{T_m^0} = \frac{R}{\Delta H_u} \frac{V_u}{V_1} \left(v_1 - \chi_1 v_1^2\right) + \frac{R N_A}{\Delta H_u} \ln(1 + K a_c) \tag{3.50}$$

For experiments carried out at fixed polymer concentration the last term on the right in Eq. (3.50) represents the depression of the melting temperature, at the given composition, due to the chemical reaction. If the factor $K a_c$ is small, then

$$T_m(v_2) - T_{m,r}(v_2) \cong \frac{R T_m^2(v_2) N_A K a_c}{\Delta H_u} \tag{3.51}$$

$T_m(v_2)$ represents the melting temperature of the mixture, which is devoid of the reactant, at the composition v_2 and $T_{m,r}(v_2)$ is that after complexing.

The melting of collagen and other fibrous proteins follows the form given by Eq. (3.51). It is found that the melting temperature is a linear function of the concentration of the binding species.(106) By assuming that binding to each repeating unit is equally probable, i.e. $N_A = 1$, and knowing the value of ΔH_u, the intrinsic binding constant K can be calculated from the experimental data by identifying the activity of the salt with its concentration. The values of the intrinsic binding constants calculated from the melting point depression are comparable to those obtained by other methods.(107) The binding constants for salts such as KCl and NaCl are relatively low. Salts that cause a significant depression of the melting temperature, such as LiCl, LiBr, KCNS, KBr, have high binding constants.(107) This latter class of salts is thought to have strong preferential interaction with the peptide or amide band. The similar actions of the salts on a variety of fibrous and globular proteins suggest a melting mechanism caused by preferential binding to the peptide, or amide, bond.

Structural transformations of globular proteins in dilute solution are well known to be induced by the action of urea. Such reactions are also found to obey Eq. (3.51). For this reactant both the carbonyl oxygen and the amino hydrogen of the peptide group are assumed to be involved. Therefore, in this case the right-hand side of Eq. (3.51) must be increased by a factor of 2.

The melting temperatures of the polyamides are also depressed in a systematic manner by the interaction of inorganic salts.(108–110) Studies with the synthetic polymers can be carried out with the polymer in the pure state. Equation (3.51) should still be applicable, with $T_m(v_2)$ being replaced by T_m^0. Figure 3.24 is a plot of the extrapolated equilibrium melting temperatures of poly(caproamide) as a function of salt concentration for three different salts.(110) A linear relation results, as would be expected from Eq. (3.51). Moreover, we note that KCl has a trivially small effect on depressing the melting temperature consistent with expected low binding constant. On the other hand the lithium salts give a relatively large melting point depression consistent with a much larger binding constant. The similarity in results between this synthetic polyamide and the fibrous and

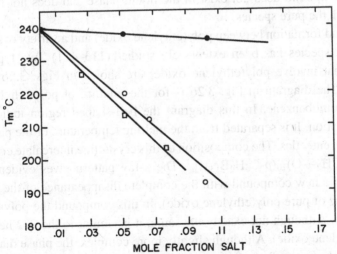

Fig. 3.24 Plot of equilibrium melting temperature against mole fraction of salt concentration. KCl ●; LiCl ○; LiBr □. (From Valenti *et al.* (110))

globular proteins is not surprising since preferential binding to the same type group is involved.

In general, and in particular for open systems, the more complete Eq. (3.50) must be used to allow for changes in both the polymer and reactant concentrations. The sensitivity of the melting temperature to the specific chemical reaction is embodied in the last term of this equation. For example, if there is a very strong affinity for complexing, i.e. if K is large, only a small change in the activity (or concentration) of the reactant will suffice to cause a marked shift in the equilibrium. Conversely, if K is small, a large value of the activity may be required to lower the melting temperature. However, in the vicinity of the melting point small changes in a_c will still drastically alter the concentration of the various species. Therefore, melting on crystallization can be carried out isothermally by this type of chemical interaction.

The detailed discussion up to now has been focused on a specific type chemical process, i.e. binding restricted to the liquid state. The consequence of this process is a reduction of the melting temperature. Other processes that affect the chemical potential of the polymer unit in either state will also influence the equilibrium. For example, binding could be restricted to the crystalline state and a similar type of analysis results.

In addition to simple binding there are many examples where a low molecular weight species enters either the crystal interior or the interlamellar space with compound formation. These situations, although not uncommon, must obviously be very specific in nature and are termed inclusion compounds or clathrates. An example is given by the phase diagram of Fig. 3.25 for polyethylene–perhydrotriphenylene mixtures.(112) A compound is formed that melts congruently at 178.2 °C. This inclusion compound does not exist in the liquid phase and does not form mixed crystals with the pure species.

Compound formation between poly(ethylene oxide) and a variety of low molecular weight species has been extensively studied.(113–121) Two typical phase diagrams that involve poly(ethylene oxide) are shown in Figs. 3.26 (113) and 3.27.(116) The diagram in Fig. 3.26 is for the mixture of poly(ethylene oxide) with p-dibromobenzene. In this diagram the bell-shaped region indicates compound formation. It is separated from the melting temperature of the pure components by two eutectics. The composition of this crystalline intercalate compound is $[-(CH_2-CH_2-O)_{10}(p-C_6H_4Br_3)_3]_n$. The x-ray pattern gives evidence for the formation of a new compound with the complete disappearance of the reflections characteristic of pure poly(ethylene oxide). In this compound the polymer adopts a 10/3 helix. Although not exactly the same, it is similar to the 7/2 helix typical of poly(ethylene oxide). Although slightly more complex, the phase diagram with p-nitrophenol again reflects compound formation. The compound has the same composition and melting temperature as the second eutectic. Analysis of the x-ray pattern and infra-red spectra indicate that the chain conformation in the compound

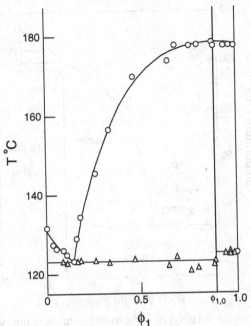

Fig. 3.25 Phase diagram of binary mixtures of polyethylene and perhydrotriphenylene. (From Farina, DiSilvestre and Grasse (112))

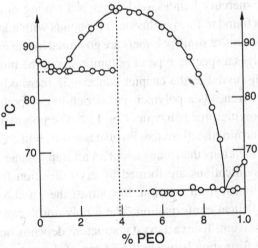

Fig. 3.26 Phase diagram of the poly(ethylene oxide)–p-dibromobenzene system. (From Point and Coutelier (113))

again departs significantly from the 7/2 helical form of the pure polymer. Interestingly, the phase diagram with either ortho or meta nitrophenol does not give any indication of compound formation.(116) Thus, we have an example that very specific structural interactions are required for compound formation. This more complex type phase diagram is found with other poly(ethylene oxide) mixtures.

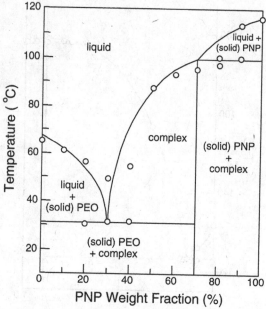

Fig. 3.27 Phase diagram for the poly(ethylene oxide)–*p*-nitrophenol system. (From Tsutsui *et al.* (116))

(119–121) Other species that form compounds with poly(ethylene oxide) include lithium salts, urea, mercury halides, and resorcinol among others.(117,118,122–124) Urea has been found to form inclusion compounds with many polymers.(125)

The ordered structures of some polymers are governed by the influence of specific diluents. This involves a specific type of polymorphism, the more general aspects of which will be discussed in the chapter concerning thermodynamic quantities. Syndiotactic poly(styrene) is a polymer that is rich in compound formation with solvent mediated polymorphic behavior.(126–130) The polymer can crystallize in four major crystalline modifications that involve two different chain conformations. In the α and β modifications the chains adopt an all trans planar zigzag conformation. These two modifications are formed by crystallization from the melt and, under special conditions, from solution. In contrast the γ and δ modifications are characterized by a helical conformation. The δ polymorph can only be prepared in the presence of solvent. Its exact crystal structure depends on the nature of the solvent. Compound formation between the δ form of the polymer and the solvent has been demonstrated. Complete elimination of the solvent results in the pure, helical γ form.

An example of a phase diagram involving a good solvent for syndiotactic poly(styrene) is given in Fig. 3.28.(130) The diagram illustrated for chlorobenzene is similar to that found with *o*-xylene, another good solvent. In this solvent the β form (planar zigzag conformation) gives a normal melting point depression at the

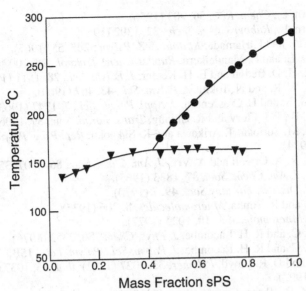

Fig. 3.28 Phase diagram for syndiotactic poly(styrene)–chlorobenzene. Melting of β phase ●; melting of δ phase ▼. (From Roels, Deberdt and Berghmans (130))

high polymer concentrations. However, at a polymer concentration of about 40% the δ polymorph is formed. Compound formation is indicated with incongruent melting. Variations in the phase diagrams are obtained, depending on the thermodynamic interaction between polymer and solvent. Compound formation has also been demonstrated with syndiotactic poly(methyl methacrylate) in different organic solvents.(131)

In this chapter we have found that for melting and phase equilibrium theory the same basic principles that are applicable to low molecular weight species also apply to polymers. In fact, a rather good measure of success is achieved. The only special treatment afforded to polymers is the formulation of the free energy of mixing of polymer and diluent. This also follows basic principles.(1) It is important to recognize that no new basic laws have had to be developed to understand the melting behavior of polymer–diluent mixtures.

References

1. Flory, P. J., *Principles of Polymer Chemistry*, Cornell University Press (1953) pp. 495ff.
2. Flory, P. J., *J. Chem. Phys.*, **17**, 223 (1949).
3. Flory, P. J., *Principles of Polymer Chemistry*, Cornell University Press (1953) pp. 508ff.
4. Mandelkern, L., R. R. Garrett, P. J. Flory, *J. Am. Chem. Soc.*, **74**, 3949 (1952).
5. Cheng, L. P., A. H. Dwan and C. C. Gryte, *J. Polym. Sci.: Pt. B: Polym. Phys.*, **32**, 1183 (1984).

6. Mandelkern, L., *Chem. Rev.*, **56**, 903 (1956).
7. Mandelkern, L., *Rubber Chem. Tech.*, **32**, 1392 (1959).
8. Nakajima, A. and F. Hamada, *Kolloid Z. Z. Polym.*, **205**, 55 (1965).
9. Roberts, D. E. and L. Mandelkern, *Rubber Chem. Technol.*, **28**, 1007 (1955).
10. Flory, P. J., H. D. Bedon and E. H. Keefer, *J. Polym. Sci.*, **28**, 1511 (1958).
10a. Krigbaum, W. R. and N. Tokita, *J. Polym. Sci.*, **43**, 467 (1960).
10b. Gechele, G. B. and L. Crescentini, *J. Appl. Polym. Sci.*, **7**, 1347 (1963).
11. Jackson, J. B., P. J. Flory and R. Chiang, *Trans. Farad. Soc.*, **59**, 1906 (1963).
12. Miyata, S., M. Sorioka, T. Arikawa and K. Sakaoku, *Rep. Prog. Polym. Phys. Jpn*, **17**, 233 (1974).
13. Flory, P. J., R. A. Orwoll and A. Vrij, *J. Am. Chem. Soc.*, **86**, 3507, 3515 (1964).
14. Flory, P. J., *J. Am. Chem. Soc.*, **87**, 1833 (1965).
15. Flory, P. J., *Discuss. Faraday Soc.*, **49**, 7 (1970).
16. Olabisi, O. and R. Simha, *Macromolecules*, **8**, 206 (1975).
17. Simha, R., *Macromolecules*, **10**, 1025 (1977).
18. Sanchez, I. C. and R. H. Lacombe, *J. Phys. Chem.*, **80**, 2352 (1976).
19. Sanchez, I. C. and R. H. Lacombe, *J. Polym. Sci.: Polym. Lett.*, **15B**, 71 (1977).
20. Kim, S. S. and D. R. Lloyd, *Polymer*, **33**, 1027 (1992); *ibid.* **33**, 1037 (1992); *ibid.* **33**, 1047 (1992).
21. Richards, R. B., *Trans. Faraday Soc.*, **42**, 10 (1946).
22. Bueche, A. M., *J. Am. Chem. Soc.*, **74**, 65 (1952).
23. Flory, P. J., L. Mandelkern and H. K. Hall, *J. Am. Chem. Soc.*, **73**, 2532 (1951).
24. Nakajima, A. and H. Fujiwara, *J. Polym. Sci. A2*, **6**, 723 (1968).
25. Frushour, B. G., *Polym. Bull.*, **1**, 1 (1982).
26. Burghardt, W. R., *Macromolecules*, **22**, 2482 (1989).
27. van Emmerik, P. T. and C. A. Smolder, *Eur. Polym. J.*, **9**, 157 (1973).
28. Chiang, R. and P. J. Flory, *J. Am. Chem. Soc.*, **83**, 2857 (1961).
29. Shultz, A. R. and C. R. McCullough, *J. Polym. Sci., Pt. A-2*, **7**, 1577 (1969).
30. Butte, W. A., C. C. Price and R. E. Hughes, *J. Polym. Sci.*, **61**, 28 (1962).
31. Flory, P. J. and R. R. Garrett, *J. Am. Chem. Soc.*, **80**, 4836 (1958).
32. Zaides, A. L., *Kolloid Z.*, **16**, 265 (1954).
33. Feughelman, M., R. Langridge, W. E. Seeds, A. R. Stokes, H. R. Wilson, C. W. Hopper, M. H. F. Wilkins, R. K. Barclay and L. D. Hamilton, *Nature*, **175**, 834 (1955).
34. Papkov, S. P., *Vysokomol Soyed*, **A20**, 2517 (1978).
35. Smith, P. and A. J. Pennings, *Polymer*, **15**, 413 (1974).
36. Smith, P. and A. J. Pennings, *J. Mater. Sci.*, **11**, 1450 (1976).
37. Hodge, A. M., G. Kiss, B. Lotz and J. C. Wittman, *Polymer*, **23**, 985 (1982).
38. Smith, P. and A. J. Pennings, *J. Polym. Sci.: Polym. Phys. Ed.*, **15**, 523 (1977).
39. Gryte, C. C., H. Berghmans and G. Smits, *J. Polym. Sci.: Polym. Phys. Ed.*, **17**, 1295 (1979).
40. Wittmann, J. C. and R. St. John Manley, *J. Polym. Sci.: Polym. Phys. Ed.*, **15**, 2277 (1977).
41. Wittmann, J. C. and R. St. John Manley, *J. Polym. Sci.: Polym. Phys. Ed.*, **15**, 1089 (1977).
42. Flory, P. J. and W. R. Krigbaum, *J. Chem. Phys.*, **18**, 1086 (1950).
42a. Hobbs, J. K., M. J. Hill, A. Keller and P. J. Barham, *J. Polym. Sci.: Pt. B: Polym. Phys.*, **37**, 3188 (1999).
43. Beech, D. R. and C. Booth, *Polymer*, **13**, 355 (1972).
44. Pennings, A. J., in *Characterization of Macromolecular Structure*, National Academy of Science (U.S.) (1968) p. 214.

45. Sanchez, I. C. and E. A. DiMarzio, *Macromolecules*, **4**, 677 (1971).
46. Prasad, A. and L. Mandelkern, *Macromolecules*, **22**, 4666 (1989).
46a. Mattice, W. L. and L. Mandelkern, *Macromolecules*, **4**, 271 (1971).
47. Flory, P. J., *Statistical Mechanics of Chain Molecules*, Interscience (1969) pp. 59ff.
48. Newman, S., W. R. Krigbaum and D. K. Carpenter, *J. Phys. Chem.*, **60**, 648 (1956).
49. Harrington, W. F. and M. Sela, *Biochem. Biophys. Acta*, **27**, 24 (1958).
50. Blout, E. R. and G. D. Fasman, in *Recent Advances in Glue Research*, G. Stainsky ed., Pergamon Press (1958) p. 122.
51. Mandelkern, L. and M. H. Liberman, *J. Phys. Chem.* **71**, 1163 (1967).
52. Bailey, F. E., Jr., G. M. Powell and K. L. Smith, *Ind. Eng. Chem.*, **50**, 8 (1958).
52a. Bailey, F. E., Jr. and R. W. Callard, *J. Appl. Polym. Sci.*, **1**, 56, 373 (1959).
53. Doty, P., A. M. Holtzer, J. H. Bradbury and E. R. Blout, *J. Am. Chem. Soc.*, **76**, 4493 (1954); P. Doty, J. H. Bradbury and A. M. Holtzer, *J. Am. Chem. Soc.*, **78**, 947 (1956); J. T. Yang and P. Doty, *J. Am. Chem. Soc.*, **78**, 498 (1956); **79**, 761 (1957).
54. Doty, P., K. Iamhori and E. Klemperer, *Proc. Natl. Acad. Sci. U.S.*, **44**, 474 (1958); P. Doty, A. Wada, J. T. Yang and E. R. Blout, *J. Polym. Sci.*, **23**, 851 (1957).
55. Doty, P., *Rev. Mod. Phys.*, **31**, 107 (1959).
56. Warner, R. C., *J. Biol. Chem.*, **229**, 711 (1957).
57. Rich, A. and D. R. Davies, *J. Am. Chem. Soc.*, **78**, 3548 (1956); G. Felsenfeld and A. Rich, *Biochim. Biophys. Acta*, **26**, 457 (1957); A. Rich, *Nature*, **181**, 521 (1958); A. Rich, *Biochim. Biophys. Acta*, **29**, 502 (1958).
58. Fresco, J. R. and P. Doty, *J. Am. Chem. Soc.*, **79**, 3928 (1957).
59. Doty, P., H. Boedtker, J. R. Fresco, B. D. Hall and R. Haselkorn, *Ann. N.Y. Acad. Sci.*, **81**, 693 (1959).
60. Cohen, C., *Nature*, **175**, 129 (1955); *J. Biophys. Biochem. Cytol.*, **78**, 4267 (1955).
61. Boedtker, H. and P. Doty, *J. Am. Chem. Soc.*, **77**, 248 (1955); **78**, 4267 (1956).
62. Crothers, D. M., N. R. Kallenbach and B. H. Zimm, *J. Mol. Biol.*, **11**, 802 (1965).
63. Rice, R. V., *Proc. Natl. Acad. Sci. U.S.*, **46**, 1186 (1960).
64. Marmur, J. and D. Lane, *Proc. Natl. Acad. Sci. U.S.*, **46**, 453 (1960); P. Doty, J. Marmur, J. Eigner and C. Schildkraut, *Proc. Natl. Acad. Sci. U.S.*, **46**, 461 (1960).
65. Schellman, J. H., *Compt. Rend. Trav. Lab. Carlsberg, Ser. Chim.*, **29**, 230 (1955).
66. Zimm, B. H. and J. K. Bragg, *J. Chem. Phys.*, **28**, 1246 (1958); **31**, 523 (1959).
67. Gibbs, J. H. and E. A. DiMarzio, *J. Chem. Phys.*, **28**, 1247 (1958); **30**, 271 (1959).
68. Peller, L., *J. Phys. Chem.*, **63**, 1194, 1199 (1959).
69. Rice, S. A. and A. Wada, *J. Chem. Phys.*, **29**, 233 (1958).
70. Hill, T. L., *J. Chem. Phys.*, **30**, 383 (1959).
71. Schellman, J. H., *J. Phys. Chem.*, **62**, 1485 (1958).
72. Zimm, B. H., *J. Chem. Phys.*, **33**, 1349 (1960).
73. Nagai, K., *J. Phys. Soc. Jpn*, **15**, 407 (1960); *J. Chem. Phys.*, **34**, 887 (1961).
74. Lifson, S. and A. Roig, *J. Chem. Phys.*, **34**, 1963 (1961).
75. Flory, P. J., *Statistical Mechanics of Chain Molecules*, Interscience (1969) pp. 286ff.
76. Poland, D. and H. A. Scheraga, *Theory of Helix-Coil Transition in Biopolymers*, Academic Press (1960).
77. Pauling, L., R. B. Corey and H. R. Bransom, *Proc. Natl. Acad. Sci. U.S.*, **37**, 205 (1951).
78. Landau, L. D. and E. M. Lifshitz, *Statistical Physics*, Pergamon Press, Ltd., London (1958) p. 482.
79. Zimm, B. H., P. Doty and K. Iso, *Proc. Natl. Acad. Sci. U.S.*, **45**, 1601 (1959).
80. Flory, P. J., *Advances in Polymer Science*, **59**, Springer-Verlag (1984).
81. Uematsu, I. and Y. Uematsu, *Advances in Polymer Science*, **59**, Springer-Verlag (1984) p. 37.

82. Papkov, S. P., *Advances in Polymer Science*, **59**, Springer-Verlag (1984) p. 75.
83. Samulski, E. T., in *Physical Properties of Polymers*, Second Edition, J. E. Mark ed., American Chemical Society (1993) p. 201.
84. Donald, A. M. and A. H. Windle, *Liquid Crystalline Polymers*, Cambridge University Press (1991).
85. Ciferri, A. ed., *Liquid Crystallinity in Polymers*, Cambridge University Press (1991).
86. Blumstein, A. ed., *Polymer Liquid Crystals*, Plenum (1985).
87. Blumstein, A. ed., *Liquid Crystalline Order in Polymers*, Academic Press (1982).
88. Ciferri, A., W. R. Krigbaum and R. B. Meyer, eds., *Polymer Liquid Crystals*, Academic Press (1982).
89. DeGennes, P. G., *The Physics of Liquid Crystals*, Oxford University Press (1974).
90. Maier, W. and A. Saupe, *Z. Naturforsch.*, **14a**, 822 (1959); *ibid.* **15a**, 287 (1960).
91. Onsager, L., *Ann. N. Y. Acad. Sci.*, **51**, 627 (1949).
92. Flory, P. J., *Proc. R. Soc. London*, **234A**, 73 (1956).
93. Flory, P. J. and G. Ronca, *Mol. Cryst. Liq. Cryst.*, **54**, 289 (1979).
94. Miller, W. G., C. C. Wu, E. L. Wee, G. L. Santee, J. H. Rai and K. G. Goekel, *Pure Appl. Chem.*, **38**, 37 (1974).
95. Herman, J., Jr., *J. Coll. Sci.*, **17**, 638 (1962).
96. Nakajima, A., T. Hayashi and M. Ohmori, *Biopolymers*, **6**, 973 (1968).
97. Gido, S., *Macromolecules*, **28**, 4530 (1995).
98. Warner, M. and P. J. Flory, *J. Chem. Phys.*, **73**, 6327 (1980).
99. Nakajima, A., T. Hinae and T. Hayashi, *Polym. Bull.*, **1**, 143 (1978).
100. Aharoni, S. M., *J. Polym. Sci.: Polym. Phys. Ed.*, **18**, 1439 (1980).
101. Aharoni, S. M., *Polym. Bull.*, **9**, 186 (1983).
102. Brian, A. A., H. L. Frisch and L. S. Lerman, *Biopolymer*, **20**, 1305 (1988).
103. Krigbaum, W. R. and A. Ciferri, *J. Polym. Sci.: Polym. Lett.*, **18B**, 253 (1980).
104. Ballbi, C., E. Bianchi, A. Ciferri, A. Tealdi and W. R. Krigbaum, *J. Polym. Sci.: Polym. Phys. Ed.*, **18**, 2037 (1980).
105. Flory, P. J., *J. Polym. Sci.*, **49**, 105 (1961).
106. Von Hippel, P. H. and K. Y. Wong, *Biochem.*, **1**, 1399 (1963).
107. Mandelkern, L. and W. E. Stewart, *Biochem.*, **3**, 1135 (1964).
108. Ciferri, A., E. Bianchi, F. Marchere and A. Tealdi, *Makromol. Chem.*, **150**, 265 (1971).
109. Frasci, A., E. Martuscelli and V. Vittoria, *J. Polym. Sci.: Polym. Lett.*, **98**, 561 (1971).
110. Valenti, B., E. Bianchi, G. Greppa, A. Tealdi and A. Ciferri, *J. Phys. Chem.*, **77**, 389 (1973).
111. Valenti, B., E. Bianchi, A. Tealdi, S. Russo and A. Ciferri, *Macromolecules*, **9**, 117 (1976).
112. Farina, M., G. DiSilvestre and M. Grasse, *Makromol. Chem.*, **180**, 104, (1979).
113. Point, J. J. and C. Coutelier, *J. Polym. Sci.: Polym. Phys. Ed.*, **23**, 231 (1985).
114. Point, J. J. and P. Damman, *Macromolecules*, **25**, 1184 (1992).
115. Damman, P. and J. J. Point, *Polym. Int.*, **36**, 117 (1995).
116. Tsutsui, K., K. Adachi, Y. Tsujita, H. Yoshimizu and T. Kimoshita, *Polym. J.*, **30**, 753 (1998).
117. Zahurak, S. M., M. L. Kaplan, E. A. Rietman, D. W. Murphy and R. J. Cava, *Macromolecules*, **21**, 654 (1988).
118. Lascaud, S., M. Perrier, A. Vallee, S. Besner, J. Prud'homme and M. Armand, *Macromolecules*, **27**, 7469 (1994).
119. Paternostre, L., P. Damman and M. Dosiére, *Macromolecules*, **32**, 153 (1999).
120. Moulin, J. F., P. Damman and M. Dosiére, *Polymer*, **40**, 5843 (1999).

121. Paternostre, L., P. Damman and M. Dosiére, *J. Polym. Sci.: Pt. B: Polym. Phys.*, **37**, 1197 (1999).
122. Chatani, Y. and S. Okamaura, *Polymer*, **28**, 1815 (1987).
123. Iwamoto, R., Y. Saito, H. Ishihara and H. Tadokoro, *J. Polym. Sci., Pt. A-2*, **6**, 1509 (1968).
124. Yokoyama, M., H. Ishihara, R. Iwamoto and H. Tadokoro, *Macromolecules*, **2**, 1841 (1969).
125. Chenite, A. and F. Brisse, *Macromolecules*, **25**, 776 (1992).
126. Vittoria, V., F. Candia, P. Ianelli and A. Immirzi, *Macromol. Chem. Rapid Comm.*, **9**, 765 (1988).
127. Vittoria, V., R. Russo and F. Candia, *Polymer*, **32**, 3371 (1991).
128. Berghmans, H. and F. Deberdt, *Phil. Trans. R. Soc. London*, **A348**, 117 (1994).
129. Deberdt, F. and H. Berghmans, *Polymer*, **34**, 2192 (1993); *ibid.* **35**, 1694 (1994).
130. Roels, D., F. Deberdt and H. Berghmans, *Macromolecules*, **27**, 6216 (1994).

4

Polymer–polymer mixtures

4.1 Introduction

This chapter is concerned with the thermodynamic aspects of the fusion of binary mixtures of two homopolymers. The structural and morphological features that result from the departure from equilibrium, and their influence on properties, will be discussed in a subsequent chapter. Binary polymer blends present several different situations. An important distinction has to be made as to whether the components are miscible, immiscible or partially miscible with one another in the molten or liquid state. Flory has pointed out (1) that the mixing of two polymeric components in the liquid state follows normal thermodynamic principles. Since the entropy change of mixing two long chain molecules is small, only a minute, positive enthalpic interaction will produce limited miscibility. It can then be expected that incompatibility of chemically dissimilar polymers should be the general rule. Experiment supports this conclusion. Compatibility, or miscibility, should be the exception. However, many polymer pairs have been found that are miscible, or partially miscible, with one another.(2) Miscibility involves very specific, favorable interactions between the two components.(3,4,5) Among the types of interactions involved are hydrogen bonding, charge transfer complexing and dipolar effects.

There are different situations within the miscible or partially miscible categories that need to be recognized and analyzed separately. The main groupings are: mixtures of two chemically different species, only one of which crystallizes; two chemically different species, each of which crystallizes independently; two chemically different species that co-crystallize (6); and mixtures of chemically identical polymer species that either do or do not co-crystallize. Each particular case must be specified *a priori* before an analysis can be undertaken. The literature concerned with the behavior of blends of crystallizable components is voluminous. Selection has, therefore, been limited to examples that illustrate the basic principles involved.

4.2 Homogeneous melt: background

The first class of blends to be analyzed is that of a homogeneous, disordered liquid phase in equilibrium with a pure crystalline phase, or phases. If both species crystallize they do so independently of one another, i.e. co-crystallization does not occur. With these stipulations the analysis is relatively straightforward. The chemical potentials of the components in the melt are obtained from one of the standard thermodynamic expressions for polymer mixtures. Either the Flory–Huggins mixing expression (7) or one of the equation of state formulations that are available can be used.(8–16) The melting temperature–composition relations are obtained by invoking the equilibrium requirement between the melt and the pure crystalline phases. When nonequilibrium systems are analyzed, additional corrections will have to be made for the contributions of structural and morphological factors.

4.2.1 Homogeneous melt: only one component crystallizes

The melting temperature–composition relation for the common situation of two dissimilar polymers, only one of which crystallizes, was formulated by Nishi and Wang.(17) This relation is based on the free energy of mixing of two dissimilar polymers in the disordered state, as given by Scott (18), within the framework of the Flory–Huggins lattice treatment.(7) The chemical potentials of each species in the binary mixture can be expressed as

$$\mu_1 - \mu_1^0 = RT\left[\ln v_1 + \left(1 - \frac{x_1}{x_2}\right)v_2 + x_1\chi_{12}v_2^2\right] \qquad (4.1)$$

and

$$\mu_2 - \mu_2^0 = RT\left[\ln v_2 + \left(1 - \frac{x_2}{x_1}\right)v_1 + x_2\chi_{12}v_1^2\right] \qquad (4.2)$$

where x_1 and x_2 are the numbers of segments per molecule for each of the chains and χ_{12} is the polymer–polymer interaction parameter. The parameter χ_{12} represents a free energy interaction. It is not limited to an enthalpic contribution, as is often assumed. It can be introduced either in the free energy of mixing expression (1,19) or directly into the respective chemical potentials.(20) The first two terms in Eqs. (4.1) and (4.2) represent ideal mixing of segments. Any deviations from this ideal will automatically be incorporated in the χ_{12} parameter. The numbers of segments are defined by

$$x_1 = V_1/V_0 \qquad x_2 = V_2/V_0 \qquad (4.3)$$

where V_1 and V_2 are the molar values of the polymer species and V_0 is the volume of a lattice cell, so chosen as to accommodate a segment from either chain.[1]

Let species 2 be designated as the crystallizing component. Dividing Eq. (4.2) by the number of structural repeating units per molecule $(V_0/V_{2u})x_2$, where V_{2u} is the volume of the repeating unit, the chemical potential per repeating unit, μ_{2u}, becomes

$$\mu_{2u} - \mu_{2u}^0 = RT\left(\frac{V_{2u}}{V_0}\right)\left[\frac{\ln v_2}{x_2} + \left(\frac{1}{x_2} - \frac{1}{x_1}\right)(1 - v_2) + \chi_{12}(1 - v_2)^2\right] \quad (4.4)$$

Equation (4.4) can be compared with the analogous equation given by Nishi and Wang(17)

$$\mu_{2u} - \mu_{2u}^0 = \frac{RT V_{2u}}{V_{1u}}\left[\frac{\ln v_2}{x_2} + \left(\frac{1}{x_2} - \frac{1}{x_1}\right)(1 - v_2) + \chi_{12}(1 - v_2)^2\right] \quad (4.5)$$

where V_{1u} is the molar volume of the repeating unit of the noncrystallizing component, species 1. The volume of a lattice cell in Eq. (4.5) has been identified with the volume of the noncrystallizing repeating unit. The volume of the segment of the crystallizing component is then defined. This procedure carries with it the implication that the repeating unit of species 1 and the segment (a defined number of repeating units) of species 2, are interchangeable within a lattice cell. An implied flexibility, or inflexibility, as the case may be is thus given to species 2, the crystallizing species.

Using Eq. (3.7), the expression for the chemical potential of a unit in the crystal relative to that in the melt, along with Eq. (4.5), the equilibrium condition yields

$$\frac{1}{T_m} - \frac{1}{T_m^0} = -\frac{R V_{2u}}{\Delta H_u V_{1u}}\left[\frac{\ln v_2}{x_2} + \left(\frac{1}{x_2} - \frac{1}{x_1}\right)(1 - v_2) + \chi_{12}(1 - v_2)^2\right] \quad (4.6)$$

for the melting temperature–composition relation of the mixture. If we let $x_1 = 1$ and $x_2 \to \infty$, the relation for a high molecular weight polymer–low molecular weight diluent, Eq. (3.2) results. For the problem at hand x_1 and x_2 will usually be very large. Therefore, Eq. (4.6) reduces to

$$\frac{1}{T_m} - \frac{1}{T_m^0} = -\frac{R V_{2u}}{\Delta H_u V_{1u}}\chi_{12}(1 - v_2)^2 \quad (4.7)$$

The melting point–composition relation of binary mixtures can also be analyzed by equation of state theories.(5,8–16) In this development, the first two terms of Eq. (4.6) again appear as the lead terms. The main difference in the two approaches is in the expression for the effective interaction parameter. The dependency of this parameter on composition, temperature and pressure is explicitly accounted for by

---

[1] This procedure is analogous to the treatment of polymer–solvent mixtures where the volume of the lattice cell is equated with the volume of the solvent. The polymer segment is thus defined since it must be able to occupy one lattice site.

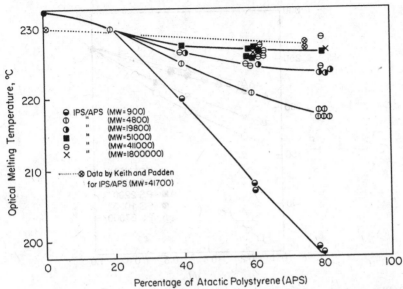

Fig. 4.1 Plot of melting temperature against percentage atactic poly(styrene) for isotactic–atactic poly(styrene) blends. Molecular weights of atactic poly(styrene) are indicated in the figure. (From Yeh and Lambert (21))

this method. The melting point depression of a mixture is then calculated in terms of composition, and reduced temperature, pressure, and the core volume of a segment characteristic of the pure polymer species.

The influence of a noncrystallizing polymeric component on the melting temperature of the crystallizing species should be relatively small, since the melting point depression is a colligative property and the added species is of high molecular weight. This conclusion does not depend on the choice of any particular expression for the free energy mixing. An example of the colligative effect can be found in mixtures of isotactic poly(styrene) with the atactic polymers of different molecular weights.(21,22) As is illustrated in Fig. 4.1(21) there is only a small decrease in T_m for up to 80% of the high molecular weight added species. However, the melting point depression progressively increases as the molecular weight of the atactic species is decreased. When the molecular weight of the atactic polymer is 900 the melting point depression is greater than 30 °C. There are other reports that indicate a somewhat larger depression for poly(styrene) blends.(23,24) However, these can be attributed to varying crystallization conditions with blend composition.

Other mixtures of chemically identical polymers such as isotactic and atactic poly(lactides) (25) and blends of bacterial poly(β-hydroxybutyrate) and its atactic counterpart (26,27,28) also show very small melting point depressions with modest to high molecular weight atactic components. Another example of the basic colligative nature of the melting point depression in such binary blends is given in

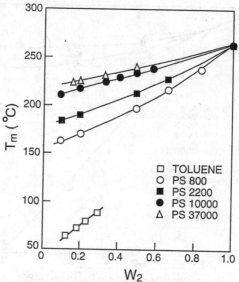

Fig. 4.2 Plot of melting temperature of poly(2,6-dimethyl 1,4-phenylene oxide) against its weight fraction, w_2, in mixtures with toluene and atactic poly(styrene). (From Kwei and Frisch (29))

Fig. 4.2.(29) Here, the melting temperature of the poly(2,6-dimethyl 1,4-phenylene oxide) is plotted against its weight fraction for mixtures with either toluene or atactic poly(styrene) of varying molecular weights. It is evident that adding toluene results in a large melting point depression. However, when poly(styrene) is the added component the decrease in melting temperature is very dependent on its molecular weight.

The expectation of a small, or negligible, depression in the melting temperature is also observed in many blends consisting of dissimilar components.(30–41a) However, in polymer–polymer mixtures, as contrasted with low molecular weight diluents, the magnitude of χ_{12} plays a decisive role in determining the melting temperature depression. The equilibrium melting temperature will be depressed only if χ_{12} is negative. The magnitude of χ_{12}, and the amount of the depression will depend on the strength of the interaction. There are examples where the observed melting point depression is greater than predicted solely on the basis of size and weak interactions. For example, the melting temperature of a poly(vinylidene fluoride)–poly(ethyl methacrylate) blend is depressed by about 15 °C for a mixture with 40% poly(vinylidene fluoride).(42) Depressions of similar magnitude are also found in blends of poly(butylene terephthalate)–poly(acrylate),(43,44) and poly(vinylidene fluoride)–poly(vinyl pyrollidone)(45,46) among others.

A literal interpretation of Eq. (4.7) leads to the expectation that an elevation of the melting temperature would occur if χ_{12} were positive. However, the condition for miscibility of the two components, based on the free energy of mixing formulation, requires that

$$\chi_{12} \le \frac{1}{2} \left[\frac{1}{x_1^{1/2}} + \frac{1}{x_2^{1/2}} \right]^2 \tag{4.8}$$

Thus, χ_{12} must be near zero or negative for miscibility of a given polymer pair. If this condition is not satisfied, then liquid–liquid phase separation will occur in the melt. There are, however, reports of the elevation of the melting temperature as concentration of the noncrystallizing component increases.[47–50] These results can probably be attributed to nonequilibrium, structural and morphological contributions.[50] When $\chi_{12} = 0$, the melting temperature will be invariant with composition, even for a homogeneous melt.

If χ_{12} is assumed to involve only enthalpic interactions, or if very small temperature interval is involved, then to a good approximation[1]

$$\chi_{12} = \frac{B V_{1u}}{RT} \tag{4.9}$$

Equation (4.7) then becomes

$$1 - T_m/T_m^0 = \frac{B V_{2u}}{\Delta H_u} (1 - v_2)^2 \tag{4.10}$$

Equation (4.10) is a consequence of the convenient identification of V_{1u} with V_0.

One of the main reasons that the melting temperatures of polymer–polymer mixtures are measured is to determine the interaction parameter χ_{12}. One can quantitatively discuss the miscibility in the melt of the two polymeric species involved with a knowledge of χ_{12}. According to Eq. (4.7), a plot of $1/T_m - 1/T_m^0$ against v_2 should result in a straight line. The value of χ_{12} can then be calculated from its slope. Appropriate plots are given in Figs. 4.3 and 4.4 for blends of isotactic poly(styrene)–poly(phenylene oxide)(30) and poly(butylene terephthalate)–poly(arylate) respectively.[43] The plot in Fig. 4.3 is almost linear, and a χ_{12} value of 0.17 is obtained from the slope. A linear plot is obtained for small concentrations of the added component in the poly(butylene terephthalate)–poly(arylate) blend.[43] However, as is shown in Fig. 4.4 significant deviations are observed when the data are extended over a wide composition range. This result indicates that in this blend, χ_{12} is concentration dependent, which is not an unexpected finding. One can assume that many other binary mixtures will display similar behavior. In practice, it is also found that the plot does not go through the origin in many cases.[5]

An additional problem is that it is mandatory that the equilibrium melting temperature for each composition be used in the analysis. A common method of obtaining the equilibrium melting temperature is by a linear extrapolation of the observed

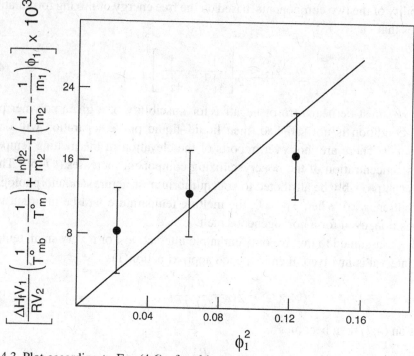

Fig. 4.3 Plot according to Eq. (4.6) of melting temperature of isotactic poly(styrene) in blends with poly(phenylene oxide). (From Plans, MacKnight and Karasz (30))

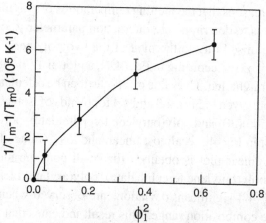

Fig. 4.4 Plot of $1/T_m - 1/T_m^0$ against square of poly(arylate) mass fractions for poly (butylene terephthalate)–poly(arylate) blends. (From Huo and Cebe (43))

melting temperature as a function of the crystallization temperature. The necessary linearity is often not found (51), since the observed melting temperature is influenced by crystallite thickness, other morphological and structural features and the crystallite reorganization that often accompanies fusion.(29,39,52) For these pragmatic reasons, and the fact that a relatively small equilibrium melting temperature range is usually involved, the melting point depression method makes difficult an accurate determination of χ_{12}. For this particular objective it is more useful to employ the equation of state approaches or other methods.(52a). To further complicate the interpretation, the actual miscibility of a given polymer pair often depends on the blending temperature (53) or the casting solvent.

One type of interaction that leads to miscibility of a chemically dissimilar polymer pair is hydrogen bonding. Painter *et al.* have expanded the Flory–Huggins mixing expression to account for this type of interaction.(54) The thermodynamic interaction parameter, χ_{12}, which appears in the free energy of mixing is divided into two parts. One, χ, represents the nonpolar interaction. The other, ΔG_H, is a composition dependent terms that represents the hydrogen bonding mixing interaction. The depression in the melting temperature can then be expressed as

$$1/T_m - 1/T_m^0 = -\left(\frac{R}{\Delta H_u}\frac{V_{2u}}{V_{1u}}\right)\left[\frac{\ln v_2}{x_2} + \left(\frac{1}{x_2} - \frac{1}{x_1}\right)(1 - v_2) + \chi(1 - v)^2 + \Delta \bar{G}_H\right]$$

(4.11)

The partial molar quantity $\Delta \bar{G}_H$ is composition dependent and can be obtained from infra-red measurements. The parameter χ can be estimated from group contributions to the solubility parameters. Better agreement between theory and experiment is found for specific hydrogen bonding systems by use of Eq. (4.11), although the observed melting temperatures involved are small.(54)

4.2.2 Homogeneous melt: both components crystallize

When two polymers crystallize independently of one another from a homogeneous melt separate crystalline domains can form. Consequently, two distinct and separate melting temperatures are observed, each of which is essentially independent of composition.(55–55c) In turn this result indicates a very small value of χ_{12} for these two components. An example of this type of behavior is illustrated in Fig. 4.5 for a pair of polyamides.(55) Similar results have been obtained with other binary blends.(56–60) The melting point depression of the higher melting component, at temperatures above that of the lower melting species, can be analyzed in the conventional manner. However, analyzing the melting point depression of the lower melting component presents some difficulty. In blends of poly(vinylidene fluoride)– poly(1,4-butylene adipate) the equilibrium melting point depression of the higher melting poly(vinylidene fluoride), which is 8.4 °C for a 20/80 mixture, can be explained conventionally with a χ_1 value of -0.19.(56) The depression of the melting point of the polyester component is indicated in Fig. 4.6. It is only slightly more than 3 °C for a 60/40 mixture. The dashed curve in the figure represents the melting temperature–composition relation that would be expected if all the

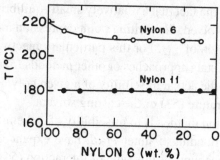

Fig. 4.5 Plot of melting temperature against weight percent nylon 6 for homogeneous blends of nylon 6 and nylon 11. (From Inoue (55))

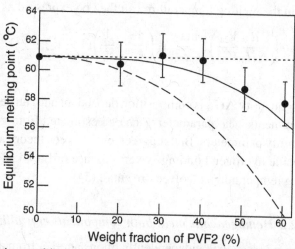

Fig. 4.6 Equilibrium melting temperature of poly(1,4-butylene adipate) in its blends with poly(vinylidene fluoride). The curves are calculated according to theory assuming that (a) all the poly(vinylidene fluoride) is availabe for mixing (– – –); (b) only the amorphous fraction of poly(vinylidene fluoride) is available for mixing (—); (c) there is no melting temperature depression (- - -). (From Penning and Manley (56))

poly(vinylidene fluoride) contributed. This assumption clearly does not represent the experimental results. However, when it is assumed that only the noncrystalline portion of the poly(vinylidene fluoride) contributes to the melting point depression then the solid curve results. This latter curve provides a rather good representation of the data. If smaller χ_{12} values were involved, as in many other blends of this type, then the melting point depression would be much smaller.

In contrast to the case of two species crystallizing independently of one another it is also possible for co-crystallization to take place. The occurrence of isomorphic blends between two polymer components is not common. There are just a few reports of this phenomenon.(6,57,61) Here, two species with slightly different crystallographic structures co-crystallize. The general requirements for the isomorphic behavior of two polymers crystallizing from a homogeneous melt have been described.(6) A simple type of co-crystallization is found in polymer pairs that have the same crystal structure. An example is found in the different structural forms of the polyethylenes.(62,62a,63) However, having the same crystal structure does not necessarily imply that co-crystallization will take place, since kinetic factors can intervene.(63,64)

Miscible binary blends of several different poly(aryl ether ketones) display isomorphic behavior.(58,61) These polymers can be considered to be phenylene units linked to one another by either an ether oxygen or a carbonyl group. The blends found to be isomorphic are miscible, while the blends that are not miscible in the melt are not isomorphic in the crystalline state. The melting temperature–composition relation of two sets of isomorphic blends of miscible poly(aryl ether ketones) are shown in Fig. 4.7.(61) The poly(ether ether ketone)–poly(ether ketone) blends and those of poly(ether ether ketone)–poly(ether ether ketone ketone) behave in a similar manner. The melting temperatures vary in a nonlinear manner from one pure species to the other. The melting temperatures are essentially the same as the higher melting component until high concentrations of the lower melting component are

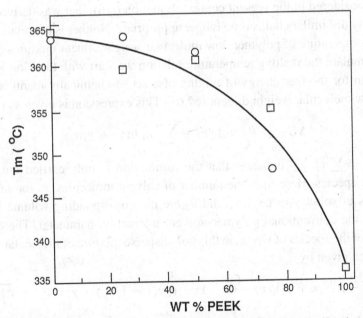

Fig. 4.7 Plot of melting temperature against composition for blends of poly(ether ether ketone)–poly(ether ketone) ○; and of poly(ether ether ketone)–poly(ether ether ketone ketone) □. (From Harris and Robeson (61))

present. The fact that only one melting temperature is observed, when the melting temperatures of the individual species are different, is indicative of isomorphism.[2] The poly(aryl ether ketone) blends have been reported to form isomorphic blends only on rapid crystallization from the melt.(58) The blends were not isomorphic for other modes of crystallization, indicating the influence of the crystallization kinetics. The melting temperature–composition curve in Fig. 4.7 is typical of isomorphic blends. Similar curves are found for poly(vinyl fluoride)–poly(vinylidene fluoride)(65) and poly(p-phenylene oxide)–poly(p-phenoxy phenyl methane)(66) blends.

4.3 Two chemically identical polymers differing in molecular weight

In analyzing blends composed of two chemically and structurally identical polymers, that differ only in molecular weight, we limit ourselves to mixtures that do not co-crystallize, i.e. each of the polymeric species crystallizes independently from a homogeneous melt. This restraint automatically limits the discussion to low molecular weight species since co-crystallization occurs between high molecular weight components.(66–68) Since the interest here is only with equilibrium conditions, the analysis is, by necessity, limited to extended chain crystals. High molecular weight polymers do not usually form extended chain crystals so that their mixtures are not considered in the present context. Equation (4.6), that was derived for two chemically dissimilar chains is no longer appropriate. Neither is the relation for the melting temperature of polymer–low molecular weight diluent mixtures (Eq. 3.2).

To formulate the melting temperature relation we start with the Flory–Huggins expression for the free energy of mixing of a set of chemically identical species with a low molecular weight diluent.(67,68) This expression is given as (7)

$$\Delta G_M = RT n_1 \ln v_1 + {\sum}' n_i \ln v_i + \chi_1 n_1 v_2 \tag{4.12}$$

where $v_2 = \sum' v_i$. $\sum'$ indicates that the summation is only carried out over all the solute species. Here n_1 is the number of solvent molecules, n_i the number of molecules of solute species i, v_1 and v_2 are the corresponding volume fractions and χ_1 is the conventional polymer–solvent interaction parameter. The chemical potential of the species of size x, in this polydisperse mixture of molecular weights, μ_{2x}, is then given by

$$\mu_{2x} - \mu_{2x}^0 = RT[\ln v_x - (x-1) + v_2 x(1 - 1/\bar{x}_n) + \chi_1 x(1-v_2)^2] \tag{4.13}$$

---

[2] The melting temperatures plotted in Fig. 4.7 do not represent equilibrium values. However, it can be expected that only single equilibrium melting temperatures will be observed for these blends and the composition dependence will be similar to that found in the figure.

The chemical potential per repeating unit of the x-mer, μ_x, for the pure polymer system, i.e. when $v_2 = 1$, is given by

$$\mu_x^{(1)} - \mu_x^0 = RT\left[\frac{\ln v_x}{x} + \frac{1}{x} - \frac{1}{\bar{x}_n}\right] \tag{4.14}$$

By expressing the chemical potential of a unit in the crystal relative to that in the melt in the standard way, and applying the conditions for phase equilibrium one finds that

$$\frac{1}{T_m^*} - \frac{1}{T_m} = \frac{R}{\Delta H_u}\left[\frac{1}{\bar{x}_n} - \frac{1}{x} - \frac{\ln v_x}{x}\right] \tag{4.15}$$

Equation (4.15) represents the equilibrium melting temperature relation for a binary mixture of two homopolymers that have the same chain repeating unit but differ in molecular weight. The analysis is predicated on the assumption that only one species crystallizes, i.e. co-crystallization does not occur. Here T_m is the equilibrium melting temperature of the pure crystallizing species of size x; T_m^* is the melting temperature of the mixture characterized by $\bar{x}_n$ and v_x. Equation (4.15) indicates that a significant change in the melting temperature will only take place with low molecular weight species in a mixture of low number average molecular weight. Hence, the restraints that were initially placed on the analysis do not have any practical significance, since interesting results are only expected in the low molecular weight range.

The validity of Eq. (4.15) has been experimentally tested with binary mixtures of low molecular weight fraction of linear polyethylenes and poly(ethylene oxides).(67) Examples of melting point–composition relation for each system are given in Figs. 4.8 and 4.9 respectively. Figure 4.8 gives the melting temperature–composition results for a mixture whose low molecular weight component is the n-alkane $C_{60}H_{122}$ while the higher molecular weight component is a linear polyethylene fraction $M_w = 1262$, $M_n = 1148$. The dashed line is calculated according to Eq. (4.15) for the higher molecular weight component. The melting temperature of the low molecular weight component is invariant with composition as expected. The agreement between theory and experiment is very good in this example. A eutectic temperature is predicted in the vicinity of $v_{1000} \simeq 0.2$ as is observed. The phase diagram for blends of $C_{10}F_{22}/C_{20}F_{42}$ is qualitatively very similar.(69)

The results for low molecular weight poly(ethylene oxide) mixtures also show good agreement with theory.(67) Figure 4.9 gives the melting temperatures of mixtures of molecular weight fractions 1500 and 3000. The melting temperature of the low molecular weight component is again found to be constant with composition. The melting point of the higher molecular weight component varies according

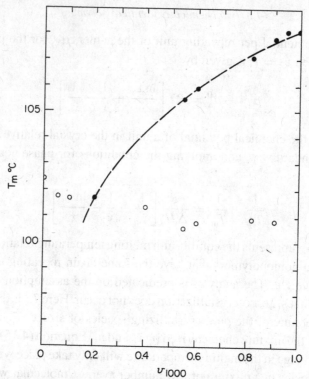

Fig. 4.8 Plot of experimentally observed melting temperature as function of composition for mixtures of $C_{60}H_{122}$ and a linear polyethylene fraction $M_n = 1148$, $M_w = 1262$. Symbols, experimental results; dashed curve, calculated from Eq. (4.15).(67)

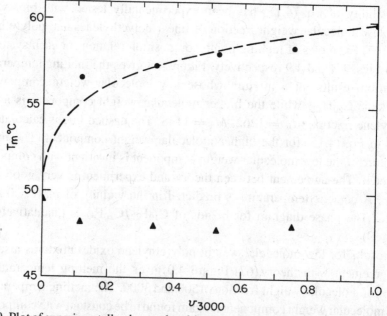

Fig. 4.9 Plot of experimentally observed melting temperatures as a function of composition for poly(ethylene oxide) mixture of molecular weight fractions $M = 1500$ ▲ and 3000 ●. Dashed curve calculated according to Eq. (4.15).(67)

to theory, as indicated by the dashed curve. Thus, basic thermodynamic theory explains in a straightforward manner the melting temperature–composition relation of mixtures of chemically identical polymers, with the same chain structure, that form extended chain crystals.

4.4 Crystallization from a heterogeneous melt

The previous discussion has been limited to blends whose components were completely miscible in the melt. Also to be considered are binary mixtures whose components are either completely immiscible or are only partially miscible. A distinction has to be made again as to whether one or both components crystallize. In order to analyze the melting temperature–composition relations in such systems it is necessary to examine some of the basic phase diagrams.(11,13,70,71) A typical set of such diagrams is given in Fig. 4.10. As a reference diagram Fig. 4.10a represents a mixture with only one component crystallizing. In this diagram the melting temperature of the crystallizing component decreases continuously as its concentration decreases. This curve can be represented either by Eq. (4.6), or the corresponding expression obtained from equation of state theory.

For a partially miscible mixture the binodial needs to be specified as well as its relative location on the melting temperature–composition curve. An upper critical solution temperature (UCST) type binodial is taken as an example. The analyses of other types diagrams, such as those with either a lower critical solution temperature or an hour-glass type follow in a similar manner. The parameters that determine the nature of the binodial are given by Scott.(18) If all else is equal, the nature of the

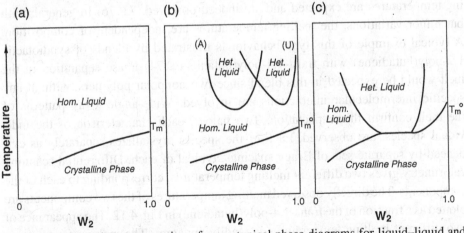

Fig. 4.10 Schematic representation of some typical phase diagrams for liquid–liquid and liquid–crystal transformations.

phase diagram depends mainly on the interaction parameter χ_{12}. In Fig. 4.10b two types of UCST binodials are illustrated. In one case the binodial is symmetric with composition while in the other it is not. Here, the melting temperature–composition diagram is so positioned that it never intersects either of the binodials. Consequently, melting takes place into a homogeneous melt. The melting temperature–composition relation, therefore, is the same as shown in Fig. 4.10a.

Of particular interest is the structure where the melting point curve actually intersects the binodial and thus traverses the two-phase melt. This case is depicted schematically in Fig. 4.10c. Here, at high polymer concentration the melt is homogeneous and a very small decrease in the melting temperature is expected with an increase in the concentration of the noncrystallizing component. In the two-phase region the melting temperature becomes invariant with composition as a consequence of the Phase Rule. At lower concentrations of the crystallizing polymer, melting again takes place into a homogeneous melt. Therefore, a decrease in melting temperature will be observed. These expectations are identical to those expected and observed for polymer–low molecular weight diluent mixtures.

An example of the schematic illustrated in Fig. 4.10c is found in mixtures of poly(vinylidene fluoride) and poly(ethyl acrylate). The phase diagram for this blend is given in Fig. 4.11.(72) Here, the melting temperatures represent equilibrium values obtained by extrapolative methods. The solid points represent the boundary of part of the heterogeneous region obtained by cloud point measurements. The expected major features of the melting point–composition curve are observed. The invariance of T_m in the two-phase melt region is apparent, as is the expected slight decrease in T_m in the two homogeneous melt regions.

When two crystallizable polymers are immiscible in the melt the crystallization of the two species should occur independently of one another. Two sets of melting temperatures are expected and are indeed observed.(73–76) In general, with but minor variations, the melting temperatures are independent of composition. A typical example of this type behavior is illustrated by blends of syndiotactic 1,2-poly(butadiene) with trans-1,4-poly(butadiene).(74) Phase separation in the melt would be expected in this blend since two nonpolar polymers, without any specific intermolecular interactions are involved. Wide-angle x-ray patterns of the melt confirm this expectation. Two haloes, each characteristic of the individual species, are observed. Each of the species crystallizes separately as evidenced by separate sets of Bragg spacings, typical of each. Differential scanning calorimetry gives two different melting temperatures, corresponding to each of the components. The directly observed melting temperatures of the two components are plotted as a fraction of the trans-1,4-poly(butadiene) in Fig. 4.12. The appearance of two different melting temperatures is readily apparent. The melting temperature of the syndiotactic 1,2-poly(butadiene) is independent of concentration, except

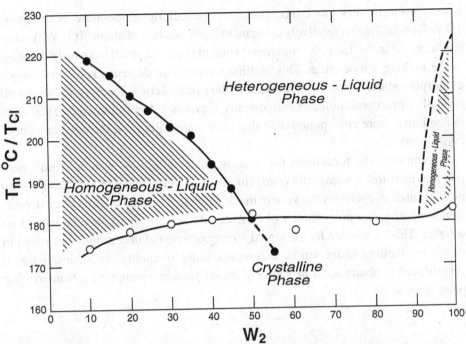

Fig. 4.11 Phase diagram for blends of poly(vinylidene fluoride)–poly(ethyl acrylate). Weight fraction, w_2, of poly(vinylidene fluoride) component. (Data replotted from Briber and Khoury (72))

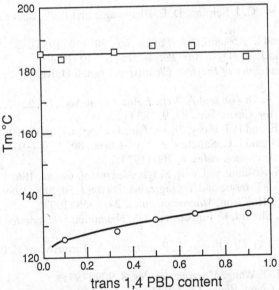

Fig. 4.12 Melting temperature of each component in blends of syndiotactic 1,2-poly (butadiene) □ with trans-1,4-poly(butadiene) ○. (From Nir and Cohen (74))

when present in very low concentrations. The melting temperature of the trans-1,4-poly(butadiene) is relatively constant to a composition of about 50%. With a further increase in the 1,2-poly(butadiene) content there is a small but steady decrease in the melting temperature. This melting temperature decrease is a consequence of morphological factors. The higher melting syndiotactic 1,2-poly(butadiene) will crystallize first upon cooling. Consequently, the trans-1,4-poly(butadiene) will crystallize into a more constrained melt that must eventually limit the development of crystallinity.

In summary, the formalism for treating the equilibrium aspects of polymer–polymer mixtures is straightforward. However, a careful and distinct classification must be made of each specific system that is studied. Since the melting point depression is a colligative property it will by necessity be very small for high molecular weights. This inexorable fact, coupled with experimental uncertainties involved in measuring melting temperatures and extrapolating to equilibrium values, makes it very difficult to obtain accurate values of the interaction parameter χ_{12} from melting point studies.

References

1. Flory, P. J., *Principles of Polymer Chemistry*, Cornell University Press (1953) p. 555.
2. Barlow, J. W. and D. R. Paul, *Ann. Rev. Mater. Sci.*, **11**, 299 (1981).
3. Wahrmund, D. C., R. E. Bernstein, J. W. Barlow and D. R. Paul, *Polym. Eng. Sci.*, **18**, 677 (1978).
4. Coleman, M. M., C. J. Serman, D. E. Bhagwagar and P. C. Painter, *Polymer*, **31**, 1187 (1990).
5. Walsh, D. J. and S. Rostami, *Adv. Polym. Sci.*, **70**, 119 (1985).
6. Allegra, G. and J. W. Bassi, *Adv. Polym. Sci.*, **6**, 549 (1969).
7. Flory, P. J., *Principles of Polymer Chemistry*, Cornell University Press (1953) pp. 495ff.
8. Flory, P. J., R. A. Orwoll and A. Vrij, *J. Am. Chem. Soc.*, **86**, 3507, 3515 (1964).
9. Flory, P. J., *J. Am. Chem. Soc.*, **87**, 9, 1833 (1965).
9a. Eichinger, B. E. and P. J. Flory, *Trans. Faraday Soc.*, **64**, 2035 (1968).
10. Lacombe, R. H. and I. C. Sanchez, *J. Phys. Chem.*, **80**, 2568 (1976).
11. Master, L. P., *Macromolecules*, **6**, 760 (1973).
12. Walsh, D. J., S. Rostami and V. B. Singh, *Makromol. Chem.*, **186**, 145 (1985).
13. Kammer, H. W., T. Inoue and T. Ougizawa, *Polymer*, **30**, 888 (1989).
14. Jo, W. H. and I. H. Kwon, *Macromolecules*, **24**, 3368 (1991).
15. Hamada, F., T. Shiomi, K. Fujisawa and A. Nakajima, *Macromolecules*, **13**, 729 (1980).
16. Chen, X., J. Yin, G. C. Alfonso, E. Pedemonte, A. Turturro and E. Gattiglia, *Polymer*, **39**, 4929 (1998).
17. Nishi, T. and T. T. Wang, *Macromolecules*, **8**, 909 (1975).
18. Scott, R. L., *J. Chem. Phys.*, **17**, 279 (1949).
19. Sanchez, I. C., *Polymer*, **30**, 471 (1989).

20. Flory, P. J., *Discuss. Faraday Soc.*, **49**, 7 (1970).
21. Yeh, G. S. Y. and S. L. Lambert, *J. Polym. Sci., Pt. A2*, **10**, 1183 (1972).
22. Keith, H. D. and F. J. Padden, *J. Appl. Phys.*, **35**, 1270 (1964).
23. Warner, F. P., W. J. MacKnight and R. S. Stein, *J. Polym. Sci.: Polym. Phys. Ed.*, **15**, 2113 (1977).
24. Martuscelli, E., G. Demma, E. Driole, L. Nicolais, S. Spina, H. B. Hopfenberg and V. T. Stannet, *Polymer*, **20**, 571 (1979).
25. Tsuji, H. and Y. Ikada, *J. Appl. Polym. Sci.*, **58**, 1793 (1995).
26. Pearce, R., J. Jesudason, W. Orts, R. H. Marchessault and S. Bloembergen, *Polymer*, **33**, 4647 (1992).
27. Kumagai, Y. and Y. Doi, *Makromol. Chem. Rapid Comm.*, **13**, 179 (1992).
28. Abe, H., Y. Doi, M. M. Satkowzki and I. Noda, *Macromolecules*, **27**, 50 (1994).
29. Kwei, T. W. and H. L. Frisch, *Macromolecules*, **6**, 1267 (1976).
30. Plans, J., W. J. MacKnight and F. E. Karasz, *Macromolecules*, **17**, 810 (1984).
31. Hsiao, B. S. and B. B. Sauer, *J. Polym. Sci.: Pt. B: Polym. Phys.*, **31**, 901 (1993).
32. Pompe, G., L. Haubler and W. Winter, *J. Polym. Sci.: Polym. Phys.*, **34**, 211 (1996).
33. Robeson, L. M. and A. B. Furtek, *J. Appl. Polym. Sci.*, **23**, 645 (1979).
34. Katime, I. A., M. S. Anasagasti, M. C. Peleteiro and R. Valenciano, *Eur. Polym. J.*, **23**, 907 (1987).
35. Lezcano, E. G., C. S. Coll and M. G. Prolongo, *Polymer*, **37**, 3603 (1996).
36. Eguiburu, J. L., J. J. Iruin, M. J. Fernandez-Beride and J. San Roman, *Polymer*, **39**, 6891 (1998).
37. Paul, D. R., J. W. Barlow, R. E. Bernstein and D. C. Wahrmund, *Polym. Eng. Sci.*, **18**, 1225 (1978).
38. Morra, B. S. and R. S. Stein, *J. Polym. Sci.: Polym. Phys. Ed.*, **20**, 2243 (1982).
39. Rim, P. B. and J. P. Runt, *Macromolecules*, **17**, 1520 (1984).
40. Jonza, J. M. and R. S. Porter, *Macromolecules*, **19**, 1946 (1986).
41. Alfonso, G. C. and T. P. Russell, *Macromolecules*, **19**, 1143 (1986).
41a. Yam, W. Y., J. Ismail, H. W. Kammer, H. Schmidt and C. Kolmmerlowe, *Polymer*, **40**, 5545 (1999).
42. Imken, R. L, D. R. Paul and J. W. Barlow, *Polym. Eng. Sci.*, **16**, 593 (1976).
43. Huo, P. P. and P. Cebe, *Macromolecules*, **26**, 3127 (1993).
44. Liu, D. S., W. B. Liau and W. Y. Chiu, *Macromolecules*, **31**, 6593 (1998).
45. Galin, M., *Makromol. Chem. Rapid Comm.*, **5**, 119 (1984).
46. Alfonso, G. C., A. Turturro, M. Scandola and G. Ceocorulli, *J. Polym. Sci.: Pt. B: Polym. Phys.*, **27**, 1195 (1989).
47. Eshuis, E., E. Roerdiuk and G. Challa, *Polymer*, **23**, 735 (1982).
48. Kalfoglou, N. K., *J. Polym. Sci.: Polym. Phys. Ed.*, **20**, 1259 (1982).
49. Rim, P. B. and J. Runt, *Macromolecules*, **16**, 762 (1983).
50. Runt, J., P. B. Rim and S. E. Howe, *Polym. Bull.*, **11**, 517 (1984).
51. Alamo, R. G., B. D. Viers and L. Mandelkern, *Macromolecules*, **28**, 3205 (1995).
52. Runt, J. and K. P. Gallagher, *Polym. Comm.*, **32**, 180 (1991).
52a. Rostami, S. D., *Eur. Polym. J.*, **36**, 2285 (2000).
53. Sauer, B. B., B. S. Hsiao and K. L. Farron, *Polymer*, **37**, 445 (1996).
54. Painter, P. C., S. L. Shenoy, D. E. Bhagwagar, J. Fischburn and M. M. Coleman, *Macromolecules*, **24**, 5623 (1991).
55. Inoue, M., *J. Polym. Sci. Pt. A2*, **1**, 3427 (1963).
55a. Azuma, Y., N. Yoshie, M. Sakurai, Y. Inoue and R. Chujo, *Polymer*, **33**, 4763 (1992).
55b. Guo, M. and H. G. Zachmann, *Macromolecules*, **30**, 2746 (1997).
55c. Matsuda, T., T. Shimomura and M. Hirami *Polymer J.*, **31**, 795 (1999).

56. Pennings, J. P. and R. St. John Manley, *Macromolecules*, **29**, 77 (1996).
56a. Escola, A. and R. S. Stein, *Multiphase Polymers*, S. L. Cooper and G. M. Este, eds., Advances in Chemistry Series, American Chemical Society (1979).
57. Guerra, G., F. E. Karasz and W. J. MacKnight, *Macromolecules*, **19**, 1935 (1986).
58. Sham, C. K., G. Guerra, F. E. Karasz and W. J. MacKnight, *Polymer*, **29**, 1016 (1988).
59. Guo, Q., *Makromol. Chem.*, **191**, 2639 (1990).
60. Zhang, H. and R. E. Prud'homme, *J. Polym. Sci.: Pt. B: Polym. Phys.*, **24**, 723 (1987).
61. Harris, J. E. and L. M. Robeson, *J. Polym. Sci.: Pt. B: Polym. Phys.*, **25**, 311 (1987).
62. Hu, S. R., T. Kyu and R. S. Stein, *J. Polym. Sci.: Pt. B: Polym. Phys. Ed.*, **25**, 71 (1987).
62a. Tashiro, K., K. Imanishi, Y. Izume, M. Kobayashi, K. Kobayashi, M. Satoh and R. S. Stein, *Macromolecules*, **28**, 8477 (1995).
63. Mandelkern, L., R. G. Alamo, G. D. Wignall and F. C. Stehling, *Trends Polym. Sci.*, **4**, 377 (1996).
64. Ueda, M. and R. A. Register, *J. Macromol. Sci., Phys.*, **335**, 23 (1996).
65. Natta, G., G. Allegra, I. W. Bassi, D. Sianesi, G. Caporiccio and E. Torti, *J. Polym. Sci., A*, **3**, 4263 (1965).
66. Montaudo, G., P. Maravigna, P. Finocchiaro and G. Centineo, *J. Polym. Sci.: Polym. Chem. Ed.*, **11**, 65 (1973).
67. Mandelkern, L., F. L. Smith and E. K. Chan, *Macromolecules*, **22**, 2663 (1989).
68. Hirami, M. and T. Matsuda, *Polym. J.*, **31**, 801 (1999).
69. Smith, P. and K. H. Garner, *Macromolecules*, **18**, 1222 (1985).
70. Burghardt, W. R., *Macromolecules*, **22**, 2482 (1989).
71. Kammer, H., *Polymer*, **37**, 1 (1986).
72. Briber, R. M. and F. Khoury, *Polymer*, **28**, 38 (1987).
73. Aret-Azar, A., J. N. Hay, B. J. Maislen and N. Walker, *J. Polym. Sci.: Polym. Phys. Ed.*, **18**, 637 (1980).
74. Nir, M. M. and R. E. Cohen, *Rubber Chem. Tech.*, **66**, 295 (1993).
75. The, J. W., *J. Appl. Polym. Sci.*, **28**, 605 (1983).
76. Thomann, R., J. Kessler, B. Rudolf and R. Mülhaupt, *Polymer*, **37**, 2635 (1996).

5

Fusion of copolymers

5.1 Introduction

The introduction into a crystallizable homopolymer chain of units that are chemically or structurally different from the predominant chain repeating unit can be expected to alter its crystallization behavior. Many different types of structural irregularities can be incorporated. There can of course be chemically dissimilar repeating units or co-units. Depending upon the chemical nature of the major chemical unit, geometric or stereo isomers, as well as regio defects can be introduced into the chain. For example, although a polymer may be termed isotactic or syndiotactic a perfectly regular structure cannot be inferred without direct structural evidence. Most often, the stereo configurations of the units are not complete and the polymer is properly treated as a copolymer. Branch points and cross-links represent other types of structural irregularities.[1] Some structural irregularities can be quite subtle in nature. In this context copolymer behavior is observed with chemically identical repeating units, as well as those that are distinctly different. Quite obviously the type and concentration of the co-units will be important. In low molecular weight binary mixtures attention is focused on the molecules. In contrast, with copolymers emphasis must be given to the sequences, their length and distribution. The manner in which the co-units are distributed along the chain, i.e. the sequence distribution, is of primary, underlying importance. The major concern of this chapter is the course of fusion of copolymers as a function of temperature and the dependence of the melting temperature on the nature, type and distribution of the co-units. The crystallization kinetics and the morphological and structural features of copolymers will be discussed in subsequent volumes.

---

[1] The influence of intermolecular cross-links needs to be treated separately. Although they are clearly structural irregularities, their influence on the crystallization behavior requires special treatment, because the properties that result depend on the physical state in which the cross-links are introduced.(1) Consequently a special chapter, Chapter 7, is devoted to this subject.

The first step in the analysis of copolymer crystallization is the development of quantitative concepts that are based on equilibrium considerations. Subsequently, deviations from equilibrium and a discussion of real systems will be undertaken. Problems involving the crystallization and melting of copolymers cannot in general be uniquely formulated since two phases and at least two species are involved. The disposition of the species among the phases needs to be specified. It cannot be established *a priori* by theory. This restraint is not unique to polymeric systems. It is a common experience in analyzing similar problems that involve monomeric components.(2) Thus, in the development of any equilibrium theory a decision has to be made prior to undertaking any analysis of the disposition of the co-units between the phases. Theoretical expectations can then be developed based on the assumptions made.

Two possibilities exist with respect to the disposition of the co-units. In one case the crystalline phase remains pure, i.e. the co-units are excluded from entering the crystal lattice. In the other, the co-unit is allowed to enter the lattice on an equilibrium basis. Typical examples of the latter would be akin to compound formation, or isomorphous replacement, where one unit can replace the other in the lattice. In either of these two main categories ideal conditions are first calculated and analyzed. Subsequently nonideal contributions to both phases can be considered while still maintaining equilibrium. There is an analogy here to solution theory and to gases, where equilibrium conditions are established first. In the next step, nonequilibrium effects in either or both phases can be brought to bear on the problem. It needs to be recognized that deviations from equilibrium in copolymers exist and are in fact important.

In general, one can expect to observe the types of phase diagrams that are found with low molecular weight systems in crystal–liquid equilibrium. For polymeric systems the liquid composition can usually be determined in a straightforward manner. However, establishing the composition in the solid state is quite difficult and presents a major problem in properly analyzing phase diagrams.

5.2 Equilibrium theory

5.2.1 Crystalline phase pure

The theory for the case where the crystalline phase remains pure has a mature development and is rich in concepts. This case will be treated first, utilizing Flory's classical work.(3,4) A model copolymer is considered that contains only one type of crystallizable unit, designated as an A unit. The noncrystallizable comonomeric unit will be designated as a B unit. In the initial molten state the A units occur in a specified distribution that is determined by the copolymerization mechanism. Upon crystallization, with the exclusion of the B units from the lattice, the sequence distribution in the residual noncrystalline melt is altered. The problem in copolymer crystallization is, thus, more complex than just having a set of isolated impurities as in the case of the crystallization of a binary mixture of low molecular weight compounds. In polymers, both the composition and sequence distribution of the residual melt is altered upon crystallization.

In this model the crystalline state is comprised of crystallites of varying lengths. The length of a given crystalline sequence is expressed by ζ, the number of A units of a given chain that traverse one end of the crystallite to the other. A sequence of A units tends to occur in crystallites which are not much shorter than themselves. The development of a crystallite in the chain direction is restricted by the occurrence of a noncrystallizing B unit. The lateral development of crystallites of length ζ is governed by the concentration of sequences of sufficient length in the residual melt and the free energy decrease that occurs upon the crystallization of a sequence of ζ A units.

Following Flory (4) these concepts can be formulated in a quantitative manner so that a description of the ideal equilibrium crystalline state results. The probability that a given A unit in the noncrystalline amorphous region is located within a sequence of at least ζ such units is defined as P_ζ. The probability that a unit chosen at random from the noncrystalline region is an A unit, and also a member of a sequence of ζ A units that are terminated at either end by B units is represented by w_j. The probability that the specific A unit selected is followed in a given direction by at least $\zeta - 1$ similar units can be expressed as

$$P_{\zeta,j} = \frac{(j - \zeta + 1)w_j}{j} \qquad j \geq \zeta$$
$$P_{\zeta,j} = 0 \qquad j < \zeta \tag{5.1}$$
$$P_\zeta = \sum_{j=\zeta}^{\infty} P_{\zeta,j} = \sum_{j=\zeta}^{\infty} \frac{(j - \zeta + 1)w_j}{j}$$

Solving Eq. (5.1) for successive values of ζ yields the difference equation

$$w_\zeta = \zeta(P_\zeta - 2P_{\zeta+1} + P_{\zeta+2}) \tag{5.2}$$

Equations (5.1) and (5.2) represent general properties of the noncrystalline region. They do not depend on the presence or absence of crystallites. In the completely molten polymer, prior to the development of any crystallinity

$$w_\zeta^0 = \frac{X_A \zeta v_\zeta^0}{N_A} \tag{5.3}$$

Here N_A is the total number of A units in the copolymer, X_A the corresponding mole fraction, and v_ζ^0 the number of sequences of A units initially present in the melt. If we assume that the probability, p, of an A unit being succeeded by another A unit is independent of the number of preceding A units then Eq. (5.3) can be expressed as[2]

$$w_\zeta^0 = \frac{\zeta X_A (1-p)^2 p^\zeta}{p} \tag{5.4}$$

Digressing for a moment, it should be recognized that the sequence propagation probability, p, in the melt can be related to the comonomer reactivity ratio for addition type copolymerization. Formulating copolymerization kinetics in the classical manner(5), we let F_A represent the fraction of monomer M_A in the increment of copolymer formed at a given stage of the polymerization. Then one can write

$$F_A = \frac{r_A f_A^2 + f_A f_B}{r_A f_A^2 + 2 f_A f_B + r_B f_B^2} \tag{5.5}$$

where r_A and r_B are the respective monomer reactivity ratios, and f_A and f_B represent the mole fractions of the unreacted monomers at this point. The sequence propagation parameter p_A can be expressed as

$$p_A = \frac{r_A f_A}{r_A f_A + f_B} \tag{5.6}$$

It is then found that

$$p_A / F_A = 1 + \left(\frac{1}{1-y}\right)^2 (r_A r_B - 1) \tag{5.7}$$

where $y \equiv r_A f_A / (1 - f_A)$. For the special case where the product of reactivity ratios $r_A r_B = 1$, $p_A = F_A$.

Returning to the main theme, one finds that by combining Eqs. (5.3) and (5.4) that

$$P_\zeta^0 = X_A p^{\zeta - 1} \tag{5.8}$$

for the completely molten polymer, represented by the superscript zero. For the crystalline polymer under thermodynamic equilibrium, the probability P_ζ^e that in the noncrystalline region a given A unit is located in a sequence of at least ζ such units is given by

$$P_\zeta^e = \exp\left(-\frac{\Delta G_\zeta}{RT}\right) \tag{5.9}$$

---

[2] In the present context the parameter p is defined in terms of Bernoullian trials. The problem can be further generalized to the case where p is influenced by the penultimate group. This situation will be discussed shortly. Since the crystalline state remains pure in the case under consideration p refers to the liquid, or amorphous state. Strictly speaking it should be designated as p_A^l, the superscript referring to the liquid state.

Here ΔG_ζ is the standard free energy of fusion of a sequence of ζ A units for a crystallite ζ units long. ΔG_ζ can be expressed as

$$\Delta G_\zeta = \zeta \Delta G_u - 2\sigma_{eq} \tag{5.10}$$

Crystallites are assumed to be sufficiently large in the direction transverse to the chain axis so that the contribution of the excess lateral surface free energy to Eq. (5.10) can be neglected. Equation (5.9) can be expressed as

$$P_\zeta^e = \frac{1}{D} \exp(-\zeta\theta) \tag{5.11}$$

where

$$\theta = \frac{\Delta H_u}{R} \left(\frac{1}{T} - \frac{1}{T_m^0} \right) \tag{5.12}$$

and

$$D = \exp\left(-\frac{2\sigma_{eq}}{RT} \right) \tag{5.13}$$

If crystallites of length ζ, $\zeta + 1$ and $\zeta + 2$ are present, and are in equilibrium, within the melt, then Eqs. (5.2) and (5.11) can be combined to give

$$w_\zeta^e = \zeta D^{-1} [1 - \exp(-\theta)]^2 \exp(-\zeta\theta) \tag{5.14}$$

Equation (5.14) expresses the residual concentration in the melt of sequences of A units that are ζ units long.

The necessary and sufficient conditions for crystallization can be stated as

$$P_\zeta^0 > P_\zeta^e \tag{5.15}$$

for one or more values of ζ. The condition $w_\zeta^0 > w_\zeta^e$, for one or more values of ζ, is a necessary but not sufficient condition for crystallization. Equation (5.4) for the completely molten polymer and Eq. (5.14) for the equilibrium crystalline polymer are functions of the sequence length ζ. Thus, for copolymers the sequence length distributions within and outside the crystallite determine the condition for phase equilibrium. For low molecular weight systems only the concentration of the species would be involved. The results for copolymers represent a more generalized case. There must therefore be a critical value $\zeta = \zeta_{cr}$ at which these two distributions are equal. This condition is given by

$$\zeta_{cr} = \frac{-\{\ln(DX_A/p) + 2\ln[(1-p)/(1-e^{-\theta})]\}}{\theta + \ln p} \tag{5.16}$$

For values of $\zeta < \zeta_{cr}$, w_ζ^e is greater than w_ζ^0; for $\zeta > \zeta_{cr}$ the converse holds. Thus, ζ_{cr} represents the limiting size above which crystallites can exist at equilibrium. Smaller values of ζ cannot be maintained at equilibrium.

The theory that has been outlined allows for an estimate of the fraction of A units that are crystalline at temperatures below the melting temperature. This estimate can be obtained by summing all the sequences of A units that participate in crystallites. This procedure leads to a slight overestimate of the degree of crystallinity since sequences greater than ζ units in length can participate in crystallites which are only ζ units long. If w_ζ^c is the concentration of sequences of ζ units involved in a crystallite then (4)

$$w_\zeta^c = w_\zeta^0 - w_\zeta^e \tag{5.17}$$

The fraction of A units in the crystalline state, w_c, is given by

$$w_c = \sum_{\zeta=\zeta_{cr}}^{\infty} w_\zeta^c = \sum_{\zeta=\zeta_{cr}}^{\infty} \left(w_\zeta^0 - w_\zeta^e\right) \tag{5.18}$$

Using the expression for w_ζ^0 and w_ζ^e from Eqs. (5.3) and (5.14) one obtains

$$w_c = \frac{X_A}{p}(1-p)^2 p^{\zeta_{cr}}\{p(1-p)^{-2} - e^{-\theta}(1-e^{-\theta})^{-2}$$
$$+ \zeta_{cr}[(1-p)^{-1} - (1-e^{-\theta})^{-1}]\} \tag{5.19}$$

Equation (5.19) is an expression for the degree of crystallinity as a function of the reduced temperature θ, as it depends on the copolymer structure embodied in the parameters p and the interfacial free energy σ_{eq}.

It is important to keep in mind that the theoretical development outlined above, and its implications, are for equilibrium conditions at, and below, the melting temperature. It requires the participation of all sequences above a critical value, particularly the very long ones. All must be in extended form. Very practical and important matters such as the kinetic barriers to the crystallization, possible folding of the chains, defects within the crystallites, as well as other nonequilibrium phenomena are not taken into account at this point. However, the ideal equilibrium requirements serve as reference base from which nonideal contributions as well as nonequilibrium behavior can be treated.

Theoretical curves of the degree of crystallinity as a function of temperature can be constructed from Eq. (5.19) by the assignment of the appropriate parameters, θ, D and p. For purposes of illustration we assume a random copolymer, where $p = X_A$ (cf. seq.). Typical plots of the degree of crystallinity as a function of temperature, for the indicated parameters, are given in Fig. 5.1. It is found, for the random copolymers being considered, that at comparable temperatures the theoretical equilibrium degree of crystallinity is severely reduced as the concentration of the noncrystallizing B units increases. The fusion of copolymers is expected to occur over a very broad temperature range. This conclusion is in marked contrast to

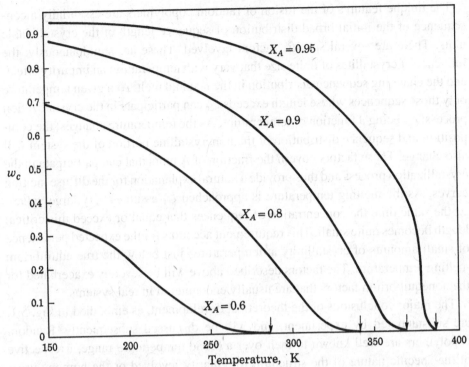

Fig. 5.1 Theoretical plot, according to Eq. (5.19) of the fraction of crystalline A units, w_c, as a function of temperature for random type copolymers of different compositions. Short vertical arrows indicate melting temperature T_m of copolymer. For case illustrated, $T_m^0 = 400$ K, $\Delta H_u = 10^3$ cal mol^{-1}, and ln $D = -1$.

what is expected, and observed, in homopolymers. For random copolymers, a small but significant amount of crystallinity persists over an appreciable temperature interval before the transformation is complete. The breadth of this interval increases substantially with the concentration of noncrystallizable co-units. Although the absolute amount of crystallinity that persists is small, it is significant from the point of view of theory and of its influence on properties. In general, therefore, random copolymers can be expected to melt broadly and to attain only low levels of crystallinity at high co-unit contents.

The diffuse nature of the fusion curves given in Fig. 5.1 does not appear to be typical of a first-order phase transition. However, these melting curves are natural consequences of the constitution of random copolymers and the requirement that the B units are restricted to the noncrystalline phase. Theoretically, at the melting temperature w_c and all its derivatives vanish. Hence a true discontinuity exists at this temperature. There is then a true thermodynamic melting temperature that is representative of the disappearance of crystallites composed of very long sequences. Its detection by experimental methods may very well be difficult.

The unique features of the fusion of random copolymers are essentially a consequence of the initial broad distribution of sequence length of the crystallizable units. There are several specific factors involved. These are simultaneously: the influence of crystallites of finite size that vary with temperature; an impurity effect; and the changing sequence distribution in the residual melt. At a given temperature only those sequences whose length exceeds ζ_{cr} can participate in the crystallization process; ζ_{cr} being a function of temperature. As the temperature changes, the composition and sequence distribution of the noncrystalline portion of the system will also change. These factors govern the fraction of A units that can participate in the crystallization process and thus provide a natural explanation for the diffuse melting curves. As the melting temperature is approached ζ_{cr} assumes very large values. At the same time the concentration of sequences that equal or exceed this critical length becomes quite small. This requirement accounts for the expected persistence of small amounts of crystallinity at temperatures just below the true equilibrium melting temperature. The factors described above will be severely exacerbated for the nonequilibrium factors that are usually encountered in real systems.

The major conclusions of the theoretical development, as embodied in Fig. 5.1, are substantiated by experiment, and will be discussed subsequently. Random copolymers are well known to melt over a broad temperature range, irrespective of the specific nature of the structural irregularity involved or the homopolymer from which they are derived. The parent homopolymers which display these features include poly(esters), poly(amides), poly(olefins), fluorocarbon polymers and crystallizable vinyl polymers to cite but a few examples. In some instances the structural irregularities can be subtle in nature so that the copolymeric character of the chain, and its subsequent crystallization behavior, is not always recognized.

From the theory outlined above it is also possible to develop melting point relations. Returning to the inequality of Eq. (5.15) and substituting the expressions for P_ζ^0 and P_ζ^e from Eqs. (5.8) and (5.11) respectively one obtains (4)

$$\frac{X_A}{p}p^\zeta > \frac{1}{D}e^{-\theta\zeta} \tag{5.20}$$

Except for the special case of copolymers that exhibit a very strong tendency for alternation, $1/D$ will be greater than X_A/p. Thus, the inequality of Eq. (5.20) becomes

$$\frac{1}{T} - \frac{1}{T_m^0} > -\frac{R}{\Delta H_u}\ln p \tag{5.21}$$

A limiting temperature must therefore exist above which the inequality of Eq. (5.21) cannot be fulfilled. Above this temperature crystallization cannot occur. This limiting temperature is the ideal equilibrium melting temperature T_m of the copolymer.

Therefore

$$\frac{1}{T_m} - \frac{1}{T_m^0} = \frac{R}{\Delta H_u} \ln p \qquad (5.22)$$

The melting temperature depends only on the parameter p and not directly on composition. The derivation of this melting point relation is such that the interfacial contribution, as was found in homopolymers, is absent.

Equation (5.22) can be derived in an alternative manner by treating the copolymer as an ideal binary mixture.(6) In analogy to the classical statistical derivation of Raoult's law in three dimensions, we seek the number of distinguishable ways, Ω, that the sequences of A and B units can be arranged in the linear chain. Orr has shown that(7)

$$\Omega = \frac{N_A! N_B!}{\prod\limits_{\zeta_A=1}^{\infty} N_{\zeta_A}! \prod\limits_{\zeta_B=1}^{\infty} N_{\zeta_B}!} \qquad (5.23)$$

Here N_A and N_B are the total number of sequences comprised of A and B units respectively; N_{ζ_A} is the number of sequences containing ζ A units and N_{ζ_B} is defined similarly. Following Orr (7), N_A, N_B, N_{ζ_A} and N_{ζ_B} can be related to the sequence propagation parameter p. For the case where the crystalline phase is pure there is only one arrangement of the A units in the crystal. The number of arrangements in the melt is given by Eq. (5.23). Since in this ideal calculation there are no enthalpic or other entropic contributions to the free energy of mixing, ΔG_M, one then finds that

$$\Delta G_M / kT = N_{0A}[(1-p)\ln(1-p) + \ln p - (1-p)\ln p] \qquad (5.24)$$

Only terms that depend on the number of A units in the chain, N_{0A} are included in Eq. (5.24). Terms which involve B units are not important for present purposes since it has been assumed that the crystalline phase is pure and we are only concerned with the melting of A units. From Eq. (5.24) it follows that the chemical potential of an A unit in the copolymer, $\mu_{A,c}$, relative to that of the parent homopolymer, $\mu_{A,h}$, is given by

$$\mu_{A,c} - \mu_{A,h} = RT \ln p \qquad (5.25)$$

Equation (5.25) is based solely on ideal mixing. This equation can then be applied to problems involving crystal–liquid phase equilibrium of copolymers. For a crystalline phase that contains only A units, $p = 1$. It immediately follows from the conditions for phase equilibrium that

$$\frac{1}{T_m} - \frac{1}{T_m^0} = \frac{-R}{\Delta H_u} \ln p \qquad (5.26)$$

Equation (5.26) is based on the ideal mixing law and the stipulation that total equilibrium prevails throughout the system. It thus represents the ideal melting point relation of copolymers under equilibrium conditions. There is an analogy here of Raoult's law for real solutions and the ideal gas law for imperfect gas theory. Alterations can be made to Eq. (5.26) by adding possible enthalpic and noncombinatorial entropic contributions to the mixing free energy while still maintaining equilibrium and a pure crystalline phase. An analogous relationship exists between ideal solutions, regular solutions and other nonideal ones. Hence, it is possible that the crystalline phase can remain pure while Eq. (5.26) is not obeyed, even though equilibrium melting temperatures are used. It cannot be overemphasized that, even under equilibrium conditions and a pure crystalline phase, observed equilibrium melting temperatures can be greater or smaller than the values specified by Eq. (5.26). Observed deviations from Eq. (5.26) do not by themselves indicate that the crystalline phase is no longer pure, i.e. the B units enter the lattice.

A major consequence of Eq. (5.26) is the expectation that the melting temperature of a copolymer, where only one type unit is crystallizable, depends only on the sequence propagation probability p and not directly on composition. This result is rather unusual, and is unique to long chain molecules. Considering the major categories of copolymer structure we find in the extremes that

for a random copolymer $\qquad p = X_A$
for a block copolymer $\qquad p \gg X_A$
and for an alternating copolymer $\qquad p \ll X_A$

Many real systems will not fit these conditions exactly but will fall between the specifications cited above. These relations of p to X_A are based on the assumption that the same crystal structure of the homopolymer is involved and the melt is homogeneous. These conditions are not always fulfilled. It is predicted that, depending on the sequence arrangement, very large differences can be obtained between the melting temperatures of copolymers of exactly the same composition. For example, for an ideally structured block copolymer there is only one arrangement of sequences. Therefore $p = 1$ and T_m will equal T_m^0. These conclusions are based on ideal equilibrium theory and can be tempered by structural and morphological factors.

The theory outlined has been predicated on the assumption that the crystallizable sequences are propagated by Bernoullian trials, i.e. the probability of a given placement is independent of the nature of the preceding unit. The probability of a given type addition can be affected by the preceding placement. When the penultimate unit (or structure) is important, the process can be treated by first-order Markovian statistics. It is of interest to apply this procedure to chains with stereochemical differences, such as isotatic and syndiotactic placements. By considering only the effect of the last placement, the following scheme is evolved.(8,9,9a) If the last two units in the chain are in isotactic placement relative to one another, α_i is defined as the conditional probability that the addition of the next unit will also result in an isotactic placement, while β_i represents the conditional probability that a syndiotactic placement will result. Similarly, α_s represents the conditional probability that a syndiotactic placement will be followed by an isotactic one, while β_s represents a syndiotactic placement following a syndiotactic one. Then

$$\alpha_i + \beta_i = 1$$
$$\alpha_s + \beta_s = 1 \tag{5.27}$$

The unconditional probability α that two adjacent monomer units selected at random are in isotactic placement with one another is obtained by summing over all the possible outcomes of the previous placement. Thus

$$\alpha = \alpha\alpha_i - \beta\alpha_s$$
$$\beta = \beta\beta_s + \alpha\beta_i \tag{5.28}$$

When stereosequences are generated by this process, X_A can be identified with α, and p with α_i, when only units in isotactic placement crystallize. By applying the necessary and sufficient conditions for the crystallization of copolymers, Eq. (5.20) can be written as

$$\alpha\alpha_i^{\zeta-i} > \frac{1}{D}e^{-\zeta\theta} \tag{5.29}$$

When a favorable correlation exists in the stereosequence generation so that $\alpha > \alpha_i$, the limiting temperature at which crystallization can occur (the melting temperature) is expressed as

$$\frac{1}{T_m} - \frac{1}{T_m^0} = -\frac{R}{\Delta H_u}\ln\alpha_i \tag{5.30}$$

Hence it does not suffice to specify solely the compositional variable α in order to express the melting point–composition relation. The conditional probability α_i is also needed. However, if there is no correlation ($\alpha = \alpha_i$) or unfavorable correlation ($\alpha < \alpha_i$) a melting point relation similar to Eq. (5.26) is obtained, with p replaced by α. Similar results are also obtained if only units in syndiotactic placement are capable of crystallizing. An equivalent analysis has also been given.(9a)

This section has been concerned solely with the ideal equilibrium conditions for a pure crystalline phase. Specifically it has been assumed that sequences of the same length form crystallites of that length i.e. the crystallites are composed of extended sequences. All the sequences crystallize according to the requirements of phase equilibrium. Very long sequences, except for block copolymers, will be scarce and few in number and must also crystallize according to these conditions. It will, therefore, not be surprising, or unexpected, to learn that these rather stringent conditions are rarely, if ever, achieved by experiment. Consequently the observed melting temperatures and crystallinity levels are affected accordingly. An important problem is to distinguish whether discrepancies are due to either shortcomings in the ideal theory, with the need to add terms, or the inability to attain equilibrium. Before discussing modifications that have been proposed to the theory described, it is important to remove the stricture that the crystalline phase is pure and allow the B units to enter the lattice and participate in the crystallization. This aspect of the problem is discussed in the following section.

5.2.2 Comonomers in both phases

When B units enter the lattice a distinction has to be made as to whether they do so as an equilibrium basis or as a set of nonequilibrium defects. This is an important distinction since the analysis of the problem is quite different in the two cases. At present we focus attention on the equilibrium case. When both comonomers are present in the crystalline and liquid phases the analysis of the equilibrium condition is more complex as compared to when the equilibrium phase remains pure. The necessary requirements, and consequences thereof, can however be stated in a formal manner.

In addition to the uniformity of the temperature and pressure two further quantities need to be satisfied. The chemical potential of each of the species, A and B, in each of the phases must be equal. Thus

$$\mu_{Al} = \mu_{Ac}$$
$$\mu_{Bl} = \mu_{Bc} \tag{5.31}$$

For monomeric systems the chemical potentials of the species in each of the phases is specified in terms of either composition, or activity. The melting temperature relations are then derived in a straightforward manner.(2) For an ideal mixture of low molecular weight species the free energy of mixing in each phase is determined by a Raoult's law type calculation, i.e. only the combinatorial entropy is considered. The composition is then expressed in terms of mole fraction. The equilibrium melting temperature in terms of the composition of each phase is then specified.

For copolymers one can in principle proceed, in analogy with Eq. (5.23), to calculate the number of distinguishable ways the different sequences in the crystalline phase can be arranged. The sequence distribution in the pure melt will be unaltered and determined by the copolymerization mechanism. The sequence distribution will depend on the concentration of the B units in this phase and the specifics of the crystal structure containing the B units. Specifically, the stoichiometric relation between the A and B units in the crystallite is required. With this information it would be possible to calculate the equivalent of Eq. (5.23). Then the ideal chemical potential of the A and B units in the crystalline phase can be obtained. By evoking Eq. (5.31) for both units in both phases the melting temperature of the ideal system will result. However, to accomplish this task requires the *a priori* specification of the number and lengths of the different sequences involved and the crystallite composition. In general, these requirements are difficult to fulfill, so that the melting temperature–composition relation is not as yet available for the ideal case.

Efforts have been made, however, to develop an equilibrium theory without consideration of the ideal contributions.(9–13) In none of these works has the importance of the sequence distribution in the crystalline state been explicitly taken into account. In some cases a distribution was assumed for ease of calculation. Since these results have been applied to experimental data, it is appropriate to consider the approaches that have been taken to describe the equilibrium condition.

In one approach it is assumed that there is a binomial (most probable) distribution of B units in the melt, i.e. only the case $p = X_A$.(10–12) All that is considered is the excess free energy that is involved for a B unit replacing an A in the crystalline lattice. This free energy is designated by ϵ. With these assumptions, and applying equilibrium conditions, the free energy of fusion of such a crystal, ΔG, is given by

$$\Delta G = \Delta G^0 + RT \ln[1 - X_B + X_B \exp(-\epsilon/RT)] \tag{5.32}$$

Here X_B is the overall, or nominal mole fraction of B units and ΔG^0 is the free energy of fusion of the pure crystallite. In deriving Eq. (5.32) the sequence distribution within the crystalline phase is not taken into account. Thus the ideal reference situation is ignored. At the equilibrium melting temperature T_m, $\Delta G = 0$. The melting point depression can then be expressed as

$$\frac{1}{T_m} - \frac{1}{T_m^0} = \frac{-R}{\Delta H_u} \ln[1 - X_B + X_B \exp(-\epsilon/RT_m)] \tag{5.33}$$

Equation (5.33) differs from that for a random copolymer (most probable or binomial distribution) with a pure crystalline phase, by the last term in the argument of the logarithm. The result embodied in Eq. (5.33) is a perturbation on the melting point equation pertinent to a pure crystalline phase. When ϵ is very large the change in free energy that is involved becomes excessive. The B units will then not enter the lattice and Eq. (5.33) becomes

$$\frac{1}{T_m} - \frac{1}{T_m^0} = \frac{-R}{\Delta H_n} \ln(1 - X_B) \tag{5.34}$$

The fact that the Flory expression is regenerated does not by itself make Eq. (5.33) valid. Since ϵ is an arbitrary parameter Eq. (5.33) will have an advantage in explaining experimental results. However, the basic assumptions that have been made in deriving Eq. (5.33) need to be borne in mind. Only a nonideal term appropriate to the crystalline phase has been added to the ideal equilibrium expression for the case of the crystalline phase being pure.

This type of analysis was extended by Wendling and Suter (13) who incorporated into Eq. (5.33) a proposal made by Killian (14) and by Baur.(15) In this concept, only sequences of length ζ are included in lamellar crystallites whose thicknesses correspond to that length. This assumption describes a particular nonequilibrium situation, and is not appropriate to an equilibrium theory. However, following this argument it is found that

$$\frac{1}{T_m} - \frac{1}{T_m^0} = \frac{-R}{\Delta H_u} \left\{ \ln\left[1 - X_B + X_B \exp\left(-\frac{\epsilon}{RT_m} \right) \right] - \langle \zeta \rangle^{-1} \right\} \quad (5.35)$$

where $\langle \zeta \rangle$ is given as

$$\langle \zeta \rangle^{-1} = 2[X_B - X_B \exp(-\epsilon/RT_m)][1 - X_B + X_B \exp(-\epsilon/RT_m)] \quad (5.36)$$

The introduction of an additional parameter can be expected to give better agreement with experimental results. However, we have been concerned in this section with equilibrium concepts. Hence comparison needs to be made with experimental data obtained under as close to equilibrium conditions as possible.

5.3 Nonequilibrium considerations

Although the subject matter of this volume is primarily concerned with equilibrium concepts it is appropriate at this point to also consider the nonequilibrium aspects of the fusion of copolymers. The reason is that for copolymer melting even the approach to equilibrium is extremely difficult, if not impossible, to attain. Relations that have been developed to analyze experimental results of copolymers have their roots in equilibrium theory. It is important to distinguish between modifying and enhancing ideal equilibrium theory, and nonequilibrium concerns. This distinction holds when the crystalline phase is pure as well as when the B units enter the lattice. When nonequilibrium situations are analyzed the restraints that have been previously imposed can be relaxed. A variety of real nonequilibrium features can be addressed. These include, among others, the formation of crystallites of small size, folded chain crystallites, the role of the interfacial free energy, σ_{ec}, characteristic of the surface normal to the chain axis, and its dependence on copolymer composition. For convenience the discussion that follows is divided into two categories. In one, the B units are excluded from the lattice; in the other they are allowed to enter.

It can be expected that for kinetic reasons crystallites smaller than predicted from equilibrium theory will usually develop. The appropriate melting temperature relation can be formulated in a straightforward manner by invoking the Gibbs–Thomson equation. The result for an ideal random copolymer is (16)

$$\frac{1}{T_m} - \frac{1}{T_m^0} = \frac{-R}{\Delta H_u} \ln X_A + \frac{2\sigma_{ec}}{T_m \Delta H_u \rho_c L_c} \tag{5.37}$$

Here T_m is the observed temperature, ρ_c and L_c are the density and thickness respectively of the crystallite. Equation (5.37) merely states how the equilibrium melting temperature is reduced by crystallites of finite size. Both L_c and σ_{ec} will be expected to depend on copolymer composition. The enthalpy of fusion, ΔH_u, results from the expansion of the free energy of fusion about the melting temperature. The temperature variation of this free energy will be more sensitive than that of a homopolymer because of the changing sequence distribution in the melt. Thus, the temperature expansion only of ΔG_u is not sufficient.

As might be anticipated the equilibrium requirement that the largest sequence of A units crystallize, and do so in extended form is extremely difficult to attain experimentally. To account for the size of the crystallites that actually form, attention is focused on the mean sequence length $\langle \zeta \rangle$, and the melting of crystallites of the same thickness. For random copolymers it is found that (15,17)

$$\frac{1}{T_m} - \frac{1}{T_m^0} = \frac{-R}{\Delta H_u} [\ln(1 - X_B) - \langle \zeta \rangle^{-1}] \tag{5.38}$$

Here $\langle \zeta \rangle = [2X_B(1 - X_B)]^{-1}$, is the average length of A unit sequence in the pure melt. This quantity is also taken to represent the thickness of an average crystallite.

A kinetic approach, based on "rough surface growth" (18) that also focuses on the finite thickness of the lamellae leads to a modification of Eq. (5.37). With a set of approximations, the melting temperature can be expressed as (19)

$$\frac{1}{T_m} - \frac{1}{T_m^0} = \frac{-R}{\Delta H_u} \left(\frac{L_c - 1}{2} \right) \ln p + \frac{2\sigma_{ec}}{\Delta H_u \rho_c L_c} \tag{5.39}$$

Equations (5.37) to (5.39) represent nonequilibrium situations where the crystalline phase remains pure. Primary attention has been given to the finite size of the crystallites through use of the Gibbs–Thomson equations and the influence of the sequence selected. One also has to consider the alternative situation where the B units enter the crystal lattice as defects.

Following the previous analysis, the melting temperature when the B units enter the lattice on a nonequilibrium basis is given by (10–12)

$$\frac{1}{T_m} - \frac{1}{T_m^0} = \frac{R}{\Delta H_u} \left[\frac{\epsilon X_{CB}}{RT_m} + (1 - X_{CB}) \ln \frac{1 - X_{CB}}{1 - X_B} + X_{CB} \ln \frac{X_{CB}}{X_B} \right] \tag{5.40}$$

Here X_{CB} is the mole fraction of B units in the lattice, and X_B is their overall composition. A random sequence distribution of B units in the crystalline phase has been assumed.(12)

When $X_{CB} = X_B$, what has been termed the uniform exclusion model results. Equation (5.40) can now be written as (12)

$$\frac{1}{T_m} - \frac{1}{T_m^0} = \frac{R}{\Delta H_u}\left[\frac{\epsilon X_{CB}}{RT_m}\right] \tag{5.40a}$$

Combining these results with those of Baur (15,17) it is found that (13)

$$\frac{1}{T_m} - \frac{1}{T_m^0} = \frac{R}{\Delta H_u}\left[\frac{\epsilon X_{CB}}{RT_m} + (1 - X_{CB})\ln\frac{1 - X_{CB}}{1 - X_{CB}} + X_{CB}\ln\frac{X_{CB}}{X_B} + \langle\zeta\rangle^{-1}\right] \tag{5.41}$$

These relations have also been extended to copolymers whose sequence distributions follow Markovian statistics.(9,9a)

In the course of analyzing experimental results of melting point depressions, recourse will be made to the different expressions that have been developed. It can be expected, however, that with the many expressions available, and the possibility of adding additional terms to the ideal Flory theory, it will be difficult to differentiate whether or not the crystalline phase is pure based solely on melting temperature–composition relations. Except in a few special cases recourse will have to be made to direct physical measurements to determine the composition of the crystalline phase.

5.4 Experimental results: random type copolymers

5.4.1 Course of fusion

Copolymers that have a random sequence distribution are characterized by the sequence propagation probability p being equal to X_A, the mole fraction of the A crystallizing units. Copolymers where p is closely related to X_A can also be considered, for all practical purposes, to be random type copolymers. We shall be concerned particularly with the course of the fusion process, the level of crystallinity that is attained, the determination of the melting temperature, the melting temperature–composition relations and related phase diagrams, as well as isomorphic and diisomorphic replacement of the repeating units, and the role of long chain branching. These aspects of melting will be analyzed in terms of the theoretical developments that have been presented. The observed fusion process will depend on the details of the crystallization mode and the subsequent heating schedule. Although this statement is true for homopolymers, fusion is more complex with copolymers.

Typical examples of the fusion of random copolymers, where the crystalline phase is pure, can be found in ethylene copolymers that contain co-units of the type $\begin{smallmatrix} R \\ | \\ -CH \end{smallmatrix}$ incorporated into the chain. These particular copolymers were prepared by the copolymerization of mixtures of diazomethane and the corresponding higher diazoalkane.(20) Special measures were adopted to assure the random distribution of the comonomer. Crystallization was allowed to occur while the temperature of the initially molten copolymer was reduced gradually by small increments in the vicinity of the melting temperature over a period of many days. This procedure provides the optimum opportunity for the crystalline copolymers to approach the equilibrium conditions. Figure 5.2 gives specific volume–temperature plots for a series of such copolymers. Slow heating rates were utilized subsequent to the crystallization, the temperature being raised 1 degree per day in the interval $T_m - 15$ to T_m. The copolymer composition indicated for each curve is presented as the ratio of CHR/100CH$_2$. Typically, sigmoidal melting curves are observed as is expected theoretically for random copolymers. The transformation occurs over

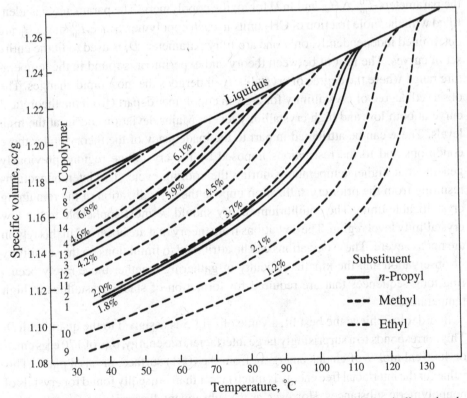

Fig. 5.2 Melting curves for polymethylene copolymers containing the indicated substituents as co-ingredients. Composition of copolymers is indicated as percentage of co-ingredient present. (From Richardson, Flory, and Jackson (20))

a wide temperature interval as compared to homopolymers. The melting range becomes broader as the concentration of noncrystallizable units is increased. Small amounts of crystallinity persist at temperatures just below T_m, in accordance with the theoretical expectations. The merging of the solidus with the liquidus curve is a gradual and asymptotic process. Careful examination of the data in the vicinity of the melting temperature gives no direct evidence of a discontinuity. This result is also consistent with theory. Although theory predicts a discontinuity, its magnitude is beyond the reach of the usual experimental observation. The theoretically desired T_m refers to this experimentally inaccessible discontinuity. The temperature at which measurable departure from the liquidus vanishes is taken to be the experimental melting temperature.

The degree of crystallinity can be calculated from the specific volume data of Fig. 5.2 and then quantitatively compared with the ideal equilibrium theory. The degree of crystallinity, $(1 - \lambda)_d$, is plotted against the temperature in Fig. 5.3 for the *n*-propyl and ethyl branched copolymers.[3] The dotted lines in Fig. 5.3 are calculated according to equilibrium theory, Eqs. (5.16) and (5.19), using the same values for the parameters T_m^0, ΔH_u, and $\ln D$ for all the copolymers. The parameter p is identified with the mole fraction of CH_2 units in each copolymer. Since T_m^0 and ΔH_u are determined independently, only one arbitrary parameter, D, is used to fit the entire set of curves. The best fit between theory and experiment is found in the temperature range where the degree of crystallinity undergoes the most rapid changes. The observed degree of crystallinity for a given copolymer departs from the theoretical curve at both low and high crystallinity levels. Major deviations occur at the high levels. These can be attributed in part to the inadequacy of the theory under these conditions and to the restrictions imposed by the crystalline regions previously generated at higher temperatures during the cooling cycle. The interconnections resulting from the prior crystallization impede the crystallization of the remaining crystallizable units. The equilibrium theory should be most applicable to the low crystallinity level region. The deviations from theory that are observed in this region are not as severe. These deviations can be attributed to limitations on the sensitivity of observation and the kinetic difficulty of gathering together those rarely occurring long sequences that are required for formation of stable crystallites at high temperatures.

In order to achieve the best fit, a value of -11.5 is assigned to the quantity $\ln D$. This corresponds to a surprisingly large interfacial free energy, σ_{ec}, of 170 ergs cm^{-2} or 4600 cal mol^{-1} of chains emerging from the (001) surface, or basal plane. This value for the interfacial free energy is much larger than is usually found for crystals of nonpolymeric substances. However, as the subsequent discussion of crystallization

---

[3] The copolymers with directly bonded methyl groups are excluded from this particular analysis since this co-unit enters the lattice on an equilibrium basis.(20,21)

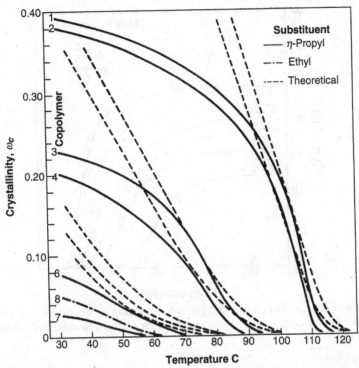

Fig. 5.3 The degree of crystallinity (calculated from the specific volume data) as a function of temperature for the copolymers with *n*-propyl and ethyl substituents. The theoretical curves are calculated from Eqs. (5.16) and (5.17). (From Richardson, Flory, and Jackson (20))

kinetics of both homopolymers and copolymers will show, this interfacial free energy is relatively high because of the necessity to dissipate the crystalline order through the depth of an interfacial layer. There is not a sharp demarcation between the crystalline and liquid-like regions.

Despite the lack of quantitative agreement between theory and experiment, much of which can be attributed to experimental shortcomings and inaccessibility of the very long sequences, the data in Figs. 5.2 and 5.3 qualitatively show all of the major characteristics of the theoretical fusion curves. They can be expected to be typical of the fusion of random type copolymers, irrespective of the chemical nature and structure of the noncrystallizing chain units. Random ethylene copolymers, prepared by a completely different method, display similar fusion characteristics.(21) It is important, however, to assess the generality of the conclusions with other copolymers, rather than just depending on the results of ethylene copolymers.

Copoly(amides) and copoly(esters) represent an important class of random copolymers. Figure 5.4 gives plots of the crystallinity level as a function of temperature for copolyamides of caprolactam with capryllactam at different compositions.(23)

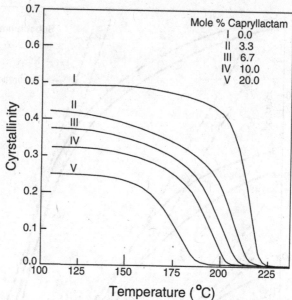

Fig. 5.4 Plot of degree of crystallinity against temperature for copolymers of caprolactam with capryllactam as comonomer. Mole percent capryllactam is indicated. (From Kubota and Newell (23))

The fusion curves are qualitatively similar to those shown in Fig. 5.3 for the ethylene copolymers. The melting of the homopolymer is relatively sharp. As the capryllactam concentration increases the level of crystallinity progressively decreases and the melting range broadens considerably. The experimental results that are illustrated in Figs. 5.3 and 5.4 indicate a universal pattern for the melting of random copolymers that is in accord with theoretical expectation.

Diene type polymers, prepared by either free radical or anionic methods, contain chain units that although chemically identical are isomeric to one another. Hence, from a crystallization point of view this class of polymers behave as copolymers. For example, polymers prepared from the 1,3-dienes are subject to several different kinds of chain irregularities. For poly(butadiene), the following structures are known to exist:

$$
\left(
\begin{array}{c}
\underset{|}{\overset{H_2}{C}} \quad \overset{H}{\underset{|}{C}} \\
C = C \\
\underset{H}{|} \quad \underset{H_2}{|}
\end{array}
\right)
\left(
\begin{array}{c}
\underset{|}{\overset{H_2}{C}} \quad \overset{H_2}{\underset{|}{C}} \\
C = C \\
\underset{H}{|} \quad \underset{H}{|}
\end{array}
\right)
\left(
\begin{array}{c}
\overset{H}{\underset{|}{C}} \\
CH_2 - C \\
| \\
CH \\
\parallel \\
CH_2
\end{array}
\right)
\quad I
$$

1,4 trans   1,4 cis   1,2 vinyl

Thus, poly(isoprene), poly(chloroprene), poly(butadiene), and other polymers in this class, contain units that can exist either in the trans-1,4 or cis-1,4 configuration, or as pendant vinyl groups that can be in either the D or L configuration, as well as regio defects.(24) The diene polymers that occur naturally, hevea rubber and gutta-percha, contain an overwhelming concentration of units that are in either the cis or trans configuration. Crystallization and fusion of these polymers are typically those of homopolymers. However, the chain composition, or microstructure, of the synthetically prepared diene polymers depends on the methods and mechanism of polymerization and it is possible to achieve a wide range in the concentration of the different structural units. For example, it is possible to synthesize poly(butadienes) with either trans-1,4 or cis-1,4 units as the predominant structure and thus the species that actually crystallize.

Figure 5.5 gives specific volume–temperature plots for three different poly-(butadienes) that were determined by dilatometric methods and utilizing slow heating rates subsequent to the development of crystallinity. The concentration of the

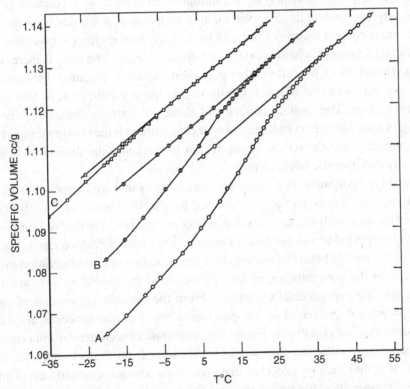

Fig. 5.5 Plot of specific volume against temperature for poly(butadienes) of varying concentrations of crystallizing trans 1,4 units. For curve A, $X_A = 0.81$; for curve B, $X_A = 0.73$; for curve C, $X_A = 0.64$. Curves B and C are arbitrarily displaced along the ordinate.(25)

trans-1,4 crystallizing unit ranges from $X_A = 0.64$ to $X_A = 0.81$.(25) These melting curves are sigmoidal and the transformation occurs over a very wide temperature interval. The melting range becomes broader with decreasing concentration of crystallizing units. Small amounts of crystallinity persist at temperatures just below the melting temperature and this final portion of the melting curve encompasses a larger temperature interval as the crystallizing unit concentration decreases. The level of crystallinity also decreases substantially with decreasing concentration of crystallizable units. All of these features are again characteristic of the fusion of random copolymers. Consequently the change that occurs on melting can be very small and can very often be undetected. In the example illustrated, recourse was again taken to establish the supercooled liquidus as an aid in the determination of T_m. There is an inherent difficulty in detecting crystallinity and in determining the melting temperature of a random copolymer which has a relatively high concentration of noncrystallizing structural units. Yet the persistence of even very small amounts of crystallinity can influence properties in a significant manner. (cf. seq.) Further decreases in the concentration of the trans-1,4 units lead to a completely amorphous polymer.(26) Butadiene–styrene copolymers show a similar effect of the trans-1,4 content.(27) For a fixed styrene content the observed melting temperature decreases by about 40 °C as the trans content decreases from 88 to 76%. A similar effect is observed when the trans-1,4 content is held constant and the stryene concentration is increased.

Poly(butadienes), as well as other poly(dienes), can be prepared that contain an adequate concentration of cis-1,4 units so that the crystallization of this species can take place. The melting behavior of these polymers is similar to when the trans-1,4 units are the crystallizing isomer.(28) Although the melting temperatures are quite different the nature of the fusion process and the diffuse melting are typically copolymeric in character.

Diene type polymers, that originally contain units that are either in the all trans or all cis state, can be partially isomerized by chemical methods.(26,29–32) The isomerization usually occurs via a free radical mechanism. The double bond is temporarily converted to a single bond of the transitory adduct. Subsequently, regeneration of the double bond occurs. Whether the same or a new configuration evolves depends on the concentration of the reactants and the equilibrium requirements of the specific experimental conditions. From the crystallization point of view a homopolymer is converted to a copolymer. If the reaction proceeds at random, a random type of copolymer results that contains a much greater concentration of structural irregularities than would be apparent from the concentration of reactant. It would then be expected that significant alterations should occur in the melting temperature, the fusion process, the crystallization kinetics, and in other properties of the polymer that are related to crystallization. These expectations are, in fact, borne out. For example, the melting temperature of gutta-percha can be

significantly lowered so that it becomes an amorphous polymer at room temperature and possesses the elastic properties associated with a noncrystalline polymer
above its glass temperature. Similarly, significant depressions of the melting temperature have been achieved in 1,4-cis-poly(butadiene), by isomerizing a portion
of the cis configurations to trans.(31)

Polypeptides are a class of polymers that can show copolymer characteristics
with either chemically different repeating units or by exhibiting a special type of
geometric isomerism. The chemical formula for a portion of a polypeptide chain
can be written as

$$II$$

with the repeating unit indicated by the chain elements contained within the bracket.
A polypeptide containing identical R groups is a homopolymer. Its crystallization
behavior depends solely on the nature of the amino acid residue of which it is
comprised. On the other hand, if units having different R groups are present, crystallization is that of a copolymer.

In addition to chemical differences among the chain repeating units, geometric
isomerism can also exist in a chemically identical polypeptide chain. According to
Pauling and Corey,(33) the bond between the carbon atom containing the carbonyl
oxygen and the nitrogen atom has about 40% double bond character because of
resonance between the structures:

$$III$$

As a result of the double bond character of this linkage, the amide group must be
nearly coplanar. Hence, a choice between either a cis or a trans configuration exists.
According to Pauling and Corey (33) and Mizushima (34) the trans configuration
of the amide group in polypeptide chains is, in general, the most stable one. If all
units in a molecule assume the same configuration, homopolymer type behavior is
expected. If, on the other hand, the two different configurations, of identical chemical repeating units, occur in the same chain, then the crystallization pattern will be
typical of a copolymer. By analogy with the results for the diene polymers the transformation of a homopolypeptide to a copolymer can, in principle, be accomplished

by the conversion of units from one configuration to another. This conversion involves rotation about the carbonyl carbon–nitrogen bond and can be induced by appropriate chemical reactions. If the reaction involves a transitory structure, where the double bond character of the peptide linkage is only temporarily lost, conversion from one configuration to another could be accomplished. The situation could also exist where certain reactions, such as the binding of specific ions, favor one of the resonance structures. The particular peptide bonds involved could completely lose their double bond character by this process. In this case the regeneration of the double bond character would require the reversion of the chemical reaction. In either case, whether geometrical isomerism develops or the peptide bond becomes more characteristically single bonded, the thermodynamic stability of the crystalline state, relative to the liquid state, will be severely reduced.

It is possible for stereoisomerism to exist among certain polymers that have chemically identical chain repeating units. The concentration and sequence distribution of the stereoisomers along the chain have an important bearing on the crystallization and melting of such polymers. An important class of polymers possessing asymmetric or pseudo-asymmetric carbon atoms are those that adhere to the general formula

$$\left(\begin{array}{ccc} & H & X \\ & | & | \\ - & C - C & - \\ & | & | \\ & H & Y \end{array} \right)_n \qquad \qquad \text{IV}$$

where X and Y represent two different substituents attached to alternate carbon atoms. Polymers of this class can be prepared from the α-olefins and appropriately substituted vinyl monomers. A more complex type of stereoisomer is formed when each of the carbon atoms contains different substituents, as in the case of a polymer prepared from a 1,2-substituted ethylenic monomer. For the simpler case illustrated, if one arbitrarily represents the chain in an extended planar zigzag form, the X or Y substituent can be located on the same or opposite side of the plane of the zigzag with respect to the same substituents of adjacent monomer units. When each of the pseudo-asymmetric carbon atoms assume identical configurations or when the configurations vary in a definite and prescribed manner throughout the molecule, homopolymer type crystallization can be expected. However, a wide variety of arrangements of the chain units in a nonregularly repeating configuration are obviously also possible. It is not surprising, therefore, that for a long time polymers of this type could not be crystallized because of the lack of sufficient stereoregularity among the chain elements.

Natta and coworkers (35) have demonstrated, in a major work, that crystallizable vinyl polymers from monomers bearing different substituents, and from α-olefins can be prepared by using suitable catalysts. These are now commonly known as Ziegler–Natta catalysts. Such polymers typically have broad molecular weight and stereoisomer distributions.(36) The lower molecular weight species of the distribution contain a much higher concentration of structural irregularities as compared to the higher molecular weights. Subsequently, using metallocene type catalysts, polymers with most probable molecular weight distribution and narrow composition distribution have been prepared.(37) Polymers prepared by this catalyst contain both regio and stereo defects in the chain. Following a suggestion by Huggins,(38) stereoregular polymers have been prepared from vinyl monomers by free-radical polymerization methods.(39–42) In this case stereocontrol is presumed to result from the directing influence of the free end of the propagating species. Variation in stereoregularity is achieved by varying the polymerization temperature, advantage being taken of the small difference in activation energy for the addition of units in the two different possible configurations.(39,43)

Two extreme conditions of chain microstructure can be envisaged for these type polymers. In one case the successive units in the chain possess identical configurations, and the resulting polymer is termed isotactic. In the other case successive alternation of the two possible configurations occurs; such a polymer is termed syndiotactic.(35) A polymer molecule need not be completely isotactic nor completely syndiotactic. In fact, they usually are not. Polymers of this type that are pure stereoisomers have yet to be prepared. A variety of intermediate chain structures can be pictured. They can range from a random sequence distribution of the two isomers with only one type participating in the crystallization, to that of an ordered copolymer where both the syndiotactic and isotactic structures can independently crystallize in the same molecule. Although a polymer may be termed isotactic or syndiotactic a completely regular structure cannot and should not be inferred. The presence of chain irregularities, stereo and regio, must always be anticipated. High resolution carbon-13 NMR has been a very valuable tool in elucidating the microstructure of these polymers.

The necessary apparatus with which to treat the fusion of stereoregular polymers is embodied in the theories already outlined. It remains to establish the sequence distribution of the structural irregularities and to question whether the crystalline phase remains pure. The sequence distribution is a reflection of the polymerization process and the probability of the addition of a particular placement. As has been indicated, addition by either Bernoullian trial or a first-order Markovian process have been treated and integrated into the melting temperature–composition theory.(8,9,9a)

Experimental results clearly indicate that stereo-irregular polymers do indeed crystallize as though they were copolymers. For example, specific volume–temperature curves for isotactic poly(propylene) display all the characteristics expected for a random type copolymer. The results of such a study by Newman (44) for different soluble portions of isotactic poly(propylene) are given in Fig. 5.6. Although the chain microstructure was not determined at the time, the extracts contain different concentrations of isotactic units. Curve A, which is for the ether extract, is linear within experimental error. This fraction is thus noncrystalline and representative of the atactic polymer. In contrast, curve F is calculated for the low temperature behavior of the hypothetical completely crystalline isotactic polymer. The melting temperatures of these fractions decrease substantially as the stereo-irregularity increases. At the same time the fusion process becomes more and more diffuse. A study where the microstructure of isotactic poly(propylene) fractions was determined indicates that there is about a 20 °C decrease in melting tempera-ture with a change in pentad concentration from 0.988 to 0.787.(45) The character of the fusion curves shown in Fig. 5.6 is similar to those observed for random copolymers having chemically different repeating units. They are also similar to those observed with the poly(butadienes) (Fig. 5.5). The principles governing the fusion are the same in all cases. They depend only on the sequence distribution of the units and not on the specific nature of the chemical or structural irregularity involved.

Stereo control can also be achieved in the homogeneous free-radical polymeriza-tion of vinyl monomers by varying the polymerization temperature. Some typical monomers that behave in this manner include methyl methacrylate,(39) vinyl acetate,(41) vinyl chloride,(42) isopropyl and cyclohexyl acrylates.(40) As the polymer temperature is lowered the crystallizability of the polymers becomes more discernible.(46) This observation can be attributed to the fact that as the tempera-ture is lowered there is a preference for units in the same configuration to be added to the growing chain. It has been found that in general there is a preference for syndiotactic sequences to develop as the temperature is lowered. As an example, the observed melting temperature of poly(vinyl chloride) increases from 285 °C to 310 °C as the polymerization temperature is lowered from −15 °C to −75 °C, with a concomitant increase in the syndiotacticity.(47)

The diffuse nature of the fusion curve makes it difficult to accurately determine the melting temperature. In Figs. 5.5 and 5.6 a first-order quantity, the specific volume, was measured. The establishment of the liquidus enabled the melting temperature to be determined in a reliable manner. It is, however, quite common to measure melting temperatures by differential scanning calorimetry (DSC). In this case a second-order quantity, in effect the derivative of a curve of the type shown in Fig. 5.6, is being measured. Because of the diffuseness of the curve it is not obvious that the maximum in the DSC thermogram corresponds to the melting temperature.(48) Crist and Howard have calculated from the Flory theory deriva-tives of curves similar to those shown in Fig. 5.1.(48) The results are shown in Fig. 5.7 for different values of the parameter p. It is clear that the maximum in the

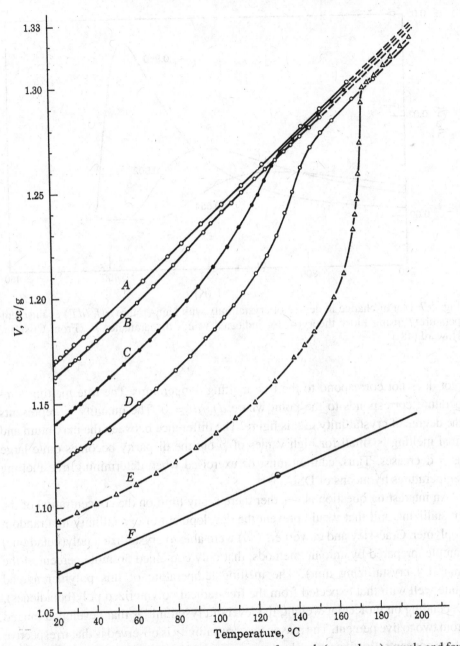

Fig. 5.6 Plot of specific volume against temperature for a poly(propylene) sample and four soluble extracts from it. A: ether extract, quenched; B: pentane extract, annealed; C: hexane fraction, annealed; D: trimethyl pentane fraction, annealed; E: experimental whole polymer annealed; F: pure crystalline polymer. (From Newman (44))

Fusion of copolymers

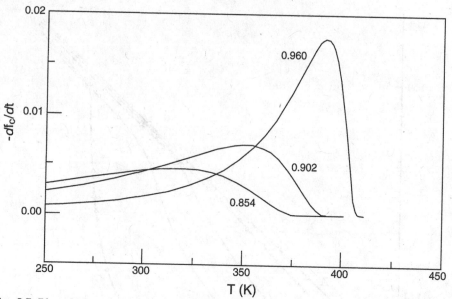

Fig. 5.7 Plot of change in degree of crystallinity with temperature (df_c/dT) against temperature T using Flory theory(4) for indicated values of parameter p. (From Crist and Howard (48))

plot does not correspond to the true melting temperature. The true melting temperature corresponds to the point where $df_c/dt = 0$. The quantity f_c represents the degree of crystallinity in this figure. The difference between the maximum and final melting is small for high values of p but the disparity becomes quite large as p decreases. Thus, caution must be exercised when determining true melting temperatures by means of DSC.

An interesting question is whether there is any limit on the concentration of the crystallizing unit that would prevent the development of crystallinity in a random coplymer. Graessley and coworkers (49) were able to crystallize a poly(butadiene) sample, prepared by anionic methods, that only contained 56 mole percent of the trans-1,4 crystallizing units. The melting temperature of this polymer agreed quite well with that expected from the free-radical polymerized poly(butadienes). Depending on molecular weight, the level of crystallinities that developed ranged from two to five percent. The reason that crystallinity is observed is that irrespective of the concentration there are always sequences of crystallizable units, albeit few in number, that are capable of crystallizing. Suitable conditions need to be found for the crystallization and detection of such sequences. At the low temperatures that were required for the small amounts of crystallinity to develop, anomalies were observed in the viscoelastic behavior of these poly(butadienes). Presumably, other physical properties will also be affected. The crystallization of the poly(butadienes) demonstrates that not only in principle but also in practice, copolymers that contain a very high noncrystallizing content can be crystallized under appropriate conditions.

A related problem of interest is when a polymer is not crystalline as prepared, but is potentially crystallizable. This situation is commonly encountered in crystallizable copolymers, and is also found in homopolymers. Some typical examples of this phenomenon are found in poly(styrene) synthesized by means of alfin type catalysts,(50) poly(methyl methacrylate), prepared by either free-radical or ionic methods,(39,51,52) and poly(carbonate).(53) Treatment with particular solvents or diluent at elevated temperatures can induce crystallinity in these polymers. The reason for the problem is kinetic restraints to the crystallization process. Treatment with appropriate diluents alleviates the problem. The principles involved, and the diluent requirements will be enunciated in the discussion of crystallization kinetics. For present purposes it should be recognized that the crystallizability of a polymer, particularly a copolymer, cannot be categorically denied unless the optimum conditions for crystallization have been investigated. Thus, in light of the previous discussion regarding the minimum concentration of chain units required for crystallization, and the need to have favorable kinetic conditions, the lack of crystallization in any given situation needs to be carefully assessed.

Certain polymers, such as poly(vinyl chloride), poly(acrylonitrile), poly-(chlorotrifluoroethylene), and poly(vinyl alcohol), are crystalline as usually prepared, despite the strong possibility of the occurrence of significant stereochemical irregularities. In many instances x-ray diffraction patterns, when used as a criterion, did not definitely support the contention of crystallinity. However, particularly in the cases of poly(vinyl chloride)(54,55) and poly(acrylonitrile),(56) solution and mechanical properties gave substantial evidence of the existence of crystallinity. Subsequent synthesis of these polymers by methods designed to impart a greater amount of chain regularity has confirmed those conclusions.(42,57)

5.4.2 *Melting temperature–composition relations: crystalline phase pure*

With the establishment of the characteristics of the fusion process of random copolymers the melting temperature–composition relation of such copolymers can be examined. To analyze the problem distinction must again be made as to whether the crystalline phase is pure or if the B units enter the crystallite, either on an equilibrium basis or as a defect. Merely establishing the liquidus is not sufficient to resolve this problem. The Flory relation is only applicable to an ideal system whose crystalline phase is pure. Deviations from this relation can be due to lack of ideality, with either the raising or lowering of the expected equilibrium melting temperature. On the other hand the B units can be entering the lattice. This possibility can really only be established by direct experimental observation. Solid-state NMR and wide-angle x-ray diffraction studies have been most useful in this connection. In some special cases indirect measurements can be helpful in resolving the problem. For example, when comonomers of different sizes and structures give the same melting temperature–composition relation it is reasonable to assume that the crystalline phase remains pure.(58) In another situation, an increase in melting temperature with comonomer concentration is suggestive of compound formation in the crystalline state.(6,21) Although conclusions from these and other indirect measurements may appear reasonable, when possible they should be substantiated by direct measurements.

When the crystalline phase only contains A units then for a random ideal copolymer

$$\frac{1}{T_{\mathrm{m}}} - \frac{1}{T_{\mathrm{m}}^0} = \frac{-R}{\Delta H_{\mathrm{u}}} \ln X_{\mathrm{A}}$$ (5.42)

The melting temperatures in this equation represent equilibrium ones. Deviations from ideality are not reflected in this equation. Additional terms can be added to the ideal mixing free energy, the basis for Eq. (5.42), and still satisfy the equilibrium requirement and the purity of the crystalline phase. As has been indicated, melting temperatures can either be raised or lowered depending on the specificity of the interaction in the melt between the two different units.

Copolymers formed by the methods of condensation polymerization are usually characterized by a sequence propagation probability parameter p that is independent of copolymer composition and the extent of conversion. Moreover, in such systems the quantity p can be equated to the mole fraction of crystallizable units. The observed melting temperature–composition relations of some representative copolyesters and copolyamides are illustrated in Fig. 5.8.(59–61) These copolymer types, whose units crystallize independently of one another, have certain characteristic features. The melting temperatures depend only on composition. They are independent of the chemical nature of the coingredient that is introduced, as is illustrated here for the copolymers of poly(ethylene terephthalate) and of poly(hexamethylene adipamide). This observation is consistent with wide-angle x-ray observations, that indicate only one of the units participates in the crystallization. As the concentration of the added ingredient is increased, a composition is reached where it can itself undergo crystallization at the expense of the other component. The melting point–composition relations for this component follow an independent curve. Thus, a eutectic type minimum results at the intersection of the two liquidus curves. This behavior is typical of random copolymers when studied over the complete composition range when each of the species is capable of crystallizing independently. Since the composition used in Fig. 5.8 is based on

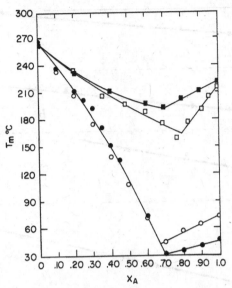

Fig. 5.8 Melting temperature–composition relations for various copolyesters and copoly-amides. ●, poly(ethylene terephthalate/adipate); ○, poly(ethylene terephthalate/sebacate); ■, poly(hexamethylene adipamide/sebacamide); □, poly(hexamethylene adipamide/capro-amide). (From Edgar and Ellery (59), Sonnerskog (60) and Izard (61))

the liquidus the shapes of the curves do not give any direct information on the composition of the crystalline phase. In these examples the fact that the same melt-ing temperature–composition relation is obtained with different comonomers gives strong evidence that the crystalline phase remains pure. However, this conclusion does not necessarily hold without independent confirmation. Other polymers, such as poly(tetrafluoroethylene) and poly(methylene oxide) behave in a similar manner for a variety of added species.(62,63) Although for some copolymers a given set of comonomers will give the same melting temperature relation, the addition of a particular co-unit will cause a different behavior.(20,64–66) Usually, the melting temperatures with such comonomers are greater at a given composition. The sur-mise is that such comonomers either enter the crystal lattice, or the parameter p increases. However, the possibility of nonideality contributions to Eq. (5.42) cannot be neglected.

According to equilibrium theory, the melting temperature–composition relations of each branch of the curves of Fig. 5.8 should be described by Eq. (5.42). Con-sequently the relations between $1/T_m$ and $-\ln X_A$ for some typical copoly(esters) and copoly(amides) are plotted in Fig. 5.9.(58,59,60,67) Each of the examples covers a wide range in copolymer composition. The melting point data in each ex-ample are well represented by a straight line which in accordance with theory extrapolates to the melting temperature of the pure homopolymer. Thus, the

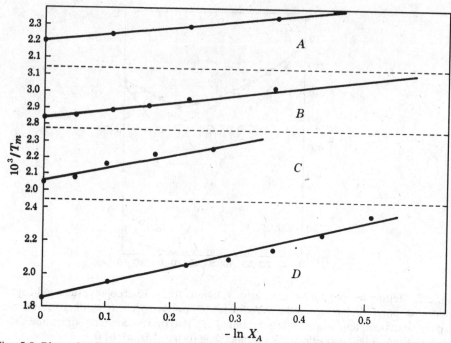

Fig. 5.9 Plot of $1/T_m$ against $-\ln X_A$ for copolyamides and copolyesters. *A*, N,N′-sebacoyl piperazine/isophthaloyl piperazine copolymer(67); *B*, decamethylene adipate/decamethylene isophthalate copolymer (from Evans, Mighton, and Flory (58)); *C*, caproamide–hexamethylene adipamide copolymer (from Sonnerskog (60)), *D*, ethylene terephthalate–ethylene adipate copolymers (from Edgar and Ellery (59)).

functional form of Eq. (5.42) is obeyed for copoly(esters) and copoly(amides) even though directly observed (nonequilibrium) melting temperatures are used. This type result is not limited to condensation type copolymers. It is found in virtually all other type copolymers that have been studied. The fact that Eq. (5.42) is obeyed when the observed melting temperatures are used leads to the expectation that the correct values of ΔH_u could be obtained from the straight lines in Fig. 5.9. This expectation is examined in Table 5.1, where a comparison is made between the ΔH_u values obtained by either the copolymer or diluent methods for some representative copolymers.

The compilation in Table 5.1 clearly indicates that although a linear relation is observed in plots suggested by Eq. (5.42), the deduced ΔH_u values are much lower than those determined by the diluent method. This is a general conclusion, being based on a widely diverse set of polymers. The differences in ΔH_u values are significant. Differences of a factor of two are observed in some cases. There are several possible reasons for this discrepancy. One is the fact that Eq. (5.42) represents equilibrium for an ideal melt. In the examples cited efforts were not made to determine

Table 5.1. *Comparison of the copolymer and diluent methods for determining* ΔH_u

Polymer	ΔH_u (cal g^{-1})	
	Copolymer method	Diluent method
Poly(decamethylene adipate)	13.4[a]	36.0[b]
Poly(decamethylene sebacate)	13.9[a]	36.0[a]
Poly(N,N'-sebacoyl piperazine)	19.8[c]	24.5[c]
Poly(decamethylene sebacamide)	23.0[a]	24.5[c]
Poly(caprolactam)	24.7[d]	37.9[e]
Poly(ethylene terephthalate)	11.5[f]	29.0[g]
Poly(methylene oxide)	49.7[h]	55.8[h]
Poly(ethylene)	41.4[i]	69.0[j]
Poly(1,4-trans-butadiene)	18.5[k]	20.4[l]
Poly(tetramethyl-*p*-silphenylene siloxane)	37.7[m]	54.5[m]

[a] Evans, R. D., H. R. Mighton and P. J. Flory, *J. Am. Chem. Soc.*, **72**, 2018 (1950).
[b] Mandelkern, L., R. R. Garrett and P. J. Flory, *J. Am. Chem. Soc.*, **74**, 3939 (1952).
[c] Flory, P. J., L. Mandelkern and H. K. Hall, *J. Am. Chem. Soc.*, **73**, 2532 (1951).
[d] Kubota, H. and J. B. Newell, *J. Appl. Polym. Sci.*, **19**, 1521 (1975).
[e] Gechele, G. B. and L. Crescentini, *J. Appl. Polym. Sci.*, **7**, 1349 (1963).
[f] Edgar, O. B. and R. Hill, *J. Polym. Sci.*, **8**, 1 (1952).
[g] Roberts, R. C., *Polymer*, **10**, 113 (1969).
[h] Inoue, M., *J. Appl. Polym. Sci.*, **8**, 2225 (1964).
[i] Phillips, P. J., F. A. Emerson and W. J. MacKnight, *Macromolecules*, **3**, 767 (1970).
[j] Quinn, F. A. and L. Mandelkern, *J. Am. Chem. Soc.*, **80**, 3178 (1958); Mandelkern, L., *Rubber Chem. Tech.*, **32**, 1392 (1959).
[k] Berger, M. and D. J. Buckley, *J. Polym. Sci., Pt. A*, **1**, 2995 (1963).
[m] Okui, N., H. M. Li and J. H. Magill, *Polymer*, **19**, 411 (1978).
[n] Natta, G. and G. Moraglio, *Rubber Plast. Age*, **44**, 42 (1963).

equilibrium melting temperatures. The theory requires that the melting temperature recorded represent the disappearance of very long crystalline sequences. This task will be difficult to fulfill under any circumstances. Such sequences will be difficult to develop for any reasonable crystallization rates and their detection will require very sensitive experimental methods. When the barrier to the crystallization of copolymers is examined it can be expected that the size of the crystallites that actually form will be significantly reduced relative to equilibrium requirements. Even under very sensitive experimental methods the recorded melting temperature of random copolymers will be less than theoretical expectation. This difference will become larger as the co-unit content increases and will result in an apparent lower enthalpy of fusion.

It has been found that using the modification proposed by Baur gives better agreement than the ideal Flory theory. For example, using extrapolated equilibrium melting temperatures gives excellent agreement with experimental results for copolymers of poly(L-lactide-meso lactide).(67a)

The effect of small crystallite thickness on the observed melting temperature–composition relation of random copolymers of vinylidene chloride and methyl methacrylate was analyzed by utilizing the Gibbs–Thomson equation.(16) However, to adapt this procedure to copolymers the dependence of both the crystallite thickness and the interfacial free energy σ_{ec} on copolymer composition needs to be specified. It was possible to explain the observed melting temperature–composition relation for this copolymer by assuming the dependence of these two quantities on composition.

Most of the experimental melting temperature–composition relations reported have involved directly observed melting temperatures. Extrapolative methods have been developed that allow for an approximation of the equilibrium temperature.[4] Several examples have been reported where extrapolated equilibrium melting temperatures were used. An example is given in Fig. 5.10 for random copolymers of syndiotactic poly(propylene) with 1-octene as comonomer.(71) In this figure the solid line represents Eq. (5.42), calculated with ΔH_u equal to 1973 cal mol^{-1}. This value was determined independently from wide-angle x-ray diffraction and differential scanning calorimetry.(72) Thus, there is excellent agreement between experiment and theory as embodied in Eq. (5.42). In contrast, $\Delta H_u = 693$ cal mol^{-1} was deduced when nonequilibrium melting temperatures were used.(64) Studies with poly(1,4 trans chloroprene), with varying concentrations of structural irregularities that utilized extrapolated equilibrium melting temperatures, also found agreement between experiment and the ideal Flory theory.(24) A ΔH_u value of 1890 cal mol^{-1} was deduced from the copolymer data as compared to 1999 cal mol^{-1} obtained by the diluent method.(72a) The agreement in ΔH_u values between the two methods is excellent. Despite the support of the Flory theory by these two investigations, studies using extrapolated equilibrium melting temperatures of copolyesters of tetramethylene terephthalate with tetramethylene sebacate gave a wide disparity in ΔH_u values.(73) There is about a factor of two between the copolymer determined ΔH_u and that obtained from the diluent depression. However, this set of copolymers

---

[4] The principles involved will be discussed later in detail when the structure and morphology of semi-crystalline polymers are considered. For present purposes it suffices to state that the method has had reasonable success when applied to homopolymers, although some important exceptions have been noted. However, major difficulties have been encountered when applied to random ethylene copolymers (68,69) as well as copolymers of isotactic poly(propylene).(70) It remains to be seen whether these methods can in fact be applied successfully to the other random copolymers. The examples that follow should be considered in this light.

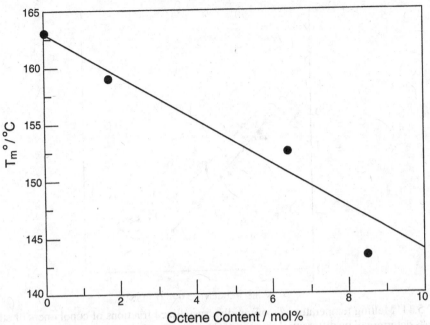

Fig. 5.10 Plots of extrapolated equilibrium melting temperature of syndiotactic poly-(propylene)–1-octene copolymers as a function of comonomer concentration. Solid line calculated according to Eq. (5.42). (From Thomann, Kressler and Mülhaupt (71))

gave four endothermic peaks on fusion. The melting temperatures used for the extrapolation were selected arbitrarily. There is, however, a fundamental problem in determining the equilibrium melting temperature of random copolymers.

The uncertainties involved in the extrapolation procedures used to obtain the equilibrium melting temperature, coupled with the limited data that is available, makes it premature to decide whether or not Eq. (5.42) holds in any meaningful way. The fact that Eq. (5.42) may not apply, even when equilibrium melting temperatures are used, does not necessarily mean that the B units enter the crystal lattice. This conclusion could be incorrect without any direct evidence for support. The addition of nonideal terms to Eq. (5.42) could also resolve the problem.

Some of the principles as well as problems involved in the melting of random copolymers are found in olefin type copolymers. The melting temperatures of a large number of random type ethylene copolymers, as determined by differential scanning calorimetry, are plotted as a function of the mole percent branch points in Fig. 5.11. The samples represented in this figure are either molecular weight and compositional fractions or those with a narrow composition distribution with a most probable molecular weight distribution.(74) These samples were crystallized and heated rapidly. In this set of data there are two different copolymers that contain ethyl

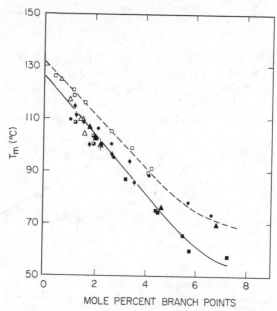

Fig. 5.11 Melting temperature T_m of rapidly crystallized fractions of copolymers of ethylene as determined by differential scanning calorimetry: hydrogenated poly(butadiene) ($\triangle$), ethylene–vinyl acetate ($\bullet$), diazoalkane copolymer with propyl side groups ($\blacktriangle$), ethylene–butene copolymer ($\square$), ethylene–octene copolymer ($\blacksquare$, $\boxdot$).(74)

branches. These are hydrogenated poly(butadienes), and a set are fractions obtained from a copolymer with broad molecular weight and composition distributions. All the copolymers represented are in a random type distribution. To analyze the melting temperature–composition relations it is convenient to divide the data of Fig. 5.11 into two regions: copolymers containing less than 3 mol % unit branch points and those which have a greater co-unit content.

A significant feature of the data in the lower concentration range of Fig. 5.11 is that except for the ethylene–butene fractions all the other copolymers give the same melting temperature–composition relation. These include the hydrogenated poly(butadienes), ethylene–hexene and two sets of ethylene–octenes prepared by two different catalysts. Similar results have been reported for other polymers of this type.(75–77) The melting temperatures, not shown, of a set of ethylene–butene copolymers, prepared with a homogeneous catalyst, that possess most probable molecular weight and narrow composition distribution fall in the solid curve, along with the other copolymers.(21) The differences between the two ethylene–butene copolymers are about 5 °C for 0.5 mol % side-groups and increase to 10 °C at about 3 mol %. These differences in melting temperatures cannot be attributed to the chemical nature of the co-units since the data for the other ethyl branched copolymers fall on the same solid line as for the other copolymers. Based on

Eq. (5.42) it can be concluded that the melting point differences are a result of a different sequence distribution between this particular ethylene–butene-1 copolymer and the hydrogenated poly(butadienes) and the other ethylene–butene-1 copolymers.

The differences in sequence distribution could be caused by different polymerization procedures, particularly the catalyst used. In the composition range of present interest, i.e. the order of just a few mole percent of co-unit, only very small differences in the sequence propagation parameter p can cause the melting point differences that are observed. For example, for a perfectly random sequence copolymer, $p = 0.9800$ for 2 mol % branch points. If we assume that deviations from equilibrium are the same for both type copolymers then from the melting temperature of the 2 mol % ethylene–butene copolymer a calculated value of $p = 0.9875$ is obtained for the higher melting ethylene–butene-1 copolymer. Thus, even in the grouping of what might be called random type copolymers small differences in the parameter p, which alter the sequence distribution, are sufficient to influence the observed melting temperatures. We have, therefore, in this set of data a striking example that for a pure crystalline phase, neither the chemical nature of the co-unit nor its nominal composition determines the melting temperature. Even small differences in sequence distribution can make significant differences in the observed melting temperature. Thus, copolymers with the same comonomer and composition can differ in melting temperatures when prepared with different catalysts.

The melting temperatures of the higher co-unit content copolymers shown in Fig. 5.11 do not give as straightforward results as found in the lower concentration range. The hydrogenated poly(butadienes) and the diazoalkane copolymers follow the same trend as in the lower concentration range. The melting temperatures of the ethylene–butene-1 fractions are still about 10–15 °C higher. On the other hand, the melting temperatures of the ethylene–vinyl acetate copolymers are beginning to deviate and also become about 10–15 °C higher. This pattern of melting points indicates a tendency for the ethylene–vinyl acetate copolymer to deviate from a completely random sequence distribution.

Figure 5.12 represents a compilation of melting temperature relations for rapidly crystallized ethylene copolymers with a set of 1-alkenes and norbornene as comonomers.(74–76,78) The melting temperatures of ethylene copolymers with bulkier side-group comonomers such as 1-decene, 4-methyl-1-pentene, cyclopentadiene and dicyclopentadiene follow the same curve as in Fig. 5.12.(78a) The plot clearly indicates that the melting points are independent of co-unit type under these crystallization conditions. Since observed melting temperatures of copolymers are known to depend on chain length the results shown have been limited to molecular weights of about 90 000.(21) Studies of ethylene–octene copolymers with much higher comonomer content indicate a continuation of the curve shown in Fig. 5.12

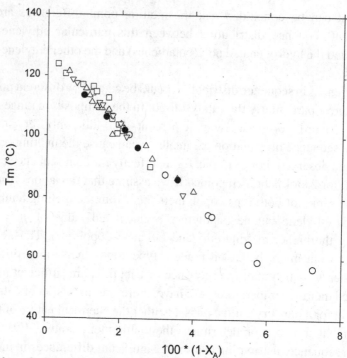

Fig. 5.12 Plot of observed melting temperature T_m against mol percent of structural irregularities in the chain. ○ HPBD; □ ethylene–butene; △ ethylene–hexene; ▽ ethylene–octene; ● ethylene–norbornene. $M_w \simeq 90\,000$. HPBD stands for hydrogenated poly-(butadiene).(74–76,78)

to much lower melting temperatures.(22) The results shown in the figure do not imply that all comonomers with the same sequence distribution give the same melting temperature–composition relations. In fact, this is not the case, as will be discussed in the next section where the melting temperatures of ethylene–propylene, ethylene–vinyl chloride and ethylene–vinyl alcohol copolymers will be analyzed.

Copolymers of syndiotactic poly(propylene) behave in a similar manner.(64) Here the copolymers with ethylene, 1-pentene, 1-hexene and 4-methyl-1-pentene as co-units obey the same melting temperature–composition relation. On the other hand, the copolymer with 1-butene gives higher melting temperatures than the others. This result will also be discussed further in the next section.

The melting temperature–composition relations that were described above were for rapidly crystallized samples. This crystallization procedure results in relatively small crystallite sizes. In an alternative procedure the crystallization can be conducted isothermally at elevated temperatures and never cooled prior to fusion. It is then found that the melting temperatures are dependent on the nature of the comonomer.(79) Ethylene butene and hexene copolymers behave similarly to one another while ethylene–octene and ethylene–norbornene give lower melting temperatures. These differences can be attributed to morphological influences, particularly the perfection in the lamellae structure.

5.4.3 Melting temperature–composition relations: crystalline phase mixed

In this section, a more detailed discussion will be given of the melting temperature–composition relation when the co-unit enters the crystalline lattice. The analysis of this situation is more difficult than the previous case. Except in special cases, a decision as to whether a partitioning of the B units occurs between the two phases cannot be made solely on the basis of the liquidus. One special case is when a variety of comonomers, each with a different size and shape, yield the same melting temperature–composition relation. Under these circumstances it is reasonable to conclude that the crystalline phase remains pure for all of the co-units. Such a situation has been found with a series of aliphatic copolyesters (58) and tetrafluoroethylene with different comonomers.(79a) On the other hand, when the melting temperature–composition relation depends on the incorporated comonomer, as in the case of the ε-caprolactams (79b), it is reasonable to assume that at least a portion of the co-units enters the lattice.

The failure of the Flory theory, even when extrapolated equilibrium melting temperatures are used, does not necessarily mean that either comonomers or structural defects are entering the lattice.(45) The melting point relation given by Eq. (5.42) is for an ideal melt. Modification of this theory can be legitimately made, while still maintaining equilibrium, without requiring that the co-unit enter the lattice.

Ideally, the solidus should be established for all cases. Except in a few rare situations a complete phase diagram, where both the liquidus and solidus are presented, is not available. Determining the solidus for polymers, even on a compositional basis, is a formidable matter. As theoretical considerations have indicated, the sequence distributions in both phases are actually required for polymers rather than the composition. This makes the task of determining the solidus a very difficult one. Moreover, if a mixed crystalline phase is observed a decision has to be made as to whether it represents an equilibrium or defected state.

One method used to probe the crystallite interior involves an appropriate chemical reaction. It is assumed that the noncrystalline region is severed from the crystalline one by the reaction. An example of this method is the selective oxidation of ethylene copolymers.(80–83) It is presumed that the crystalline core remains behind. The residue can then be analyzed by several different methods. The problem here, as well as with other chemical methods, is in establishing the reactivity, and thus the contribution, of the interfacial region. The concentration of B units in this region will be relatively high. Therefore, if this region is not completely removed by the reaction it would be construed to be part of the crystalline core. It is not surprising, therefore, that conflicting results have been obtained by this and other chemical methods.(84,85)

Physical methods have also been used to probe the composition of the crystalline phase. These methods include: wide- and small-angle neutron and x-ray scattering, vibrational spectroscopy, carbon-13 solid-state NMR and the determination of unit cell dimensions by wide-angle x-ray scattering. The last cited method has a minimal interference from the interfacial region, as long as the Bragg spacings are well-defined. Despite the difficulty in quantitatively establishing the composition and sequence distribution of the crystalline phase, adequate evidence has been developed that demonstrates that co-units enter the lattice of many copolymers. The co-crystallization of the A and B units manifests itself in several different ways, similar to what is observed in binary monomeric systems. These include among others, compound formation, isomorphism and isodimorphism. The determination of the sequence distribution in the crystalline phase is an important and worthy challenge.

In the discussion of the melting temperature–composition relation of the ethylene–1-alkene random copolymers (Figs. 5.11 and 5.12) ethylene–propylene copolymers, with directly bonded methyl groups were not considered. The reason that the discussion of these copolymers was postponed is that they have significantly higher melting temperatures than those with either large alkyl branches or bulker side-groups.(6,20) Detailed studies have given a strong indication of a maximum at low branch point content in the melting temperature–composition relation for these copolymers.(6,20) The maximum in the liquidus suggests compound formation, as is observed in many binary mixtures of metals and other monomeric substances. It reflects the fact that the methyl group enters the lattice on an equilibrium basis. In contrast, co-units that enter the crystal lattice as nonequilibrium defects will invariably cause a lowering of the melting temperature. This result represents another case where the liquidus alone is strongly suggestive of the character of the crystalline phase. The melting temperature–composition relations for ethylene–vinyl chloride copolymers are virtually identical to those of ethylene–propylene.(86) Hence we can also surmise that the Cl atom enters the lattice on an equilibrium basis. Similar studies have shown that the smaller side-groups such as CH_3, Cl, OH and O can enter the lattice of ethylene copolymers.(87) Whether they all do so on an equilibrium basis has not been established. In contrast, the melting temperature relations for the other ethylene–1-alkene copolymers and ethylene–vinyl acetate are the same. It can be concluded that they act in a similar manner with respect to incorporation into the lattice. They are excluded because of their size. Other physical-chemical measurements support this conclusion.(21) However, direct determination of the sequence distribution in the crystalline phase is eventually required.

It should be noted that the dimensions of lattice parameters, determined by x-ray diffraction, have been commonly used to establish the purity of the crystalline phase. Extensive studies of this kind have been carried out with polyethylene copolymers.(21,88–94) The basic assumption is made that the expansion of the lattice reflects the inclusion of the co-unit. However, Bunn has pointed out that this interpretation is not unique.(95) The crystallite thicknesses of such copolymers are relatively small, being less than 100 Å, depending on the composition.(74) The strain that develops in the thick interfacial region of such thin crystallites could easily cause the lattice expansion. Hence, the analysis of lattice parameters does not necessarily yield definitive information with respect to the issue of interest. In some cases this analysis has led to incorrect conclusions.

A different type of pseudo-phase diagram based on the liquidus, and involving ethylene, is found in ethylene–vinyl alcohol random type copolymers.(96) These copolymers are prepared by the saponification of ethylene–vinyl acetate copolymers. Since the latter are in random sequence distribution the ethylene–vinyl alcohol copolymers have the same distribution. However, the crystallinity levels and melting points between the two are quite different. The level of crystallinity of the ethylene–vinyl acetate copolymer decreases continuously with co-unit content, as was illustrated in Fig. 5.11. The crystalline phase remains pure for this copolymer. The copolymer becomes completely noncrystalline at ambient temperature, when the co-unit content reaches about 20 mol %. The ethylene–vinyl alcohol copolymer, on the other hand, gives quite different results as is shown in Fig. 5.13.(96) This rather unusual diagram for a random type copolymer requires a more detailed

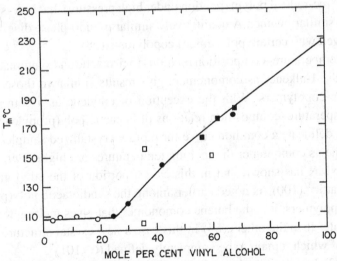

Fig. 5.13 Composite plot of melting temperature against mole percent vinyl alcohol for ethylene–vinyl alcohol copolymer.(96)

examination of the crystalline phase. The x-ray diffraction patterns, and thus the unit cell structures, depend on the co-unit content. The repeat distance along the chain axis is always found to be 2.5 Å, which is also the same for both parent homopolymers. The unit cell structure, however, varies from orthorhombic to hexagonal to monoclinic as the vinyl alcohol content increases from 20 to 55 mol %. Starting with pure poly(vinyl alcohol) there is a monotonic decrease in the observed melting temperature with increasing ethylene content until a co-unit content of about 25 mol % vinyl alcohol is reached. The portion of the phase diagram up to this point is consistent with the crystalline phase being pure in either a monoclinic or hexagonal form. Atactic poly(vinyl acetate) does not crystallize. However, poly(vinyl alcohol) prepared from the same polymer does, although hydroxyl substituents are still randomly placed on either side of the chain. From an analysis of the x-ray diffraction pattern Bunn concluded that co-crystallization occurs, i.e. the C—OH unit enters the crystal lattice.(97) Although hydroxyl groups are much larger than hydrogen atoms, neither of them are very large when compared with the space available in the crystal structure. Evidently the random removal of the H and OH positive attached to alternate carbon atoms does not lead to serious enough strain to prevent crystallization. Intermolecular hydrogen bonds between hydroxyl groups will also aid in stabilizing the crystalline structure. The co-crystallization is reflected in the basic thermodynamic properties of these copolymers. For the copolymers containing 75 mol % of ethylene and greater, the melting temperature becomes invariant with composition. This result suggests that the sequence distribution is the same within the crystal and liquid phase. The melting temperature that is extrapolated to pure polyethylene is consistent with that of a branched polyethylene (formed by free-radical polymerization under high pressure) that is crystallized and melted in a similar manner. A qualitatively similar pseudo-phase diagram has also been observed with certain poly(imide) copolymers.(98)

Melting temperature–composition relations for isotactic and syndiotactic copolymers with the 1-alkenes as comonomers give results similar to those found with the ethylene copolymers. With the exception of ethylene as a comonomer, the melting temperature–composition relations of isotactic poly(propylene)–1-alkene copolymers delineate a common curve for rapidly crystallized samples. However, with ethylene as comonomer the melting temperatures are higher.(99) Solid-state carbon-13 NMR has shown that in this case a portion of the ethyl groups enter the crystal lattice.(100) As noted earlier among the syndiotactic poly(propylene)–1-alkene copolymers, it is the butene comonomer that gives melting temperatures that are greater than the others.(64) In this case a new crystal structure is formed, the details of which remain to be completely defined.(64,101)

Natta (102) has described two types of isomorphism. In isomorphism itself, both units participate in the same crystal structure over the complete composition range. In the other type, termed isodimorphism, the system consists of two different crystalline structures. The formation of one or the other depends on the sequence distribution (composition) of the crystalline phase. Examples of these types of replacements are found in virtually all copolymer types including copolyamides, (103–109), synthetic and natural copolyesters (110–115), vinyl copolymers (102,116–118), diene polymers (119), poly(olefins) (120–123), poly(aryl ether ether ketones) (124) and poly(phenyls) (125). A detailed summary of other copolymers where co-crystallization occurs can be found in Ref. (126). There appear to be two underlying principles that govern isomorphic replacement.(126) These are that the two repeating units should have the same shape and volume and that the new ordered chain conformation be compatible for both types. The principles that are involved can best be illustrated by examining a few examples.

In a formal sense the crystallization of poly(vinyl alcohol) can be considered to be the result of isomorphic replacement.(97) Similarly, the crystallization of a poly(vinyl fluoride) with an essentially atactic structure has been reported.(117) In this case the individual chains adopt a planar zigzag conformation in the crystal structure. Here the randomly placed atoms that replace each other are hydrogen and fluorine. Their van der Waals radii of ~ 1.25 Å and 1.35 Å respectively are close enough for their substitution. Poly(trifluorochloroethylene) can achieve a relatively high degree of crystallinity, despite its stereoirregularity. The small difference in the van der Waals radii of chlorine and fluorine is such as to allow a substitution and thus a high level of crystallinity, despite the irregular distribution of two kinds of atoms.

Copolymers of isotactic poly(styrene) with either *o*-fluorostyrene or *p*-fluorostyrene have been shown, by wide-angle x-ray diffraction, to be crystalline over the whole composition range.(102) All of these copolymers have the same crystalline structure. The lattice constants in the direction of the chain axis are also the same and there are only slight deviations in the perpendicular directions. The melting temperatures vary continuously from that of the pure isotactic poly(styrene) to that of poly(*o*-fluorostyrene). This type isomorphism results from the fact that the corresponding homopolymers have the same repeat distance. In addition fluorine and hydrogen atoms have similar sizes so that fluorine can replace a hydrogen so that there are no critical van der Waals contacts with neighboring atoms. Consequently the two units can be substituted for one another on the same lattice site. This substitution causes only small variations in the lattice constant in direction normal to the chain axis.

The copolymers of isotactic poly(styrene) and *p*-fluorostyrene are also crystalline over the complete composition range. However, in this case the two corresponding homopolymers have different crystal structures and symmetries. Isotactic poly(styrene) has a threefold-helical structure while poly(*p*-fluorostyrene) has

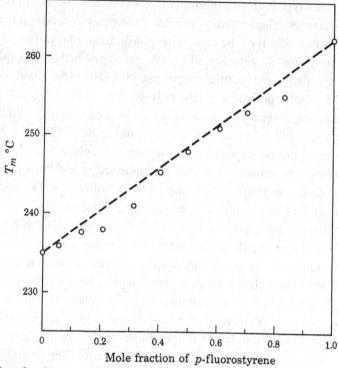

Fig. 5.14 Plot of melting temperatures of isotactic styrene–p-fluorostyrene copolymers as a function of mole fraction of p-fluorostyrene. (From Natta (102))

a fourfold one. The copolymers are formally classified as being isodimorphic. The melting temperatures are essentially a linear function of composition as is illustrated in Fig. 5.14.(102) Copolymers richer in styrene have the former structure, thus those richer in p-fluorostyrene have the latter one. The melting temperatures of poly(aryl ether ketone ketone) comprised of terephthalic and isophthalic units show a similar composition relation.(124)

Copolymers of 4-methyl-1-pentene with 4-methyl-1-hexene and ispropyl vinyl ether with sec butyl vinyl ether have also been shown to be isomorphic.(126) The melting points are not always a linear function of composition but the levels of crystallinity are relatively high, consistent with co-crystallization over the complete composition range.

Random copoly(esters) and copoly(amides) provide a set of polymers that are fruitful in yielding information about co-crystallization and isomorphic replacement. In particular the role of the distance between the carbonyl groups in the diacids can be explored in detail. As an example, Edgar and Hill (103) pointed out that the distances between the carbonyl groups in terephthalic and adipic acid are almost identical. Therefore, it could be anticipated that copoly(amides) and

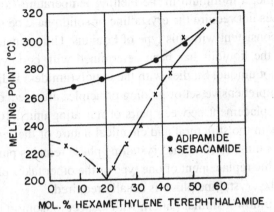

Fig. 5.15 Plot of melting temperature against composition of hexamethylene adipamide/terephthalamide (●) and hexamethylene sebacamide/terephthalamide (X) random copolymers. (From Edgar and Hill (103))

copoly(esters) of these co-units would be isomorphic. This expectation is reached for these copoly(amides), as is illustrated in Fig. 5.15.(103) Here the melting temperature–composition relation for the copolymer hexamethylene terephthalamide with adipamide is a smooth monotonic function consistent with isomorphic replacement over the complete composition range. No minimum in the melting point–composition relation is observed. This curve can be compared with that for copolymers of hexamethylene terephthalamide with sebacamide. This system gives a minimum in the melting temperature. It is consistent with the crystalline phase remaining pure and each component crystallizing separately. This is the expected result since the hexamethylene terephthalamide unit is not accommodated within the hexamethylene sebacamide lattice. Having comonomeric units of identical length is not however a sufficient condition for co-crystallization. This is evident from the observation that co-crystallization is not observed in the corresponding copoly(ester) of ethylene adipate–terephalate.(103) It was postulated that the interchain hydrogen bonding in copoly(amides) is necessary to maintain the *p*-phenylene linkage parallel to the chain axis to ensure the co-crystallization.

Tranter(107) has studied a series of copoly(amides) based on hexamethylene diamine and dibasic acids. Only one of the copolymers, hexamethylene diamine and *p*-phenylene dipropionic acid, gives a linear softening point–composition curve typical of isomorphic replacement. All the other copoly(amides) studied gave a minimum (eutectic type) softening point–composition diagram. However, from x-ray diffraction studies it was concluded that despite this type liquidus the second components dissolved in the lattice of the first until a critical concentration was reached. At this point the lattice structure changed rather abruptly. Now the second component was dissolved in the first, so that isodimorphism occurred. As a matter of general principle, a minimum in the melting temperature–composition of the liquidus curve does not require the crystalline or solid phase be pure. Several different solidi are consistent with this type of liquidus. Unfortunately, with but few exceptions, only the liquidus has been determined with copolymers, so that its interpretation is not unique. Studies with the copoly(amides) indicate that a more definitive and comprehensive set of guiding principles are needed in order to predict isomorphic replacement between pairs of repeating units.(104–106,108,127) Subtle differences in the structural and chemical nature of the repeating units are involved in determining whether the crystalline phase remains pure or if complete or partial isomorphic replacement of one type or the other takes place. Ultimately, a direct study of the crystalline phase is usually required.

Copoly(esters) follow a similar pattern with respect to isomorphic replacement. Again, specific examples have to be examined. Evans, Mighton and Flory studied a series of copolymers based on either decamethylene adipate or decamethylene sebacate.(58) The melting point depression of copolymers of poly(decamethylene adipate), with methylene glycol as comonomer is substantially less than observed for other co-units where the crystalline phase is pure. It was concluded that the situation was analogous to solid solution formation, but that the B units did not replace A units indiscriminately. The melting temperatures of copolymers that contained bulkier cyclic co-ingredients are in close agreement with expectations for a pure crystalline phase, consistent with structural considerations.

Copolymers of hexamethylene sebacate with decamethylene adipate and decamethylene sebacate with hexamethylene adipate show eutectic type minima in their respective melting temperature–composition relations.(128) However, high levels of crystallinity, characteristic of the respective homopolymers, are formed over the complete composition range. This result is not characteristic of a random copolymer with a pure crystalline phase. In the latter case a significant reduction in crystallinity level and marked broadening of the fusion range is expected and is observed. It can be concluded that in each of these copoly(esters) both repeating units participate in a common lattice.

A rare example of where both the liquidus and solidus, and thus the complete phase diagram, were determined can be found in the work of Hachiboshi *et al.* who crystallized random copolymers of ethylene terephthalate with ethylene isophthalate over the complete composition range.(110) The wide-angle x-ray patterns of these copolymers change systematically with co-unit content. It was concluded that the two units can co-crystallize and form a new unit cell. The complete phase diagram is shown in Fig. 5.16.(110) The solidus was determined by assuming the additivity of the lattice spacings. The phase diagram is a classical one. It even contains an azeotropic point. Polymer crystallization, therefore, is not atypical. For low molecular weight systems the liquid and solid must have the same composition, or

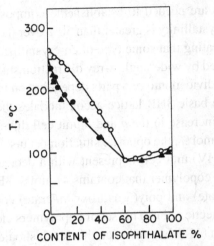

Fig. 5.16 Complete phase diagram for ethylene terephthalate–ethylene isophthalate random copolymers. (From Hachiboshi, *et al.* (110))

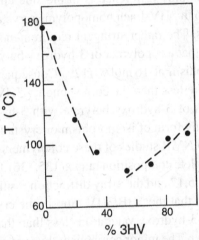

Fig. 5.17 Plot of melting temperature against the melt composition for random copolymers of 3-hydroxy butyrate and 3-hydroxy valerate. (From Scandala, *et al.* (131))

activity at the azeotropic point. For random copolymers, the comparable requirement would be that the sequence propagation probability be the same in both phases. With the utilization of advanced techniques to probe the structure and composition in the solid state, presentations of complete phase diagrams can be anticipated in the future.

The crystallization of bacterial synthesized random copoly(esters) of 3-hydroxy butyrate (3HB) with 3 hydroxy valerate (3HV) has been extensively studied.(114,115,129–135) The melting temperature–composition relation, based on the liquidus, is given in Fig. 5.17.(131) A pseudo-eutectic point is found at 41 mol %

3HV. Further studies are needed to establish the composition of the crystalline phase. The level of crystallinity is greater than 50% over the complete composition range (131,134) indicating that some type of co-crystallization has occurred. This conclusion is supported by wide-angle x-ray diffraction studies.(114,129,130) The x-ray patterns can be divided into two parts depending on the 3HV content. Below 41 mol % of 3HV the basic 3HB lattice accommodates the 3HV comonomer, as demonstrated by the increase in the *a* and *b* unit cell dimensions. At higher concentrations, above 55 mol %, the opposite situation occurs. The x-ray patterns now show that the poly(3HV) unit cell is present with a decrease in the *b* dimension in the unit cell. In the copolymer that contains 41 mol % 3HV, crystallites of both poly(3-hydroxy butyrate) and poly(3-hydroxy valerate) coexist, as would be expected at a pseudo-eutectic point. This set of copolymers clearly shows a classical case of isodimorphism.(114) In retrospect, this result should not be too surprising. The two comonomers are chemically and geometrically similar as are the crystal structures of their respective homopolymers. Both of the monomers have the same backbone structure. The only difference is in the side chains; a methyl group in 3HB and an ethyl group in 3HV. Each homopolymer crystallizes as a 2_1 helix with similar repeat distances. The rather stringent requirements for isodimorphism are accentuated by the fact that a copolymer of 3-hydroxy butyrate 4-hydroxy butyrate does not show isomorphism at 16 mol %.(129) Consequently, the crystallite size and crystallinity level are less than the corresponding 3HV copolymers. In a similar manner copolymers of 3-hydroxy butyrate with 3-hydroxy hexamate do not give any indication of any form of isomorphism or crystallization.

Solid-state carbon-13 NMR studies of these copolymers also demonstrate isodimorphism over the complete composition range.(135,136) These studies are consistent with the plot in Fig. 5.17 and the x-ray diffraction results. Moreover, the NMR studies have also shown that the 3HB/3HV ratio in the crystalline phase of poly-(3-hydroxy butyrate-co-3-hydroxy valerate) is less than that for the nominal composition of the copolymer. The minor component thus enters the lattice at a smaller concentration than the composition of the pure melt. The ratio of 3HV in the crystal to that in the melt increases with increasing 3HV content. In principle, a complete conventional type phase diagram based on composition could be obtained by this method. We should recall, however, that when treating copolymers the important quantity is the sequence distribution within the crystal relative to that of the melt.

A similar type of pseudo-phase diagram is shown in Fig. 5.18 for random copolymers of ethylene terephthalate (PET), and ethylene naphthalene 2,6-decarboxylate (PEN).(112,137) The data points corresponding to the different crystallization procedures are very close to one another. Wide-angle x-ray diffraction studies have shown that significant crystallinity is present in fibers of these random copolymers.

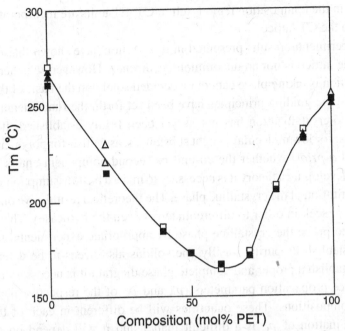

Fig. 5.18 Plot of melting temperature against melt composition for random copolymers of ethylene terephthalate and ethylene naphthalene 2,6-dicarboxylate for different crystallization procedures. △ dynamic crystallized sample; ▲ annealed sample; □ fiber sample, ■ annealed fiber sample. (From Lu and Windle (112))

The lattice parameters vary with composition in a way that indicates that the crystals are composed of both types of units. The crystallites are made up of both ethylene terephthalate and ethylene naphthalate units. A transition in crystal structure occurs at 70 mol % ethylene terephthalate, corresponding to a composition slightly to the right of the minimum in Fig. 5.18. These features are reminiscent of those found in the 3HB/3HV copolymers.

Random copolymers of ethylene 2,6 naphthalate and hexamethylene 2,6 naphthalate (PHN) illustrate the subtleties involved for co-crystallization to occur. Upon crystallization this system gives a typical eutectic type pseudo-phase diagram based on the liquidus.(113) However, there is no indication of co-crystallization occurring anywhere in the composition range. The copolymers rich in ethylene units only form PEN crystals; while those rich in hexamethylene only form PHN crystals. The importance of the lengths of the repeating units for co-crystallization is emphasized here, as in the following example. Ethylene terephthalate (ET) and 1,4-cyclohexene dimethylene terephthalate (CT) have similar chemical structures but their repeat distances are different. Despite this, co-crystallization to a limited extent is found in copolymers of the two monomers.(138,138a) In the ET rich composition range only poly(ethylene terephthalate) type crystallites are formed.

However, in the composition region rich in CT, ET units are incorporated to some extent into the CT lattice.

The experimental results presented in this section have shown that for co-units to enter the lattice is not an uncommon occurrence. However, whether or not co-crystallization is taking place cannot be decided solely on the basis of the liquidus. Although some guiding principles have been set forth, the fundamental structural basis for co-crystallization has not as yet been firmly established. It should be recalled that for low molecular weight substances, as well as for polymers, it cannot be decided *a priori* whether the co-unit or second component enters the crystal lattice. In developing a theory it is necessary to make a basic assumption with regard to the constitution of the crystalline phase. The theoretical results give one guidance as to what to seek in order to differentiate between the two cases. Therefore, it is necessary to probe the crystalline phase by appropriate experimental methods in order to establish its purity. Ideally, the solidus also needs to be determined. In order to establish a proper and complete phase diagram it is necessary to stipulate the sequence propagation parameters, p_A and p_C of the respective phases rather than the compositions. These quantities will be different in each of the phases. The determination of p_C is a difficult matter and it will depend on the details of the isomorphic structure. As has been noted earlier, several equilibrium and nonequilibrium theories have been developed to account for co-crystallization. These theoretical developments have all been based on the composition in the crystalline phase.

Despite the shortcomings that have been described, efforts have been made to explain isomorphism and isodimorphism based on the theories that have been outlined. Equilibrium theories as embodied in Eqs. (5.33) and (5.35) will require equilibrium melting temperatures, or approaches thereto. Observed melting temperatures can be used with nonequilibrium theories. Unfortunately, for cases where it has been established that co-units enter the lattice, the proposed equilibrium theories have been tested with directly observed melting temperatures. In some instances, attempts to account for the small crystallite thicknesses have been made by invoking the Gibbs–Thomson equation.(16) However, there are other factors that cause the observed melting temperatures to be reduced from the required equilibrium ones.

As an example of the problems involved consider the melting temperature results for copolymers of ethylene terephthalate–ethylene naphthalene 2,6-decarboxylate.(13,112,137) This set of copolymers is isodimorphic over the complete composition range. An analysis has been given that includes the Flory ideal model given by Eq. (5.42) and the equilibrium Sanchez–Eby model Eq. (5.33). The former assumes that the crystalline phase is pure; the latter that the B units enter the lattice on an equilibrium basis. Also, the Baur analysis, Eq. (5.38) has been invoked. The latter, a modification of the Flory theory, assumes that the average sequence

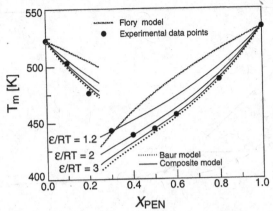

Fig. 5.19 Plot of melting temperatures of ethylene terephthalate–ethylene naphthalene 2,6-decarboxylate copolymer as functions of melt composition for different theoretical models. Flory model Eq. (5.24), Baur model Eq. (5.38); composite model Eq. (5.41). Solid points: experimental results(112,137). (From Wendling and Suter (13))

length dominates the melting. A composite model represented by Eq. (5.35) was also considered. This model combines the Sanchez–Eby (equilibrium, B units in the lattice) with the Baur theory for the crystalline phase being pure. The results of the analyses are summarized in Fig. 5.19.(13) Directly observed, nonequilibrium melting temperatures are used.

The Flory model predicts higher melting points, as would be expected, and should not even be considered here. The basic tenets of the theory do not apply to this set of experimental data since it has been established that the crystalline phase is not pure. Equation (5.38) gives a reasonable agreement with the experimental data over a major portion of the composition range. However, the theory again requires a pure crystalline phase. The hybrid, composite theory, Eq. (5.41), can be made to fit the experimental data. However, it requires large variations in the parameters ϵ/RT with composition. Thus, even with the assumptions that have been made, and the use of diverse physical situations, none of the theories that have been proposed satisfactorily explain this data set. It is not difficult to fit this type of data since ϵ is treated as an arbitrary parameter and is allowed to vary with composition. The physical significance is of concern for this reason and for combining two different theories: in one the crystal phase is pure, in the other it is not. A similar fit has also been obtained in the analysis of the melting temperature of the 3-hydroxy butyrate–3-hydroxy valerate copolymers.(138b)

It is possible to grow large copolymer crystals by taking advantage of the simultaneous polymerization and crystallization of certain monomers.(139–141) Wegner and collaborators have taken advantage of this technique to prepare well-defined, large crystals of poly(methylene oxide) from trioxane as well as

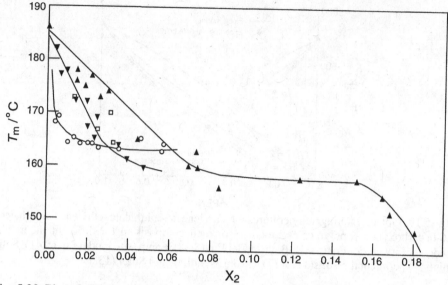

Fig. 5.20 Plot of melting temperatures against mole fraction of comonomers for copolymers of poly(methylene oxide) formed by solid-state polymerization. Comonomers: ▲ —CH$_2$—CH$_2$—O—; ▼ —(CH$_2$)$_3$—O—; □ —(CH$_2$)$_4$—O—; ○ —(CH$_2$—CH$_2$—O—)$_2$. (From Dröscher, *et al.* (144))

copolymers from trioxane with other cyclic ethers as comonomers.(142–147) The reaction proceeds until chemical equilibrium between the solid, crystalline polymer and the residual monomer is reached. Moreover, if the crystals are annealed in these closed systems, smaller crystals dissolve in favor of further growth of the already existing larger ones. Crystals as large as 10 μm in diameter and 1 μm in the chain direction can be obtained by this method.(141) Random copolymers are formed by this polymerization procedure so that the comonomer units are now randomly distributed within the crystal.(143) This unique and well-defined sequence distribution within the crystal, is a consequence of equilibrium polymerization.

The melting temperature–composition relations of such copolymers are quite interesting. Figure 5.20 is a plot of the melting temperatures of the nascent, as-polymerized copolymers, of poly(methylene oxide) against the mole fraction, X_2, of incorporated comonomer.(144) The comonomers used are indicated in the legend. The mole fraction at which extended chain crystals could be maintained ranges from 0.057 to 0.18. This is a very unique plot in that at low concentrations of co-units there is a monotonic depression of the melting temperature with mole fraction of the co-unit. However, in all cases there is a plateau region that appears at approximately the same melting temperature. At this point the melting temperature is independent of composition. Similar results have also been observed with epichlorohydrin (146) and phenyl glycidyl (147) as comonomers. Among these copolymers only the one with 1,3-dioxolane as comonomer could be prepared at concentration beyond the plateau. At $X_2 \simeq 0.15$ a monotonic decrease in the melting temperature is again observed. The interpretation of Fig. 5.20 presents a unique and complex situation. Because extended chain crystals are involved it might be expected that close to equilibrium conditions would prevail. Under these circumstances the sequence distributions will be the same in the crystal and the melt at the melting temperature. Since the parameter p is the same in both states, to a first approximation the melting temperature should be independent of composition. Although this condition is fulfilled in the plateau region it is clearly not at the low comonomer concentration. The problem is that although the distribution in the nascent crystallite is a consequence of polymerization equilibrium, it does not follow that starting with the melt the same distribution would result under equilibrium crystallization conditions. It is conceivable that at the low comonomer concentration the sequence distribution is disturbed so that either the crystalline phase remains pure or smaller than equilibrium values of the co-units enter the lattice. The results embodied in Fig. 5.20 are interesting and important. However, they are not subject to an obvious interpretation over the complete composition range. Once melted, and allowed to recrystallize, conventional behavior is observed.(143) Melting temperatures are depressed by about 10 °C for the homopolymers to about 30 °C at $X_2 = 0.1$ for 1,3-dioxalane as comonomer.(146)

It should be noted in concluding this section that the fibrous and globular proteins, as well as the nucleic acids, possess crystal structures that allow different chemical repeating units of the same general type to enter the crystal lattice. Crystallographic analysis indicates a stereochemical identity among many of the amino acid residues and the nucleotides. Under favorable circumstances, the simpler synthetic copolymers behave in a similar manner.

5.5 Branching

Another type of structural irregularity that influences melting and crystallization is long chain branching. The reason is that the branch points are structurally different from other chain units. The role of short chain branches of regular length, as for example the random ethylene–1-alkene copolymers, has already been discussed. In this case the melting temperature relations can be expressed in a formal manner. Long chain branches, on the other hand, are not usually of uniform length. Most often the branches are of sufficient length so that they can also participate in the crystallization.

Long chain branched polyethylene, commonly termed low density polyethylene typifies this class of polymers. Thermodynamic measurements, such as heat capacity (148,149) and specific volume,(150,151) indicate that long chain branched

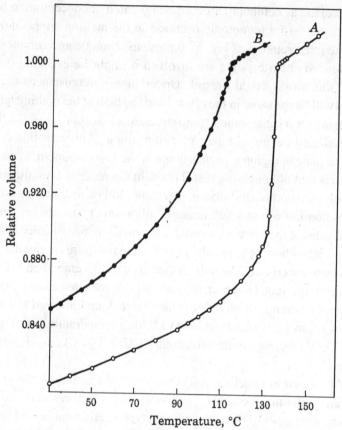

Fig. 5.21 Plot of relative volume against temperature. Curve A for linear polyethylene
(polymethylene); curve B for long chain branched polyethylene.(151)

polyethylene melts vary broadly. In typical copolymer fashion approximately half
of the crystallinity disappears over a 40 °C temperature interval. A comparison of
the course of fusion between linear polyethylene (polymethylene) and a long chain
branched polyethylene is shown in Fig. 5.21.(151) The differences between the two
polymers are readily apparent. The linear polymer melts relatively sharply, 70% of
the crystallinity disappears in only a 3 to 4 °C interval. In comparison the fusion
of the branched polymer takes place over the complete temperature range of study.
There is also about a 20 °C difference in the observed melting temperatures.

Qualitatively similar melting point depressions are observed in long chain
branched poly(ethylene terephthalate)(152) and poly(phenylene sulfide).(153) The
extrapolated equilibrium melting temperatures of poly(phenylene sulfide) decrease
by 11 °C with a modest concentration of long chain branches. Coincidentally, the
extrapolated equilibrium melting temperatures of poly(ethylene terephthalate) also
decrease by 11 °C for the range of branch concentrations studied.

Multi-arm, or star polymers represent model branched systems. Such polymers have been synthesized from ε-caprolactam. Extrapolated equilibrium melting temperatures were obtained for the linear, three-arm and six-arm polymers of comparable molecular weights.(154) The directly observed melting temperatures were systematically lowered as the branching content increased. This was reflected in the equilibrium melting temperature which decreased by 8 °C from the linear polymer to the six-arm star one. This decrease in the extrapolated equilibrium melting temperature is greater than would be expected based solely on the concentration of branch points and their disruption of the structural regularity of the chain. The concentration of arms from a common branch point plays an important role in this regard.

5.6 Alternating copolymers

When there is a strong tendency for the comonomeric units to alternate, i.e. when $p \ll X_A$, a large depression of the melting temperature is predicted by Eq. (5.26). This expectation is based on the assumption that only the A units crystallize and the crystal structure corresponding to that of the homopolymer forms over the complete composition range. This condition is usually difficult to fulfill.

An example of the melting temperature–composition relation for an alternating copolymer is that of ethylene and chlorotrifluoroethylene, shown in Fig. 5.22.(155) The observed melting temperatures are plotted against the mole fraction of the ethylene units. A maximum in the melting temperature is observed at equal molar ratios of the two components. This temperature, 264 °C, corresponds to the

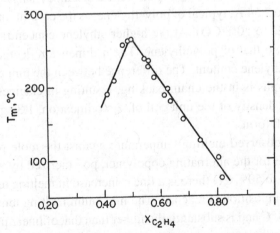

Fig. 5.22 Plot of melting temperature against mole fraction of ethylene units for alternating copolymers of ethylene–chlorotrifluoroethylene. (From Garbuglio, *et al.* (155))

melting of a sequence of $C_2H_4C_3F_3Cl$ repeating units and is much higher than that of the corresponding homopolymers. Compositions above and below the equimolar concentration represent incomplete alternation. The plot in Fig. 5.22 is obviously different from that expected for a random copolymer. Enthalpies of fusion, ΔH_u, per repeat unit of this comonomer pair have been reported to be 3175 and 4500 cal mol^{-1}.(155,156) It does not necessarily follow from these results that the melting temperature of the perfectly alternating copolymer will always be greater than the corresponding homopolymer.

A classic example of an alternating type copolymer is found in ethylene–carbon monoxide. Copolymerization of this copolymer pair by either free-radical methods or by γ-radiation at low temperature does not lead to perfect alternation.(157,158) In contrast, perfectly alternating copolymers of ethylene–carbon monoxide, as well as carbon monoxide with other olefins have been prepared by the use of homogeneous palladium catalyst systems.(159–161) The alternating ethylene–carbon monoxide copolymer is polymorphic. The α form is stable at low temperatures and transforms to the β form at about 140 °C. The melting temperature of the β form is approximately 255 °C, for the palladium catalyzed polymer (162), and is about 10 °C higher than the corresponding copolymer prepared by free-radical polymerization.(162) There is a systematic decrease in the melting temperature with increasing carbon monoxide concentration.(158,162,163) The melting temperature of the alternating copolymer is much greater than that of linear polyethylene and any of its random copolymers. This is a consequence of the high extent of alternation and the formation of a different crystal structure.

The unit cell of the α polymorph is orthorhombic, similar to that of polyethylene, and the chains have a planar zigzag conformation. However, the repeat distance in the chain direction is 7.60 Å.(162,164) The repeat distance changes discontinuously from this value to 2.54 Å, typical of polyethylene, as the co-unit ratio changes from 1:1 to 1.3:1 (44% to 50% CO). At the higher ethylene concentrations the crystal structure reverts to that of polyethylene. The a dimension decreases while the b increases with ethylene content. The difference between the unit cell structures of the two polymorphs is in the chain packing, resulting in changes in the a and b dimensions. The density of the unit cell of α modification, 1.39 g cm^{-3}, is higher than that for the β form.

A plot of the observed melting temperature against the mole percent of CO is given in Fig. 5.23 for the alternating copolymer, polymerized by γ-radiation.(158) From about 39% to 50% CO there is a linear increase in melting temperature until the equimolar composition is reached. The maximum melting temperature of this copolymer is 244 °C and is substantially higher than that of linear polyethylene and any of its radom copolymers.

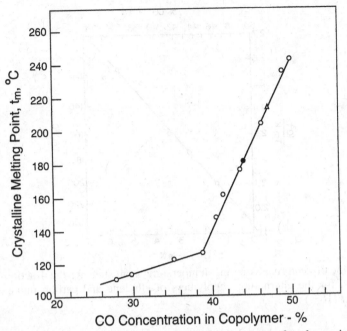

Fig. 5.23 Plot of melting temperature against CO concentration for alternating ethylene–carbon monoxide copolymers. (Adapted from Colombo, *et al.* (158))

To analyze the melting temperature–composition relation in more detail we assume, following Starkweather, that the new AB type crystal structure can be treated as the crystallizing unit in a random copolymer.(163) Thus, if y represents the fraction of CO, the concentrations of crystallizable CH_2CH_2 CO and noncrystallizable CH_2CH_2 units are proportional to y and $1 - 2y$ respectively. The fraction of crystallizable units X is then given by

$$X = \frac{y}{1 - y} \tag{5.43}$$

The Flory relation, for random sequence distribution, then becomes

$$\frac{1}{T_m} - \frac{1}{T_m^0} = -\left(\frac{R}{\Delta H_u}\right) \ln X \tag{5.44}$$

The melting point data of Fig. 5.23 are plotted in Fig. 5.24 according to Eq. (5.44). Good agreement is obtained between experiment and theory substantiating the analysis given above. Similar results are obtained when the melting temperature relation of alternating copolymers of ethylene–chlorotrifluoroethylene is treated in the same manner.

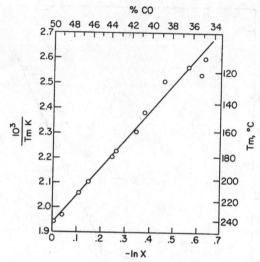

Fig. 5.24 Plot of reciprocal melting temperature against ln X, the fraction of crystallizable units, for the alternating copolymers of ethylene and carbon monoxide. (From Starkweather (163))

The question arises as to the reason for the high melting temperatures of the ethylene–carbon monoxide copolymers relative to that of linear polyethylene. There is about a 100 °C temperature difference. It has been proposed that this difference is due to the packing energy in the crystal and is thus reflected in ΔH_u.(162,165) On the other hand it has been thought to be due to a low value of ΔS_u, caused by disorder in the crystal and a preference for extended chain conformation in the melt.(163) The latter idea is, however, not supported by rotational isomeric state calculations.(165) The ΔH_u value for this poly(ketone) was determined to be 225 J g^{-1}, or 12.6 kJ mol^{-1}, by the diluent melting point depression method.(166) This result compares favorably with estimates from model compounds.(161) The corresponding value of ΔS_u of 5.3 e.u. mol^{-1} is not unduly low. The packing energy appears to play a major role in determining the high melting temperature.

Linear alternating copolymers of CO with either propylene, 1-butene, 1-hexene, norbornene or styrene have also been prepared using palladium type catalysts.(161) The melting temperatures of the crystalline propylene–carbon monoxide copolymers are much lower than those of ethylene–carbon monoxide. A melting temperature of 165 °C has been obtained for a highly regioregular chain. Alternating copolymers of carbon monoxide with either butene or hexene have not been crystallized.(167) Syndiotactic styrene–carbon monoxide copolymers have melting temperatures as high as 280 °C. A melting temperature for isotactic, optically active alternating copolymers of styrene–carbon monoxide has been reported as 353 °C.(168)

Pairs of olefins can also form alternating copolymers. By hydrogenating syndiotactic poly(cis-1,4-pentadiene) an alternating syndiotactic ethylene–propylene copolymer is formed.(169) The melting temperature of the copolymer, 39 °C, is well below the melting temperature of the respective homopolymers. The Bragg reflections characteristic of the copolymer are quite different from either linear polyethylene or syndiotactic poly(propylene), indicating the formation of a new crystal structure. Perfectly alternating copolymers of olefin pairs have also been prepared using metallocene type catalysts.(170,171) These include ethylene–propylene and ethylene–1-octene, among others.

Copolymers of tetrafluoroethylene show a strong tendency for alternation with either ethylene, propylene or isobutylene as comonomers.(172–174) The alternating copolymers with ethylene have a crystal structure that is quite different from either polyethylene or poly(tetrafluoroethylene).(175) The form of the melting temperature–composition relation is similar to that shown in Fig. 5.22 for the alternating ethylene–chlorotrifluoroethylene copolymers. The maximum melting temperature at the equimolar composition, where the extent of alternation is greater than 90%, is 285 °C. This melting temperature is substantially greater than that of linear polyethylene, and about 60 °C lower than that of poly(tetrafluoroethylene).

Copolymers of tetrafluoroethylene with isobutylene are crystalline in the equimolar range.(174) The maximum melting temperature of 203 °C corresponds to the equimolar composition. The x-ray diffraction pattern did not show any correspondence with the homopolymer of poly(tetrafluoroethylene). The prepared copolymers were not crystalline outside a narrow composition range. It is unresolved whether crystallinity could be induced over a wider composition range, or if the sequence distribution is such as to preclude the development of crystallization. In contrast, the copolymers containing propylene did not display any crystallinity over the complete composition range. This result appears to indicate a low level of stereoregularity for the propylene comonomer.

Some general features have emerged from the discussion of the crystallization and melting behavior of highly alternating copolymers. Almost invariably a new crystal structure is formed that is different from that of the corresponding pure homopolymers. Thus, structural similarity is not a requirement for alternating copolymers to crystallize, or to be crystallizable. This conclusion reflects one of the unique properties of alternating copolymers. Consequently, it is found that crystallization can occur with unlikely, or unexpected, comonomer pairs. Melting temperatures of the copolymers can be higher than either of the respective homopolymers with some pairs, lower than both in others and in between in some cases. Detailed analysis of the melting point relations in terms of the structure of the alternating crystalline sequence is hampered by lack of knowledge of the appropriate thermodynamic parameters that govern fusion.

5.7 Block or ordered copolymers

A class of copolymers of particular interest are block or ordered copolymers. They are also known, in special situations, as multiblock or segmented copolymers. In this type copolymer identical chain units are organized into relatively long sequences. The sequence propagation probability parameter p is, therefore, much greater than X_A and approaches unity in the ideal case. Consequently the equilibrium melting temperature T_m is expected to be very close to that of the pure homopolymer, provided that the melt is homogeneous and the crystalline phase is pure and devoid of any permanent built-in morphological constraints. Extensive research concerned with the crystallization of block copolymers has been reported. Thus, the literature on the subject is voluminous. Therefore, no effort will be made here to give a complete compilation of the literature. Rather, the usual procedure will be followed of seeking out examples that illustrate the basic principles that are involved.

The long sequences of A and B units of a block copolymer can be arranged in several different ways. A diblock copolymer, schematically represented as

$$AB$$

is characterized by the length of, or number of repeats in each of the sequences. A triblock copolymer has two junction points of dissimilar units and can be represented as

$$ABA \quad or \quad BAB$$

It is also characterized by the molecular weight of each block. A multiblock, or repeating copolymer can be represented in general as

$$(A-A---A-A-A)_n (B-B-B---B-B-B)_m$$

where the sequence lengths of each type sequence can be either constant or variable. Block copolymers do not have to be linear. For example star-shaped or "comb" polymers can be composed of distinct blocks. For illustrative purposes we shall limit ourselves to just two different types of repeating units. The principal interest is in crystalline, or crystallizable, blocks. There are two possible situations. In one, both blocks are potentially crystallizable. In the other only one block is able to participate in the crystallization. The case where both blocks are amorphous, or noncrystallizable, is not of interest in the present context.

Before examining the crystallization behavior of block copolymers it is necessary to first understand the nature of the melt. This is an important concern since it is from this state that crystals form and into which they melt. For reasons that will become apparent this is a particularly important consideration in understanding the crystallization and melting of block copolymers. The melt of a block copolymer is not necessarily homogeneous, even under equilibrium conditions. The melt can be heterogenous with a definite supermolecular, or domain structure. This melt structure is unique to ordered copolymers and represents an equilibrium property of the melt. It can be expected that a heterogeneous melt will affect the crystallization kinetics, because crystallization will occur in a constrained space. The thermodynamic properties should also be altered relative to the homogeneous melt. In turn the equilibrium melting temperature will be influenced.

The basis for understanding the structure of block copolymers in the liquid state is related to the problem of mixing two chemically dissimilar polymers, as was discussed in Chapter 4. To review briefly, two chemically dissimilar homopolymers will be homogeneous when the free energy of mixing is negative. The entropy to be gained by mixing two such homopolymers is very small owing to the small number of molecules that are involved. Therefore, only a small positive interaction free energy is sufficient to overcome this inherent mixing entropy. Immiscibility thus results. It can be expected, in general, that two chemically dissimilar polymers will be incompatible with one another and phase separation will occur. As was pointed out when discussing miscible binary blends, exceptions will occur between polymer pairs that display favorable interactions.

Consider now a block copolymer composed of two chemically dissimilar blocks each of which is noncrystalline. The same factors that are involved in homopolymer mixing will still be operative so that phase separation would be *a priori* expected. However, since the sequences in the block copolymer are covalently linked, macrophase separation characteristic of binary blends is prevented. Instead, microphase separation and the formation of separate domains will occur. The linkages at the A–B junction points further reduce the mixing entropy. There has to be a boundary between the two species and the junction point has to be placed in this interphase. The interphase itself will not be sharp and will be composed of both A and B units. Mixing of the sequences, and homogeneity of the melt, will be favored as the temperature is increased. There is then a transition temperature between the heterogeneous and homogeneous melt, known as the order–disorder transition.

The details of phase separation in block copolymers depend on the chain lengths of the respective blocks, their interaction and the temperature and pressure. Microphases will tend to grow in order to reduce the surface to volume ratio and hence reduce the influence of the interfacial free energy associated with the boundary between the two domains. However, the restriction on the localization of the A–B junction point is important and acts to restrain the growth. These opposing effects will produce a minimum in the mixing free energy that will depend on the size and shape of the domain for a given composition and molecular weight of the species. Depending on the composition and molecular weight of the blocks, phase separation is favored by specific domain shapes. The simplest shapes calculated, as well

Fusion of copolymers

as observed, are alternating lamellae of the two species; cylinders (or rods); spheres of one species embedded in a continuous matrix of the other. Phase diagrams in the melt, involving the different possible microphases have been calculated.(176–184) Interaction with a solvent, prior to microphase separation, can exert a profound influence on the size and shape of the domain. This is due to specific interaction with a particular block. In solution various micellar type structures are found, the specifics of which are dependent on the nature of the solvent.

For most of the AB, ABA and $(A_n B_m)$ systems which have been studied the domain diameters for spheres and cylinders, and the thickness of lamellae, are usually in the range of 50–1000 Å.(183) Details will depend, among other factors, on the molecular weight and block lengths. However, the length of cylindrical domains and the breadth and length of the lamellae can approach macro dimensions when the morphology is well developed. The boundary between the two microphases is not infinitely sharp. Rather there is a concentration gradient across the boundary where the mixing of the two species occurs. Typically the thickness of the interphase is estimated to be about 20–30 Å.

A schematic illustration of the major domain structures that are found in pure amorphous block copolymers is illustrated in Fig. 5.25.(183) Here the diblock copolymer poly(styrene)–poly(butadiene) is taken as an example. In (a) poly-(styrene) spheres are clearly seen in a poly(butadiene) matrix; the spheres change to cylinders with an increase in the poly(styrene) content, as in example (b). With a further increase in the poly(styrene) concentration, alternating lamellae of the two species are observed (c). At the higher poly(styrene) contents, (d) and (e), the situation is reversed. Poly(butadiene) cylinders, and then spheres, now form in a poly(styrene) matrix. More quantitative descriptions of the domain structures have been given.(184,186,187) Crystallization and melting often occur to or from heterogenous melts with specific microphase structures.

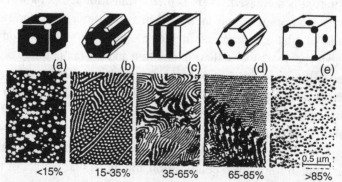

Fig. 5.25 Schematic representation of domain structures in amorphous diblock styrene–butadiene copolymers. Percentages show poly(styrene) content. (Brown *et al.* (183))

With this brief outline of the structural features of the melt of block copolymers, the equilibrium aspects of the crystalline state can be examined. Since the crystallization and melting of block copolymers will depend on the melt structure, a clear distinction has to be made as to whether it is homogeneous or heterogeneous. The type and size of the melt domains, as well as the associated interfacial structure are also important. Several studies have indicated that the rate of transformation from a homogeneous to two-phase melt takes place at a measurable rate in many block copolymers. Consequently, the crystallization can in principle be conducted from either of these melt states. It is reasonable to expect that the properties of the crystalline state would be affected accordingly. Therefore, the pathway for the crystallization needs to be specified.

An important issue is whether each of the components that comprise the copolymer can crystallize. If one cannot, it is important to specify whether it is in either the glassy or rubber-like state. The nature of one species, even if it does not crystallize, will influence the crystallization of the other. If the glass temperature of the noncrystallizing species is greater than the melting temperature of the crystallizing component, then restraints will be imposed on the crystallization process. Similar effects could also occur by very rapid cooling and vitrification or by having a highly entangled crystallizing component. The importance of these effects needs to be explored.

The interest at this point is to analyze the melting temperature–composition and melting temperature–block length relations of some typical ordered copolymers. Although the primary concern is the equilibrium condition, it can be anticipated, from the above discussion, that there could very well be complications in achieving this state. For an ideal, ordered copolymer of sufficient block length the parameter p will approach unity. Therefore, T_m should be invariant with composition. This expectation is drastically different from what is predicted for and observed with other copolymer types. This expectation is unique to chain molecules. It emphasizes the key role of the arrangement of the chain units in governing crystallization behavior. For a given composition, with the same co-units, major differences are to be expected in both melting temperatures and level of crystallization between random and block copolymers. These differences should in turn be reflected in a variety of properties. It should be recalled, however, that the free energy of fusion that was used to derive Eq. (5.26) is based on the premise that the melt is homogeneous. Even at equilibrium the melts of ordered copolymers are not necessarily homogeneous, and quite commonly are not. Therefore, the presence of the domain structures, and the interfacial region between them, could alter the conclusions reached with respect to the melting temperature relations.

The fusion of block copolymers is sharp and comparable to that of a homopolymer. This point is illustrated in Fig. 5.26 where the specific volume is plotted

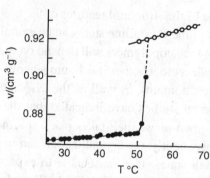

Fig. 5.26 Plot of specific volume against temperature for diblock copolymer of styrene and ethylene oxide. (From Seow, *et al.* (188))

against the temperature for a diblock polymer of poly(styrene)–poly(ethylene oxide).(188) The M_n of the crystallizing ethylene oxide block is 9900 and its weight percent in the sample is 67%. The melting range is clearly very narrow. All of the fusion characteristics are reminiscent of a well-fractionated linear homopolymer. This behavior is theoretically expected for a block copolymer with long crystallizable sequences, when there is no intervention of any morphological complications.

A striking example of the importance of sequence distribution on the melting of copolymers is shown in Fig. 5.27.(189) A compilation of the melting temperatures of random and block copolymers of ethylene terephthalate, with both aliphatic and aromatic esters as comonomers is given. The basic theoretical principles with respect to composition are vividly illustrated here. The chemical nature of the co-unit plays virtually no role in the melting temperature–composition relations. Of paramount importance are the sequence distributions. The difference in melting point between the random and block copolymers is apparent. In the block copolymers, the melting temperature is essentially invariant until a composition of less than 20% of ethylene terephthalate is reached. At this point there is a precipitous drop in the melting temperature. On the other hand, as is expected, there is a continuous, monotonic decrease in the melting temperature of the corresponding random copolymers. As a consequence, in the vicinity of 40–60 mol % of ethylene terephthalate there is more than 100 °C difference in the melting temperature of the two types of copolymers. There will of course also be major differences in the levels of crystallinity between the two. In turn, these differences in crystallinity properties will influence many other properties.

The important principle that the melting temperature of a copolymer depends on the sequence distribution of the co-units, and not directly on the composition, is also illustrated by ester interchange, that can take place in the melt. In Fig. 5.28 the melting temperature is plotted as a function of time as a 50/50 poly(ethylene

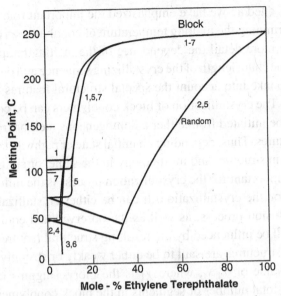

Fig. 5.27 Melting temperature against composition for block copolymers of poly(ethylene terephthalate) with ethylene succinate(1); ethylene adipate(2); diethylene adipate(3); ethylene azelate(4); ethylene sebacate(5); ethylene phthalate(6); and ethylene isophthalate(7). For comparative purposes, data from random copolymers with ethylene adipate and with ethylene sebacate also are given. (From Kenney (189))

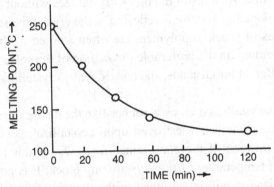

Fig. 5.28 Plot of melting temperature as a function of time for a poly(ethylene adipate), urethane linked, block copolymer heated at 250 °C. (From Iwakura, *et al.* (190))

terephthalate)–poly(ethylene adipate) urethane linked, block copolymer is heated at 250 °C.(190) The copolymer composition remains fixed during the heating. However, the melting temperature decreases with time because of ester interchange and the randomization of the copolymer. Concomitantly, the parameter p decreases with time and eventually approaches X_A. During the course of the 120 minute heating the melting temperature decreases from 250 °C for the block copolymer, to 120 °C for the random one.

The examples cited above have emphasized the important role of sequence distribution in determining the melting temperature of copolymers. However, in order to understand in more detail the dependence of the melting temperature of block copolymers on the chain length of the crystallizing sequence and on the composition it is necessary to take into account the special structural features that are inherent to such systems. The crystallization of block copolymers can be complicated since the process can be initiated from either a homogeneous melt or from different microdomain structures. Thus, depending on initial state or pathway taken, differences can be expected in structure and morphology in the same, or similarly constituted polymers. Also important for the crystallization process is the influence of the second component on the crystallization. It can be either crystallizable, rubber-like, or a glass. The fusion process, as well as the observed and equilibrium melting temperatures, will be influenced by the resulting structural features.

Microdomain structures are said to be either weakly or strongly segregated, depending on the value of $\chi_1 N_t$, where χ_1 is the Flory–Huggins interaction parameter and N_t the total number of segments in the block copolymer. When the microdomains in the melt are weakly segregated, crystallization in effect destroys the structure and a conventional lamellar type morphology results. When the molecular weight of the copolymer increases, then according to theory, the stability of the microdomain in the melt is enhanced and the structure is maintained during subsequent crystallization. As a result the block crystallizes without any morphological change, i.e. the domain structure is reflected in the crystalline state that results.

The properties of block copolymers are often studied in the form of solvent cast films. Depending on the preferable interaction of the solvent with each of the components, different initial states, and consequently crystallization pathways, can be established.

When the noncrystallizing block is rubber-like the distinct possibility exists that the domain structure will be destroyed upon crystallization.(191) The situation would be quite different if the glass temperature, T_g, of this block were greater than the melting temperature of the crystallizing block. It is possible in this case that the crystallization will be confined to the domains formed in the melt, with vitrification occurring in the noncrystallizable block. In this case, the junction points between the blocks are localized at the interface between the two components. The segmental motion involved in crystallization will be retarded and the crystallization will be confined. The influence of these factors can be better understood by studying selected examples and comparing the crystallization of di- and triblock copolymers that have the same components.

The fusion of the triblock copolymer, hydrogenated poly(butadiene–isoprene–butadiene) 27 wt % hydrogenated poly(butadiene), crystallized under two different conditions, is one example of the role of the initial melt structure or crystallization

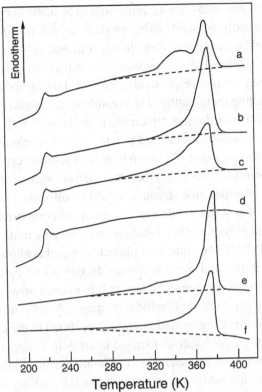

Fig. 5.29 Thermogram from differential scanning calorimetry. (a) isothermally crystallized HBIB; (b) solution crystallized HBIB; (c) melt crystallized HBIB slow cooled at −5 K/min; (d) amorphous HPI; (e) solution crystallized HPB; (f) melt crystallized HPB slow cooled at −5 K/min. HPB, hydrogenated poly(butadiene); HPI, hydrogenated poly(isoprene); HBIB, hydrogenated poly(butadiene)–poly(isoprene)–poly(butadiene). (From Séquéla and Prud'homme (192))

pathway.[5](192) The crystallizations were carried out either by slow cooling from the pure melt or from a benzene solution. The crystallization from the pure melt took place from a microphase separated domain structure assumed to be composed of hexagonally packed cylinders. In contrast, the small-angle x-ray scattering of the solution crystallized polymer indicated that the crystallization occurred without any microphase separation in the melt. This type of path dependence has been stated to be a general feature of block copolymer crystallization.(193) A comparison between the fusion of the hydrogenated poly(butadiene) in the block copolymer, and by itself, as a random copolymer, is given in Fig. 5.29 for both modes of crystallization.(192)

---

[5] Hydrogenated poly(butadiene), an ethylene–butene random copolymer, is often used as the crystallizing block, in di- and triblock copolymers. In this context the copolymer is commonly referred to as polyethylene. This nomenclature can be misleading since it carries the connotation that hydrogenated poly(butadiene) behaves as a homopolymer with respect to crystallization. In fact, it behaves as a typical random copolymer that is located within the structure of an ordered copolymer.

These thermograms also allow for a comparison to be made between the melting of the respective homopolymers. For either mode of crystallization, the melting temperature of the hydrogenated poly(butadiene) component is the same in the block copolymer or when isolated. A comparison of thermograms (b) and (c) for solution and bulk crystallized samples respectively, does not show any perceptible difference in the observed melting temperature. The breadth of melting is broader for the polymer crystallized from the domain structured melt. However, the recorded melting temperature of the solution crystallized polymer could be obscured by a melting–recrystallization process, a quite common feature of solution crystallized polymers. If this process was in fact occurring, then the interpretation of the thermograms in Fig. 5.29 in terms of initial melt structure would be difficult.

The properties of a series of diblock copolymers composed of hydrogenated poly(butadiene) and poly(3-methyl-1-butene) with varying molecular weights have also been reported.(194) The change in molecular weights allows for different degrees of incompatibility and melt structures. In this set of copolymers, the melt structure ranges from being homogeneous at low molecular weights to a strongly segregated hexagonally packed cylindrical morphology at the higher ones. Crystallization from the strongly segregated melts was confined to the cylindrical domain and was essentially independent of thermal history. In contrast, the morphology that results from either weakly segregated or homogeneous melts is dependent on the thermal history. In weakly segregated systems fast cooling from the melt confines the crystallization to the cylindrical domain; slow cooling leads to complete disruption of the cylindrical melt. Concomitantly, thermodynamic properties are altered. The lowest molecular weight samples, where crystallization proceeds from a homogeneous melt, develop the highest level of crystallinity and melting temperature. The crystallization from the strongly segregated melt results in a lower level of crystallinity, about 10%, and melting temperature reduction of about 4 °C. Although these differences are small on a global scale, they are important and emphasize the influence of the melt structure.

In a comparable study with block copolymers of poly(styrene) and poly(ε-caprolactone) the molecular weight was also varied.(195) Consequently the melt varied from being homogeneous to one that was strongly segregated. The melting temperatures of the copolymers with a homogeneous melt, M_n poly(styrene) = 6000 and varying low molecular weights of poly(ε-caprolactone) were close to that of the corresponding homopolymers of the same chain length. At most, the melting temperature of the homopolymer is 1–2 °C higher. Crystallinity levels are also comparable to one another and are in the range of 50–60%. The melting temperatures of the copolymers with a domain structured melt are also comparable to the homopolymers of corresponding molecular weights. However, the crystallinity levels are appreciably less.

Booth and coworkers have studied an interesting set of block copolymer fractions based on ethylene oxide, E, as the crystallizing sequence and propylene oxide, P, as the noncrystallizing sequence.(196–200) All of the crystallizing blocks had narrow molecular weight distributions. Studies of the mixing behavior of low molecular weight fractions of poly(ethylene oxide) and poly(propylene oxide) indicate that the two components are compatible in the melt. This observation leads to the conclusion that the corresponding block copolymers do not exhibit microphase separation in the melt. This set of copolymers then provide a good reference for melting point studies. Different types of copolymer architecture were studied. A comparison can be made of the thermodynamic behavior between the diblock PE, two triblocks PEP and EPE, as well as the multiblock copolymers $P(EP)_n$.

A set of PE type block copolymers, with E fixed at 40 units and P increasing from zero to 11 units, were studied.(196) The thickness of the crystalline portion of the lamellar structures that formed was about 25 ethylene oxide units. The crystallites are, therefore, close to extended form, but not completely so. A small, but significant, portion of the ethylene oxide units are noncrystalline and are intermixed with those of propylene oxide. The crystallinity level of the homopolymer with 40 repeating units is about 70%. This level of crystallinity is maintained by all of the diblock copolymers studied, irrespective of the length of the P blocks. The observed melting temperature of the corresponding homopolymer was 50–51 °C, depending on the crystallization temperature. There is a decrease of about 3.5 °C between the melting temperature of the homopolymer and the copolymer with 11 propylene oxide units. This small melting point depression can be attributed to interfacial effects caused by the increasing length of the noncrystallizing sequences. The basic equilibrium requirements appear to be applicable to this series of diblock copolymers.

A set of triblock copolymers, with the sequence PEP, were also studied. The length of the E block ranged from 48 to 98 repeating units and the P blocks from 0 to 30 units.(198) When E was equal to 48, either extended or folded crystallites were formed, depending on the length of the P block. This result demonstrates an important principle that extended chain crystallites can form in the central block of an ordered copolymer. This result is important since it demonstrates that folded structures that form at larger block lengths are a consequence of kinetic factors, rather than from any equilibrium requirement. Appropriate equilibrium theory must then allow for extended chain crystallites. For E blocks, whose lengths were greater than 48 units only folded type crystals formed, irrespective of the lengths of the P blocks. For the extended crystallite (E = 48) there is a 1 °C depression in T_m, relative to that of the pure homopolymer (P = 1). However, when P = 2, there is a 6 °C depression in the melting temperature. When P is increased to 5 or more, only folded chain crystallites are formed. The melting temperatures are now depressed about 15 °C relative to that of the homopolymer. The crystallinity levels remain constant at about 70% for the extended chain conformation, but increase slightly for the folded chains. As the length of the central E is increased only folded chain crystallites are observed. Concomitantly, there is a decrease in the observed melting temperature relative to that of the corresponding homopolymer. This melting point depression becomes accentuated as the length of the P end blocks increases. The formation of folded chain crystallites precludes analysis in terms of equilibrium theory. The decrease in melting temperature of the extended chain crystallites with increasing size of the end-groups is somewhat unexpected. However, the lengths of the crystallizing sequences involved here are relatively small. The melting temperature of such a sequence will be influenced by the end interfacial free energy, that in turn will be governed by the size of the end-group.

In contrast the crystallite chain structure, and melting temperatures, of the EPE type block copolymers are quite different.(197) In the EPE copolymers the P block lengths ranged from 43 to 182 units while the crystallizing E blocks contained from 18 to 69 units. If any chain folding occurs at all in this system, it only does so at the higher E block lengths. In contrast, it was found for the PEP copolymers that folding is already observed at E = 48. Consequently, the melting points of EPE would be expected to follow a different pattern. The melting temperatures of the copolymers and homopolymers of corresponding block lengths are given in Table 5.2. The melting temperatures of the block copolymers and the corresponding homopolymers are essentially identical except at the highest chain lengths. Even here, the differences are small. These results stand in sharp contrast to the melting temperatures of the PEP blocks, even for the extended chain structures. The

Table 5.2. *Comparison of the melting temperatures of poly(ethylene oxide) homopolymers with those in EPE copolymers of same chain length (197)*

Chain Length[a]	T_m (°C)	
	Homopolymer	Copolymer
43	53	52
55	56	55
59	57	57
80	61	57
132	64	60
177	65	63
182	65	61

[a] Chain length given in terms of number of repeating units.

position of the crystallizing block, in a symmetrical triblock copolymer makes a difference in the observed melting temperatures as a consequence of interfacial and morphological factors.

The studies that have been described for the model block copolymers composed of P and E units have yielded some important and interesting results. However, the block lengths, particularly of the crystallizing component, are relatively short. It is important to establish whether these results can be applied to other systems with high molecular weight crystallizing blocks. The results just described are in fact different from those for di- and triblock copolymers of styrene and tetrahydrofuran.(201) In these copolymers the crystallizing component is in the center of the triblock. Despite the inhomogeneous melt structures of these copolymers, the melting temperatures are invariant with the block type. They decrease only slightly with increasing styrene content. On the other hand there is a marked decrease in crystallinity level. The differences between the two triblock copolymers are in the molecular weights of the crystallizing blocks and the fact that crystallization occurs below the glass temperature of poly(styrene). The molecular weights of the poly(ethylene oxides) in the copolymers with propylene oxide are in the range of a few thousand g/mol. On the other hand the molecular weight of tetrahydrofuran is 60 000 or greater. In contrast to these results, in di- and triblock copolymers of ethylene oxide and isoprene neither the observed melting temperatures nor levels of crystallinity change much with composition except at high isoprene content.(202) Because of the low molecular weights of the ethylene oxide blocks, in the PEP copolymers, the interfacial free energy influences the chain structure within the crystallite. The observed melting temperature is thus affected. This effect will not be significant when the chain length of the crystallizing component is large.

Studies of block copolymers of hydrogenated poly(isoprene) and hydrogenated poly(butadiene) also addressed the role of molecular weight and character of the noncrystallizing block.(203) In these copolymers the hydrogenated poly(butadiene), B, is the crystallizing block while the hydrogenated poly(isoprene), I, is rubber-like. Di- and symmetric triblock arrangements, IB, BIB and IBI were studied. The molecular weights of the copolymers were all about 200 000 with narrow molecular weight distributions and long block lengths. The observed melting temperature of 102 °C was independent of the butadiene concentration and molecular architecture. The same melting temperature was also observed for the random copolymer, hydrogenated poly(butadiene), by itself. These results further support the basic principle that for sufficiently long chain lengths the melting temperature of the crystallizing component is independent of molecular weight and its arrangement within the copolymer. Furthermore, there is no restraint to crystallization in these copolymers by the vitrification of one of the blocks. As is illustrated in Fig. 5.30 the crystallinity levels are dependent on the composition of

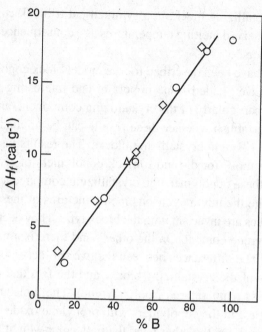

Fig. 5.30 Plot of the enthalpy of fusion against butadiene content for different block copolymers of hydrogenated poly(butadiene) HB and hydrogenated poly(isoprene) HI. ○ HBIB; ◇ HIBI; △ HIB. (From Mohajer, *et al.* (203))

the crystallizing block but independent of the chain architecture. It is evident from this plot that although at a given composition the measured enthalpy of fusion depends on composition it is independent of the sequence arrangement. Similar melting temperature results have been reported for di- and triblock copolymers of either styrene or butadiene with the crystallizing component, ε-caprolactone.(204) Except for the low chain lengths, and a slight effect of the styrene block, the observed melting temperatures are close to that of the homopolymer.

An informative study concerned with the thermal behavior of di- and triblock copolymers of hydrogenated butadiene, HB, with vinyl cyclohexane, VC, has been reported.(191) In these copolymers the 145 °C glass temperature of the poly(vinyl cyclohexane) block, is much higher than the crystallization range of the hydrogenated poly(butadiene) component. A wide range of domain structures were developed in the melt by varying the molecular weights of each block. The structures included hexagonally packed cylinders, lamellae, gyroids and spheres. The order–disorder transition of each of the copolymers was more than 60 °C greater than T_g of the poly(vinyl cyclohexane) block. Therefore, the domains in the melt are well established, or segregated, prior to the vitrification of the poly(vinyl cyclohexane) block. Crystallization in these copolymers is thus restricted by the glassy VC block. Small-angle x-ray scattering measurements showed that the domain structure of the

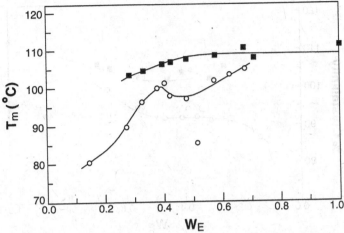

Fig. 5.31 Plot of melting temperatures, T_m for HBVC diblock (■) and VCHBVC triblocks (○) as functions of weight fraction, w_E, of HB componoent. (From Weimann, *et al.* (191))

melt was preserved upon crystallization. Crystallization was therefore restricted to lamellae, cylinders, gyroids or spheres as the case might be.

The melting temperature–composition relations for the diblock and triblock, VCHB and VCHBVC, are shown in Fig. 5.31.(191) The melting temperatures of the diblock copolymers are essentially constant for w_E values equal to, or greater than 0.5. They are only 1–2 °C lower than that of pure hydrogenated poly(butadiene). There is just a small continuous decrease in T_m as the poly(butadiene) content decreases. Thus, the constraints placed on the crystallization by the vitrification of the VC blocks are small for the diblock copolymers. More striking is the observation that at the same composition the melting temperatures of the triblock copolymers are lower than those of the diblocks. At the high butadiene concentrations the melting temperatures are relatively close to one another. However, there is a significant difference in melting temperatures at the lower butadiene compositions. The glassy nature of the end blocks places a major constraint on the crystallization.

The crystallinity levels of these di- and triblock copolymers are plotted in Fig. 5.32 against the weight fraction of hydrogenated poly(butadiene) for samples that were cooled from the melt to 25 °C at 20 °C min^{-1}. The crystallinity levels in the diblock copolymers are slightly less than that of pure hydrogenated poly(butadiene). Although fairly constant at low hydrogenated butadiene concentrations, there is a steady increase as the pure crystallizing species is being approached. The crystallinity levels of the triblock copolymers are lower than the comparable diblock ones. At hydrogenated poly(butadiene) concentrations greater than $w_E = 0.50$, a steady increase in crystallinity level is observed that approaches the value for the diblock copolymers. At lower compositions, the crystallinity level is essentially

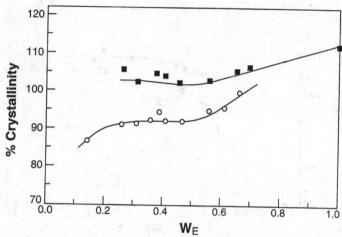

Fig. 5.32 Plot of crystallinity levels of HBVC diblocks (■) and VCHBVC triblocks (○) as functions of weight fraction w_E of HB component. (From Weimann, *et al.* (191))

constant, with a decrease occurring at the lowest concentrations. These results again illustrate the constraint that has been imposed. However, the effect is much greater for the triblock copolymers, where the crystallizing component is flanked by two glassy blocks. These results contrast with those for the di- and triblocks of poly(ethylene oxide) and poly(styrene), where the glassy component is in the center of the triblock. For these copolymers the same crystallinity level is observed for both type blocks.(205) The results in Fig. 5.32 are quite different from the hydrogenated poly(butadiene)–hydrogenated poly(isoprene) system given in Fig. 5.30. In the latter case, since the noncrystallizing component is rubber-like, the crystallinity level at a given composition is the same for the blocks of different molecular architecture. Confinement of the crystallization has also been observed if the amorphous, noncrystallizable block is highly entangled.(194,206)

A more detailed analysis of the influence of constrained crystallization on the melting temperature can be made by examining the effect of the domain width.(191) This width, w, is defined as the thickness of lamellar domain, or the diameter of the cylindrical, spherical and gyroid structures. Figure 5.33 is a plot of T_m against $1/w$. Both the di- and triblock copolymers give linear relations that extrapolate to melting temperatures that are very close to that of pure hydrogenated poly(butadiene). The difference in melting temperatures between the di- and triblock copolymers is still maintained. There is no difference in melting temperature between the domain structures in either category. The important factor here is their domain size. The lower melting temperatures of the triblock copolymers, at a constant value of w, indicate that their crystallites could be smaller than those in the diblock. Another possibility is that the constraints at the end of the crystallizing block could be

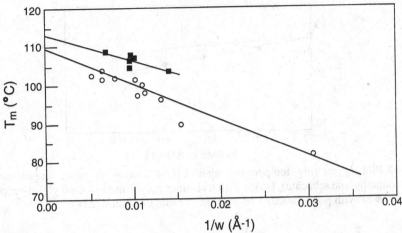

Fig. 5.33 Plot of melting temperatures for HBVC diblocks (■) and VCHBVC triblocks (○) against reciprocal of domain width, w. (From Weimann, *et al.* (191))

acting as an effective interfacial free energy and cause a lowering of the melting temperature.

Another example of constrained crystallization is when both blocks can crystallize. Di- and triblock copolymers of poly(ϵ-caprolactone) and poly(ethylene oxide) have been extensively studied in this connection.(207–211) A feature here is that the two blocks crystallize independently of one another. The block that crystallizes first influences the crystallization kinetics and morphology of the other block so that its crystallization occurs in a confined space.(212) Consequently, thermal and thermodynamic properties are in turn affected. The equilibrium melting temperatures, T_m^0, of the corresponding two homopolymers differ by only about 13 °C. Therefore, the crystallization kinetics and the component that crystallizes first will be dependent on the composition as well as the crystallization temperature. The block that crystallizes first develops a relatively high level of crystallinity that is comparable to that attained by the corresponding homopolymer. On the other hand, the block that crystallizes subsequently only attains relatively low levels of crystallinity. For example, when a given block concentration is 25% or less, its crystallinity level decreases to zero. When the concentrations of both blocks are comparable to one another, they still both crystallize but the order of crystallization depends on the crystallization temperature. The melting of each component is clearly observed irrespective of composition. The difference in melting temperatures is in the order of 4–5 °C. Copolymers, where both blocks can crystallize, offer interesting possibilities relating to structure and thermodynamic properties. These remain to be investigated.

The crystallization of multiblock copolymers has also been extensively studied. Poly(esters) have been widely used in this connection.(213–216) As an example,

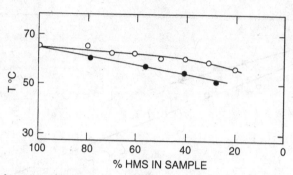

Fig. 5.34 Plot of melting temperature against composition of block copolymers of poly(hexamethylene sebacate), HMS, with its isomer poly(2-methyl-2-ethyl-1,3-propylene sebacate), ● or with poly(dimethyl siloxane) ○. (From O'Malley (214))

the melting point–composition relations for the crystallizable poly(hexamethylene sebacate), HMS, with either its noncrystallizable isomer poly(2-methyl-2-ethyl-1,3-propylene sebacate), MEB, or poly(dimethyl siloxane) are plotted in Fig. 5.34.(214,215) The melting temperatures were obtained after isothermal crystallization. The block copolymers of the two poly(esters) were prepared by coupling the hydroxy terminated polymers with hexamethylene diisocyanate. There is a decrease in the directly measured melting temperature of about 8 °C over the wide composition range studied. These results suggest that an even smaller decrease in the melting point would be found if the initial crystallization temperature was increased. A significant portion of the melting point depression can be attributed to the influence of the coupling agent. A similar effect of the urethane linkage in lowering the melting temperature has been found in other block copolymers.(217,218) When the directly coupled poly(dimethyl siloxane) is the second component the depression of the melting temperature is much smaller. Only a 3.5 °C depression is observed over the complete range. The melting temperature is essentially constant up to a composition of 50% HMS. Even though minor deviations from equilibrium theory are observed, the basic principles involved are supported by these results.

Support of the conclusions reached above is found in the melting temperature–composition relations of block copolymers composed of poly(ethylene sebacate) and poly(propylene adipate), that were also coupled with hexamethylene diisocyanate.(216) In this case extrapolation methods were adapted to approach equilibrium melting temperatures. The results are summarized in Table 5.3. Over the wide composition ranges that were studied, there is, within the experimental error of ±1 °C, virtually no change in the extrapolated equilibrium melting temperatures. The invariance of the equilibrium melting temperatures holds for a mole fraction of crystallizing units as low as 0.2. These melting temperatures are in good agreement

Table 5.3. *Extrapolated equilibrium melting temperatures of block copolymers of ethylene sebacate and propylene adipate (216)*

X_{es}^a	T_m (°C) (by extrapolation)
0.2	84.5 ± 1
0.4	85.5 ± 1
0.6	87.0 ± 1
0.8	86.3 ± 1

[a] Mole fraction of ethylene sebacate.

with that of the crystallizing homopolymer. The results agree with the pioneering work of Coffey and Meyrick (213) on similar block copolymers where, however, the melting temperatures were deduced by indirect methods. Similar melting point–composition relations are found among virtually all of the block copoly(ester) systems that have been studied when the block lengths are sufficiently long. There will be exceptions, if transesterification takes place during the copolymer synthesis.

A class of polymers, known as thermoplastic elastomers, possess the characteristics of a cross-linked rubber. These are copolymers that consist of two different block types. One of these is an amorphous, or liquid-like block that has a relatively low glass temperature. This block is often referred to as the soft segment since it imparts the rubber-like behavior to the copolymer. The other component can be either glass-like or crystalline. It is termed the hard block since it maintains dimensional stability. At sufficiently high temperature, however, this stability is lost so that the copolymer behaves as a true thermoplastic material. There are many examples of segmented block copolymers where the hard and soft segments alternate along the polymer chain. Of interest here are those in which the hard segment is crystalline.

The properties of a series of model segmented poly(urethanes), represented by the structural formula

has been reported.(219) This block copolymer was synthesized in such a manner that the crystallizable, or hard segment is monodisperse. In this formula $G = (OCH_2CH_2CH_2)_xO$ and $B = OCH_2CH_2CH_2O$. The lengths of the blocks are represented by the parameter n, that was varied from 1 to 4. These polymers give sharp endothermic melting peaks that follow a simple relation that is illustrated in Fig. 5.35. In this limited range of block chain lengths $1/T_m$ is linearly related

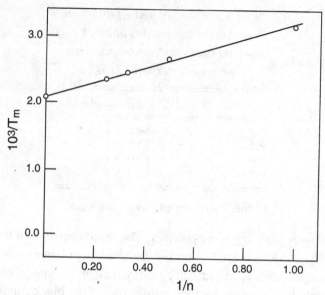

Fig. 5.35 Plot of reciprocal melting temperature, $1/T_m$, against reciprocal block length, $1/n$, for hard segment of block copolymer. (From Harrell (219))

to $1/n$. The straight line drawn extrapolates to the melting temperature of the homopolymer that corresponds to the hard segment. In this model system the crystalline blocks behave independently of the co-units. This result is supported by the fact that the enthalpies of fusion, based on the crystallizable segment content, are similar for the different copolymers. The measured enthalpy of fusion, and thus the level of crystallinity, is only slightly less than the pure homopolymer.

Similar results have also been obtained with segmented block copoly(esters). The most popular systems studied have been based on poly(tetramethylene terephthalate) as the crystallizable block and various low molecular weight poly(glycols), that do not crystallize, as the other block.(220–224) The melting temperatures of these copolymers increase with increasing poly(ester) content and approach 230 °C, the melting temperature of the pure homopolymer. These results are illustrated in Fig. 5.36 where the observed melting temperatures are plotted against the average block length. This behavior reflects the role of the increasing average block length of the crystallizable units. Following theoretical expectations the melting temperature at a given composition is independent of the chemical nature of the poly(ester) block.(221) Characteristic of these, as well as other block copolymers, is the fact that the crystallization is not complete. The noncrystalline hard segments mix in the amorphous phase with those of the soft component.

The melting temperatures of multiblock copolymers of ethylene oxide with propylene oxide, $P(EP)_m$, can be compared with the triblock polymer PEP.(198,200) The ethylene oxide and propylene oxide sequences have discrete lengths that range

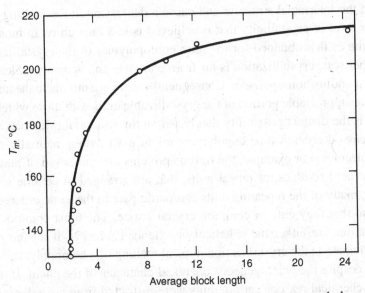

Fig. 5.36 Dependence of the melting temperature on the average tetramethylene terephthalate block length in its copolymer with poly(oxytetramethylene glycol). (From Cella (220))

from 45 to 136 for E and from 4 to 12 for P. The value of m varies from 1 to 7. The level of crystallinity in these multiblock copolymers is only about 60% of that observed for comparable PEP copolymers. The melting temperatures of the $P(EP)_m$ and PEP copolymers with the same E length sequence are, however, comparable to one another. The differences in melting temperatures being about 1–3 °C. Thus, the morphological and interfacial contributions are about the same in both copolymer types. Similarly, the crystallinity level of a multiblock poly(styrene)–poly(ethylene oxide) copolymer is less than that of the di- or triblock ones.(205) For example, the fraction of crystallinity has been reduced to 0.50 at 20% styrene and is only 0.20 at 50%. The lower levels of crystallinity are probably a consequence of less perfect microphase separation in the melt and the impeded crystallization caused by the glassy poly(styrene).

To summarize the major experimental findings, the melting temperature–composition relations of multiblock copolymers are similar to one another, irrespective of the chemical nature of the co-unit. When the sequence length of the crystallizing block is sufficiently long, the melting temperature is independent of composition. In accord with theory, it is either identical or very close to that of the corresponding homopolymer. This theoretical expectation has been found in many of the examples already cited as well as in others.(225–227) There are reasons why in some cases slightly lower melting temperatures are observed than is expected. The use of an external coupling agent can cause a lowered melting temperature at all sequence lengths. At lower chain lengths the noncrystallizing sequences can

influence the interfacial structure and cause a depression in the melting temperature. The level of crystallinity that is achieved is less than unity. In most cases it
is the same as that obtained for the pure homopolymer of the crystallizing units.
Put another way, crystallization is far from complete and is comparable to that of
the corresponding homopolymer. Consequently, there is a mixing in the amorphous
phase of an appreciable portion of the crystallizable units with those which are not.

There is the distinct possibility that based on the structural principles that have
been discussed ordered type copolymers might exist among naturally occurring
macromolecules. For example, the fibrous proteins are composed of many different amino acid residues, or repeat units, that are arranged in definite sequences.
However, many of the repeating units can participate in the same ordered conformation, so that they enter a common crystal lattice. The most common ordered
conformations are either the α-helical or extended β-forms. If all the repeating
units participate in the crystallization, typical melting of a homopolymer would be
expected despite the heterogeneous chemical character of the chain. If, however,
for stereochemical reasons certain units were restricted from crystallizing, then a
fusion process typical of a copolymer would result. Therefore, it is not required
a priori that a stoichiometric identity of repeating units be maintained between the
overall composition of the protein or nucleic acid and those involved in crystallization. This concept has important bearings on the interpretation of physical-chemical
processes involving the crystal–liquid transformation in such systems. It is a significant factor in interpreting x-ray diffraction patterns of fibrous proteins, since
only those units that crystallize contribute. For these relatively complex polymers,
the nature and concentration of the chemical units that actually participate in the
crystallization, and their sequence distribution, needs to be specified in order to
understand properties and phenomena related to crystallization.

It can be envisaged that in the fibrous proteins where the amino acid residues that
crystallize (comparable to the A units) are arranged in one block, those that do not
crystallize are present in another. The two differing blocks would then alternate
along the chain. This arrangement would be formally equivalent to that of an ordered
copolymer. It is also possible to have a random sequence distribution between the
crystallizable and noncrystallizable units. Distinction between these possibilities
involves structural determinations, thermodynamic studies, and an assessment of
physical and mechanical properties.

Silk fibroin has been recognized to be a semi-crystalline polymer.(228,229) Analyses of the small peptide fragments found in partial hydrolyzates of silk fibroin are
not in accord with the concept of a regular chemical repeating sequence throughout the molecule.(230) Rather, a structure in which certain types of residues occur
in particular portions of the chain is suggested. Specific fission of the poly(peptide) chain at tyrosine has allowed for the isolation of two major portions.(231)

Sixty percent of the chain contains only glycine, alanine, and serine residues and gives a powder x-ray diffraction pattern that is very similar to that of the native fibroin. The other portion contains all the bulky amino acid residues as well as small concentrations of glycine, alanine, and serine. It is highly doubtful whether the latter portion participates in the crystallization. These analytical results are in accord with the suggestion that has been made that the glycine, serine, and alanine residues form the crystalline regions of the polymer.(232) It has been shown further that there is a predominant sequence of six amino acid residues, glycine–alamine–serine–glycine–alanine–serine, that gives rise to the x-ray diffraction pattern and the crystal structure.(233) The other residues are relegated to the amorphous, or non-crystalline region, since they cannot be accommodated within the three-dimensional ordered structure. Other isolated residues from the six-member repeat can also be found in this region. Thus, the constitution of silk is analogous to that of a block copolymer and has been recognized as such.(234,235) Different classes of amino acids reside in each block. One group can participate in the ordered crystalline array, while the other cannot.

The concept of a partially crystalline structure for silk fibroin is further enhanced by studies of mechanical properties. The elastic properties of fibers, derived from different species of silk worms, that contain varying proportions of amino acid residues with long or bulky side chains, have been studied.(236) Fibers containing 90% of glycine and alanine residues are relatively inextensible. This behavior is expected since the fibers are highly crystalline. On the other hand, as the content of amino acid residues with more bulky side-groups increases, the fibers become more elastic. This observation is consistent with the presence of a significant number of chain units in disordered conformations. The mechanical properties, together with the sequence distribution of repeating units, give strong support to the concept that silk fibroin is properly considered to be a block copolymer.

Other fibrous proteins, particularly those in the keratin and collagen class, behave as copolymers from a crystallization point and appear to have the characteristics of ordered copolymers. For example, as has been pointed out previously, the fi-brous protein collagen has a unique amino acid composition. About one-third of the residues are glycine(G) and about 20% are imino acids(I), either proline or hy-droxyproline. The remainder are distributed among the other amino acid residues. The overall composition varies among the different kinds of collagens. There is a basic triplet repeat, G–X–I, where X is one of the other amino acid residues. It was recognized early on that the unique ordered conformation of collagen arose from a repeat of the glycine leading triplets and the role of the imino acid residues.(237) Only sequences of the type G–X–I and G–I–I can be accommodated within the ordered crystalline structure. Since all the sequences cannot participate in the or-dered structure, the melting and crystallization behavior of collagen will be typical of a copolymer. The sequence distribution of such triplets then becomes a matter of prime importance. It is interesting to note that the melting temperature of the collagen depends on the concentration of imino acid residue.

An unusual naturally occurring block copolymer has been identified with mussel byssal threads.(238) The threads are stiff at one end, the region that tethers the mussel to its target, and are extensible at the other end. It has been shown that this naturally occurring fiber is in fact a block copolymer, with three major domains. There is a central collagen domain, flanking elastic blocks and a histidine-like terminal region. The elastic domains strongly resemble the noncrystalline rubber-like protein, elastin.

Block copolymers, composed of different polypeptide sequences have also been studied.(239–241) In general the ordered conformation of a given block, and its thermodynamic stability is similar to that of the corresponding homopolypeptide. In particular, when α-helical conformations are formed in the copolymers similar helix–coil transitions are observed.

Several theories have been proposed to define the equilibrium structure of di- and triblock copolymers, one of whose components crystallizes.(242–244) Such theories should properly predict thermodynamic properties as well as equilibrium structure. However, common and central to all the theories is the basic assumption that the chains in the crystalline block are regularly folded in an adjacent re-entry array that leads to a smooth interface. The validity of this assumption for crystalline block copolymers needs to be carefully examined, in view of the experimental work that has been summarized earlier. This assumption is in contrast to homopolymers, where it has been established that the equilibrium condition requires extended chain type crystallites.(3) (See Chapter 2)

In another approach Ashman and Booth extended previous work with polydisperse homopolymers to diblock copolymers.(196,245) The analysis was specifically directed to the thermodynamic and structural properties of PE diblock with E length fixed at 40 units and the P block lengths varying from 1 to 11. The crystalline copolymer was assumed to consist of stacked lamellae of alternating crystalline and amorphous regions. Each of the lamellae were of uniform thickness, l_a and l_c, for the amorphous and crystalline ones respectively. The interface between the two lamellae was taken to be discrete. The molecular weight distribution of the low molecular weight poly(ethylene oxide) was represented by a Schulz–Zimm distribution. It was further assumed that the ends of the molecule were excluded from the crystalline lamellae. An important assumption was that the molecules fold to their maximum extent. In effect the calculation assumes folding by adjacent re-entry, with the folds included in the crystalline lamellae. With these assumptions, the melting temperature is expressed as

$$T_m = T_m^0[1 - 2\sigma_e/\Delta H_u l_c]/[1 - RT_m^0 \ln \phi_e I/\Delta H_u t l_c] \qquad (5.45)$$

The numerator in this equation is just the Gibbs–Thomson equation. It accounts for the finite crystallite thickness. The quantity ϕ_e is a reduced volume fraction. The parameter I accounts for the state of order in the crystalline sequence. It is in effect essentially the probability that a sequence of given length does not contain a chain end. The average number of times a chain traverses the crystalline lamellae is given by t. The specific relations appropriate to these quantities are given in Ref. (196). The interfacial free energy σ_e is taken to be the sum of three terms expressed as

$$\sigma_e = \sigma_o + \sigma_m + \sigma_a \tag{5.46}$$

Here σ_o is the free energy of forming the crystal–amorphous interface; σ_m is the non-combinatorial term due to the chain ends in the amorphous region; σ_c is the free energy increase due to conformational restrictions caused by restraints of the interface.

Returning to the experimental results, it should be recalled that many of the block copolymers that have been studied have hydrogenated poly(butadiene) as the crystalline core. Since hydrogenated poly(butadiene) is a random ethylene–butene copolymer, chain folding with adjacent re-entry is untenable for this component of the block copolymer. A fundamental factor that defines the interfacial structure of a crystalline lamella is the dissipation of the chain flux emanating from the basal plane.(246–250) This problem can be alleviated by regular chain folding. However, cognizance must be taken of the fact that for all the polymers that have been studied a large increase in free energy is required to accomplish folding with adjacent re-entry. Therefore, a compromise must be reached between a regular folded structure, and one in which the interfacial zone contains a significant amount of disorder.(246–248) Another way in which the chain flux can be reduced is by chain tilting, as has been observed in block copolymers.(251) A more detailed discussion of the problems involved in the dissipation of the chain flux, in both homopolymers and copolymers, will be given in Volume 3.

Experimental evidence has shown that the levels of crystallinity, based on the crystallizable blocks, are usually well below unity and are comparable to the values found with the corresponding homopolymers. The chain units of the crystallizable and noncrystallizable blocks mix in the noncrystalline phase. For sufficiently long chain lengths the observed melting temperatures of the crystalline components in block copolymers are comparable to and approach those of the corresponding homopolymers. These widespread observations are not consistent with nor can they be explained by regularly folded chain structures.(252)

Despite these concerns, the theories have apparently been successful in predicting how the domain size depends on the chain lengths of the crystalline and amorphous components. The size of the domains can be transformed into crystalline, and amorphous, thicknesses. According to one theory(242)

$$l_a \sim N_a^{2/3} \tag{5.47}$$

and

$$l_c \sim N_c N_a^{-1/3} \tag{5.48}$$

where N_a and N_c are the number of repeating units in the amorphous and crystalline lamellae respectively. In another theory(243)

$$l_a \sim N_a^{7/12} \tag{5.49}$$

and

$$l_c \sim N_c N_a^{-5/12} \tag{5.50}$$

The expectations from the two theories are similar. Although experiment has not be able to distinguish between the two there has been a general verification of this aspect of the theory.(253–256) However, the range of chain lengths studied has usually been less than a factor of ten. A much greater range in chain length is needed to verify Eqs. (5.47) to (5.50).

5.8 Copolymer–diluent mixtures

The crystallization of a copolymer from its mixture with a low molecular weight diluent occurs over the complete concentration range as in homopolymers. The melting temperature–composition relation will depend on the reduction in the free energy of the melt as a consequence of the added diluent. It can be expected that the reduction will depend on the sequence distribution in the copolymer, the specific interactions of the two co-units with the diluent and the purity of the crystalline phase. When the crystalline phase is pure, i.e. only A units crystallize and the diluent does not enter the lattice, the melting temperature reduction is derived by calculating the free energy of mixing in the melt and applying the conditions of phase equilibrium to the crystallizing repeating unit.

The Flory–Huggins free energy of mixing, appropriately modified, is conveniently used for this purpose.(3,257) There are essentially two major contributions to the mixing free energy, the combinatorial entropy and the net interaction free energy between the polymer and diluent. The latter can be expressed as (258,259)

$$\chi_1 = v_A \chi_{1A} + v_B \chi_{1B} - v_A v_B \chi_{AB} \tag{5.51}$$

When both co-units and solvent have the same volume, Eq. (5.51) reduces to

$$\chi_1 = x_A \chi_{1A} + x_B \chi_{1B} - x_A x_B \chi_{AB} \tag{5.52}$$

Here χ_1 is the interaction parameter of a binary copolymer with the pure solvent. The interaction parameters of the corresponding homopolymers in the same solvent are χ_{1A} and χ_{1B}. The interaction between the A and B units in the chain is given by χ_{AB}. The volume fractions and mole fractions of the comonomers in the copolymer molecule are v_A, v_B, x_A and x_B respectively. Equation (5.51) or (5.52) should hold for all types of copolymers. However, the entropy of mixing will depend on the copolymer sequence distribution.(3)

If the steric structures of the two units of random compolymers are not too different from one another, the mixing entropy will be similar to that of a homopolymer. The melting point depression equation will then be of the same form as Eq. (3.2) with χ_1 being expressed by either Eq. (5.51) or (5.52). For copolymers with chemically similar co-units, as stereo-irregular copolymers, χ_{1A} and χ_{1B} will be close to one another and x_{AB} will be essentially zero. Under these conditions the melting point depression will also be similar to that of a homopolymer. Experimentally, the validity of this relation has been verified for a number of stereo-irregular polymers such as poly(chlorotrifluoroethylene)(260), poly(acrylonitrile)(261), isotactic poly(propylene)(262,263) and isotactic poly(styrene).(264) For chemically dissimilar co-units Eq. (5.51) or (5.52) has to be used for χ_1. In this case the functional form of the melting point depression relation for a given copolymer will be unaltered. However, the actual depression will be different for copolymers having the same co-units, but with different compositions, because of the change in χ_1.

The entropy of mixing of a graft copolymer has been calculated using the Flory–Huggins type lattice.(257,259) This result also applies to block copolymers, since the entropy of mixing on a lattice is the same for linear and branched polymers. For example an ABA type block copolymer is a special type graft copolymer in which the grafts are located at the ends of the chain. The free energy of mixing of such a copolymer can be expressed as

$$\Delta G_M = kT[n_2 \ln(v_A + v_B) + n_1 \ln v_1 + \chi_1 n_1 v_2] \tag{5.53}$$

where χ_1 is again given by Eq. (5.51). Equation (5.53) is formally identical to the free energy of mixing of a homopolymer with solvent.(265) Here v_A and v_B are the volume fractions of the A and B co-units and thus equal to v_2; n_2 and n_1 are the number of polymer and solvent molecules respectively. Thus, the expectation is that the same melting temperature–composition relations will be observed. Inherent in the derivation of Eq. (5.53) is the implicit assumption that the melt is homogeneous and no specific account is taken of the restraint placed by the A–B junction.

The specific interactions of the A and B units with the solvent not only affect the free energy function but can also influence the morphological structures that are formed. The solvent can accentuate phase separation and modify the domain structure in the melt. Consequently, the crystallization process will be affected.(266,267) Of particular importance is the nature of the interaction of the solvent with each of the blocks. There are two extreme situations to be considered. One is where the solvent is thermodynamically a good one for both of the blocks. In the other extreme, the solvent is selective, being a good solvent for one of the blocks and a poor solvent for the other. For nonselective solvents, three distinct type domain structures, spheres, rods and lamellae, develop above a relatively low critical concentration.(266)

The crystallization of a homopolymer from dilute solution results in two distinct phases that can usually be separated by mechanical means. In contrast, when a random copolymer crystallizes from dilute solution this separation often cannot be made. The polymer molecule pervades the complete volume and the very highly fluid dilute solution is converted to a rigid medium of essentially infinite viscosity. This process is popularly termed thermoreversible gelation and is a manifestation of crystallization. Crystallization of random copolymers from dilute solution is not the only mechanism by which thermoreversible gelation can occur. It is, however, a very common occurrence. Other important gel forming mechanisms of polymers have been described.(268–270) Thermoreversible gels can also be formed by homopolymers, under appropriate conditions(270), and also by *n*-alkanes.(271)

Thermoreversible gelation, as a consequence of crystallization from dilute solutions of random copolymers has been observed in a variety of mixtures. These include, among others, poly(vinyl chloride) in dioctyl phthalate,(55) poly-(acrylonitrile) in dimethyl formamide,(56) nitrocellulose in ethyl alcohol,(272) methyl cellulose in water,(273) ethylene copolymers,(274) syndiotactic isotactic and atactic poly(styrene),(275–279) and random copolymers of ethylene terephthalate with isophthalate.(280) Flory and Garrett (281) have shown that the classical thermoreversible gelation system, gelatin in water, is the result of a crystal–liquid transformation. The gelation or dissolution can be treated as a first-order phase transition.

The important role of stereoregularity is demonstrated by the gelation of atactic poly(styrene). Although atactic poly(styrene) is generally considered to be a non-crystallizable polymer, thermoreversible gelation can be observed with this polymer in dilute solution.(279,282) Infra-red studies have demonstrated that gelation is a consequence of local conformational ordering of short syndiotactic sequences.(282) The interaction of the polymer with the solvent appears to play a crucial role in the local ordering and the resulting gelation.

The inherent copolymeric character of chain molecules is conducive for gel formation. Even if the equilibrium requirements were fulfilled, not all the chain units could participate in the crystallization. Therefore, only a small fraction of them would be transformed. The large number of chain elements that do not crystallize are interconnected by means of the ordered crystalline sequence. They pervade the entire volume, and impart the characteristic rigidity and high viscosity characteristic of gels. The accretion of chains in the lateral direction is severely restricted in the crystallization of a copolymer from dilute solution. This effect, accompanied by the retardation in longitudinal development must necessarily limit the number of chain elements that participate in the crystallization. Thus, a large number of long ordered sequences is not necessary for gel formation. However, as has been reported, homopolymers that form lamella-like crystals, can also participate in thermoreversible gelation.(270,283)

When a block copolymer is dissolved in a solvent that is a good one for one set of units and a poor one for the other a micellar structure forms.(183,284) The ability to form micelles is a distinguishing feature of block and graft copolymers. Homopolymers and random type copolymers do not form micellar structures in solution. A micelle usually consists of a swollen core of the insoluble block connected to and surrounded by the soluble blocks. As the copolymer concentration is increased the micelles aggregate and organize into structures that have been termed mesomorphic gels. It is from this organized structure, where the chains themselves are in nonordered conformation, that crystallization takes place.

References

1. Flory, P. J., *J. Am. Chem. Soc.*, **78**, 5222 (1956).
2. Adamson, A. W., *A Textbook of Physical Chemistry*, Academic Press (1973).
3. Flory, P. J., *J. Chem. Phys.*, **17**, 223 (1949).
4. Flory, P. J., *Trans. Faraday Soc.*, **51**, 848 (1955).
5. Flory, P. J., *Principles of Polymer Chemistry*, Cornell University Press (1953) pp. 178f.
6. Baker, C. H. and L. Mandelkern, *Polymer*, **7**, 7 (1966).
7. Orr, R. J., *Polymer*, **2**, 74 (1961).
8. Coleman, B. D., *J. Polym. Sci.*, **31**, 155 (1958).
9. Allegra, G., R. H. Marchessault and S. Bloembergen, *J. Polym. Sci.: Polym. Phys.*, **30**, 809 (1992).
9a. Allegra, G. and S. V. Meille, *Crystallization of Polymers*, NATO ASI Series, vol. 405, Kluwer Academic (1993) p. 226.
10. Helfand, E. and J. I. Lauritzen, Jr., *Macromolecules*, **6**, 631 (1973).
11. Sanchez, I. C. and R. K. Eby, *J. Res. Nat. Bur. Stand.*, **77A**, 353 (1973).
12. Sanchez, I. C. and R. K. Eby, *Macromolecules*, **8**, 638 (1975).
13. Wendling, J. and U. W. Suter, *Macromolecules*, **31**, 2516 (1998).
14. Kilian, H. G., *Thermochimica Acta*, **238**, 113 (1994); *Koll. Z.-Z. Polym.*, **202**, 97 (1965).
15. Baur, H., *Macromol. Chem.*, **98**, 297 (1966).
16. Obi, B. E., P. DeLassur and E. A. Gruelbe, *Macromolecules*, **27**, 5491 (1994).
17. Baur, H., *Koll. Z.-Z. Polym.*, **212**, 97 (1966).
18. Sadler, D. M. and G. H. Gilmer, *Phys. Rev. Lett.*, **56**, 2708 (1986); *Phys. Rev. B*, **38**, 5686 (1988).

19. Goldbeck-Wood, G., *Polymer*, **33**, 778 (1992).
20. Richardson, M. J., P. J. Flory and J. B. Jackson, *Polymer*, **4**, 21 (1963).
21. Alamo, R. G. and L. Mandelkern, *Thermochim. Acta*, **238**, 155 (1994).
22. Vander Eynde, S., V. B. F. Mathot, M. H. J. Koch and H. Reynaers, *Polymer*, **41**, 4889 (2000).
23. Kubota, H. and J. B. Newell, *J. Appl. Polym. Sci.*, **19**, 1521 (1975).
24. Garrett, H. R. R., C. A. Hargreaves, II. and D. N. Robinson, *J. Macromol. Sci.*, **A4**, 1679 (1970).
25. Mandelkern, L., M. Tryon and F. A. Quinn, Jr., *J. Polym. Sci.*, **19**, 77 (1956).
26. Berger, M. and D. J. Buckley, *J. Polym. Sci., Pt. A.*, **1**, 2945 (1963).
27. Fabris, H. J., I. G. Hargis, R. A. Livigni and S. L. Aggarwal, *Rubber Chem. Tech.*, **60**, 721 (1987).
28. Lanzavecchih, G., *Ind. Chim. Belge*, **Suppl. 2**, 347 (1959).
29. Cuneen, J. I. and F. W. Shipley, *J. Polym. Sci.*, **36**, 77 (1959).
30. Cuneen, J. I. and W. F. Watson, *J. Polym. Sci.*, **38**, 521 (1959).
31. Golub, M., *J. Polym. Sci.*, **25**, 373 (1957).
32. Cuneen, J. I., G. M. C. Higgins and W. F. Watson, *J. Polym. Sci.*, **40**, 1 (1959).
33. Pauling, L., R. B. Corey and H. R. Branson, *Proc. Natl. Acad. Sci. U.S.*, **37**, 205 (1951).
34. Mizushima, S., *Adv. Protein Chem.*, **9**, 299 (1954).
35. Natta, G., *J. Polym. Sci.*, **16**, 143 (1955); G. Natta, P. Pino, P. Carradini, F. Danusso, E. Mantica, G. Mazzanti and G. Moraglio, *J. Am. Chem. Soc.*, **77**, 1708 (1955).
36. Horton, A. D., *Trends Polym. Sci.*, **2**, 158 (1994).
37. Brintzinger, H. H., D. Fischer, R. Mülhaupt, B. Rieger and R. M. Waymouth, *Angew. Chem. In. Ed., Engl.*, **34**, 1143 (1998).
38. Huggins, M. L., *J. Am. Chem. Soc.*, **66**, 1991 (1944).
39. Fox, T. G., W. E. Goode, S. Gratch, C. M. Huggett, J. F. Kincaid, A. Spell and J. D. Stroupe, *J. Polym. Sci.*, **31**, 173 (1958).
40. Garrett, B. S., W. E. Goode, S. Gratch, J. F. Kincaid, C. L. Levesque, A. Spell, J. D. Stroupe and W. H. Watnabe, *J. Am. Chem. Soc.*, **81**, 1007 (1959).
41. Fordham, J. W. L., G. H. McCain and L. E. Alexander, *J. Polym. Sci.*, **39**, 335 (1959).
42. Fordham, J. W. L., P. H. Burleigh and C. L. Sturm, *J. Polym. Sci.*, **41**, 73 (1959).
43. Fordham, J. W. L., *J. Polym. Sci.*, **39**, 321 (1959).
44. Newman, S., *J. Polym. Sci.*, **47**, 111 (1960).
45. Cheng, S. Z. D., J. J. Janimak, A. Zhang and E. T. Hsieh, *Polymer*, **32**, 648 (1991).
46. Manson, J. A., S. A. Iobst and R. Acosta, *J. Polym. Sci., Pt. A-1*, **10**, 179 (1972).
47. Nakajima, A., H. Hamada and S. Hayashi, *Makromol. Chem.*, **95**, 40 (1966).
48. Crist, B. and P. R. Howard, *Macromolecules*, **32**, 3057 (1999).
49. Colby, R. H., G. E. Milliman and W. W. Graessley, *Macromolecules*, **19**, 1261 (1986).
50. Williams, J. L. R., J. Van Den Berghe, W. J. Dulmage and K. R. Dunham, *J. Am. Chem. Soc.*, **78**, 1260 (1956); J. L. R. Williams, J. Van Den Berghe, K. R. Dunham and W. J. Dulmage, *J. Am. Chem. Soc.*, **79**, 1716 (1957).
51. Fox, T. G., B. S. Garrett, W. E. Goode, S. Gratch, J. F. Kincaid, A. Spell and J. D. Stroupe, *J. Am. Chem. Soc.*, **80**, 1768 (1958).
52. Korotkov, A. A., S. P. Mitsengendlev, U. N. Krausline and L. A. Volkova, *High Molecular Weight Compounds*, **1**, 1319 (1959).
53. Blakey, P. R. and R. P. Sheldon, *Nature*, **195**, 172 (1962).
54. Doty, P. M., H. L. Wagner and S. F. Singer, *J. Phys. Coll. Chem.*, **55**, 32 (1947).
55. Walter, A. T., *J. Polym. Sci.*, **13**, 207 (1954).
56. Bisschops, J., *J. Polym. Sci.*, **12**, 583 (1954); **17**, 89 (1955).

57. Etlis, V. S., K. S. Minskev, E. E. Rylov and D. N. Bort, *High Molecular Weight Compounds*, **1**, 1403 (1959).
58. Evans, R. D., H. R. Mighton and P. J. Flory, *J. Am. Chem. Soc.*, **72**, 2018 (1951).
59. Edgar, O. B. and E. Ellery, *J. Chem. Soc.*, 2633 (1952).
60. Sonnerskog, S., *Acta Chem. Scand.*, **10**, 113 (1956).
61. Izard, E. F., *J. Polym. Sci.*, **8**, 503 (1952).
62. Guerra, G., V. Venditto, C. Natale, P. Rizzo and C. DeRosa, *Polymer*, **39**, 3205 (1998).
63. Inoue, M., *J. Appl. Polym. Sci.*, **8**, 2225 (1964).
64. Naga, N., K. Mizunama, H. Sadatoski and M. Kakugo, *Macromolecules*, **20**, 2197 (1987).
65. Kyotskura, R., T. Masuda and N. Tsutsumi, *Polymer*, **35**, 1275 (1994).
66. Okuda, K., *J. Polym. Sci., Pt. A*, **2**, 1719 (1964).
67. Mandelkern, L., *Rubber Chem. Technol.*, **32**, 1392 (1959).
67a. Huang, J., M. S. Lisowski, J. Runt, E. S. Hall, R. T. Kean, N. Buehler and J. S. Lin, *Macromolecules*, **31**, 2593 (1998).
68. Alamo, R. G., E. K. M. Chan, L. Mandelkern and I. G. Voigt-Martin, *Macromolecules*, **25**, 6381 (1992).
69. Kim, M. H., P. J. Phillips and J. S. Lin, *J. Polym. Sci.: Pt. B: Polym. Phys.*, **38**, 154 (2000).
70. Isasi, J. R., L. Mandelkern and R. G. Alamo, unpublished results (1999).
71. Thomann, R., J. Kressler and R. Mülhaupt, *Polymer*, **39**, 1907 (1998).
72. Haftka, S. and K. Könnecke, *J. Macromol. Sci.*, **B30**, 319 (1991).
72a. Krigbaum, W. R. and J. H. O'Mara, *J. Polym. Sci.: Polym. Phys. Ed.*, **8**, 1011 (1970).
73. Marrs, W., R. H. Peters and R. H. Still, *J. Appl. Polym. Sci.*, **23**, 1063 (1979).
74. Alamo, R., R. Domszy and L. Mandelkern, *J. Phys. Chem.*, **88**, 6587 (1984).
75. Alamo, R. G. and L. Mandelkern, *Macromolecules*, **22**, 1273 (1989).
76. Alamo, R. G., B. D. Viers and L. Mandelkern, *Macromolecules*, **26**, 5740 (1993).
77. Alizadeh, A., L. Richardson, J. Xu, S. McCartney, H. Marand, Y. W. Cheung and S. Chum, *Macromolecules*, **32**, 6221 (1999).
78. Isasi, J. R., J. A. Haigh, J. T. Graham, L. Mandelkern, M. H. Kim and R. G. Alamo, *Polymer*, **41**, 8813 (2000).
78a. Simanke, A. G. and R. G. Alamo, private communication.
79. Haigh, J. A., R. G. Alamo and L. Mandelkern, unpublished results.
79a. Guerra, G., V. Venditto, C. Natale, P. Rizzo and C. DeRosa, *Polymer*, **39**, 3205 (1998).
79b. Jo, W. H. and D. H. Baik, *J. Polym. Sci.: Pt. B: Polym. Phys.*, **27**, 673 (1989).
80. Culter, D. J., P. J. Hendra, M. E. A. Cudby and H. A. Willis, *Polymer*, **18**, 1005 (1977).
81. Bowmor, T. N. and J. H. O'Donnell, *Polymer*, **18**, 1032 (1977).
82. Holdsworth, P. J. and A. Keller, *Macromol. Chem.*, **125**, 82 (1969).
83. Vile, J., P. J. Hendra, H. A. Willis, M. E. A. Cudby and G. Gee, *Polymer*, **25**, 1173 (1980).
84. Vonk, C. G., *Polyethylene Golden Jubilee Conference*, Plastics and Rubber Institute, London, 1983, vol. D2.1.
85. Séquéla, R. and F. Rietsch, *J. Polym. Sci.: Polym. Lett.*, **24B**, 29 (1986).
86. Bowmer, T. W. and A. E. Tonelli, *Polymer*, **26**, 1195 (1985).
87. Mandelkern, L., in *Comprehensive Polymer Science: Polymer Properties*, vol. 2 Pergamon Press (1989) p. 363.

88. Voigt-Martin, I. G. and L. Mandelkern, in *Handbook of Polymer Science and Technology*, N. P. Cheremisinoff ed., Marcel Dekker Publishers (1989) p. 1.
89. Vonk, C. G. and A. P. Pijpers, *J. Polym. Sci.: Polym. Phys. Ed.*, **23**, 2517 (1985).
90. Vonk, C. G., *J. Polym. Sci.*, **38C**, 429 (1972).
91. Swan, P. R., *J. Polym. Sci.*, **56**, 409, (1962).
92. Kortleve, G., C. A. F. Tuijnman and C. G. Vonk, *J. Polym. Sci.*, *Pt. A2*, **10**, 123 (1972).
93. Walter, E. R. and F. P. Reding, *J. Polym. Sci.*, **21**, 501 (1956).
94. Eichhorn, R. M., *J. Polym. Sci.*, **31** (1958).
95. Bunn, C. W., in *Polyethylene*, A. Renfrew and P. Morgan, eds., Illife (1975) Chapter 5.
96. Voigt-Martin, I. G. and L. Mandelkern, in *Handbook of Polymer Science and Technology*, N. P. Cheremisinoff ed., Marcel Dekker Publishers (1989) pp. 99ff.
97. Bunn, C. W. and H. S. Peiser, *Nature*, **159**, 161 (1947).
98. Kreuz, J. A., B. S. Hsiao, C. A. Renner and D. L. Goff, *Macromolecules*, **28**, 6926 (1995).
99. Isasi, J. R., R. G. Alamo and L. Mandelkern, unpublished results.
100. Alamo, R. G., D. L. VanderHart, M. R. Nyden and L. Mandelkern, *Macromolecules*, **33**, 6094 (2000).
101. DeRosa, C., G. Talarico, L. Caporaso, F. Auriemma, M. Galinibertis and O. Fusco, *Macromolecules*, **31**, 9109 (1998).
102. Natta, G., *Makromol. Chem.*, **35**, 93 (1960).
103. Edgar, O. B. and R. Hill, *J. Polym. Sci.*, **8**, 1 (1952).
104. Yu, A. J. and R. D. Evans, *J. Am. Chem. Soc.*, **81**, 5361 (1959).
105. Yu, A. J. and R. D. Evans, *J. Polym. Sci.*, **20**, 5361 (1959).
106. Levine, M. and S. C. Temin, *J. Polym. Sci.*, **49**, 241 (1961).
107. Tranter, T. C., *J. Polym. Sci. Pt. A*, **2**, 4289 (1964).
108. Prince, F. R., E. M. Pearce and R. J. Fredericks, *J. Polym. Sci., Pt. A-1*, **8**, 3533 (1970).
109. Prince, F. R. and R. J. Fredericks, *Macromolecules*, **5**, 168 (1972).
110. Hachiboshi, M., T. Fukud and S. Kobayashi, *J. Macromol. Sci. Phys.*, **35**, 94 (1960).
111. Yoshie, N., Y. Inoue, H. Y. Yoo and N. Okui, *Polymer*, **35**, 1931 (1994).
112. Lu, X. and A. H. Windle, *Polymer*, **36**, 451 (1995).
113. Park, S. S., I. H. Kim and S. S. Im, *Polymer*, **37**, 2165 (1996).
114. Bluhm, T. L., G. K. Hamer, R. S. Marchessault, C. A. Fyfe and R. P. Veregin, *Macromolecules*, **19**, 2871 (1986).
115. Doi, Y., *Microbial Polyesters*, VCH Publishers (1990).
116. Natta, G., P. Corradini, D. Sianezi and D. Morero, *J. Polym. Sci.*, **54**, 527 (1961).
117. Natta, G., G. Allegra, I. W. Bass, D. Sianesi, G. Caporiccio and E. Torti, *J. Polym. Sci., Pt. A*, **3**, 4263 (1965).
118. Okuda, K., *J. Polym. Sci., Pt. A*, **2**, 1749 (1961).
119. Natta, G., L. Porni, A. Carbonaro and G. Lugli, *Makromol. Chem.*, **53**, 52 (1960).
120. Reding, F. P. and E. R. Walter, *J. Polym. Sci.*, **37**, 555 (1959).
121. Jones, A. T., *Polymer*, **6**, 249 (1965); *ibid.*, **7**, 23 (1965).
122. Campbell, T. W., *J. Appl. Polym. Sci.*, **5**, 184 (1961).
123. Natta, G., G. Allegra, I. W. Bassi, C. Carlini, E. Chielline and G. Montagnol, *Macromolecules*, **2**, 311 (1969).
124. Gardner, K. H., B. S. Hsiao, R. M. Matheson, Jr. and B. A. Wood, *Polymer*, **33**, 2483 (1992).

125. Montaudo, G., G. Bruno, P. Maravigna, P. Finocchiaro and G. Centineo, *J. Polym. Sci.: Polym. Chem. Ed.*, **11**, 65 (1973).
126. Allegra, G. and I. W. Bassi, *Adv. Polym. Sci.*, **6**, 549 (1969).
127. Carmer, F. B. and R. G. Beaman, *J. Polym. Sci.*, **21**, 237 (1956).
128. Howard, G. J. and S. Knutton, *Polymer*, **9**, 527 (1968).
129. Kunioka, M., A. Tamaki and Y. Doi, *Macromolecules*, **22**, 694 (1989).
130. Bloembergen, S., D. A. Holden, T. L. Bluhm, G. K. Hamer and R. H. Marchessault, *Macromolecules*, **22**, 1663 (1989).
131. Scandala, M., G. Ceccoralli, M. Pizzoli and M. Gazzani, *Macromolecules*, **25**, 1405 (1992).
132. Mitomo, H., N. Morishita and Y. Doi, *Macromolecules*, **26**, 5809 (1993).
133. Barker, P. A., P. J. Barham and J. Martinez-Salazar, *Polymer*, **38**, 913 (1997).
134. Kamiya, N., M. Sakurai, Y. Inoue, R. Chugo and Y. Doi, *Macromolecules*, **24**, 2178 (1991).
135. VanderHart, D. L., W. J. Orts and R. H. Marchessault, *Macromolecules*, **28**, 6394 (1995).
136. Doi, Y., S. Kitamura and H. Abe, *Macromolecules*, **28**, 4822 (1985).
137. Lu, X. and A. H. Windle, *Polymer*, **37**, 2027 (1996).
138. Yoo, H. Y., S. Umemoto, S. Kikutani and N. Okui, *Polymer*, **35**, 117 (1994).
138a. Yoshie, N., Y. Inoue, H. Y. Yoo and N. Okui, *Polymer*, **35**, 1931 (1994).
138b. Orts, W. J., R. B. Marchessault and T. L. Bluhm, *Macromolecules*, **24**, 6435 (1991).
139. Wunderlich, B., *Angew. Chem.*, **80**, 1009 (1968).
140. Wunderlich, B., *Adv. Polym. Sci.*, **5**, 568 (1969).
141. Wegner, G., *Faraday Discuss Chem. Soc.*, **68**, 494 (1979).
142. Mateva, R., G. Wegner and G. Lieser, *J. Polym. Sci.: Polym. Lett.*, **11B**, 369 (1973).
143. Dröscher, M., G. Lieser, H. Reimann and G. Wegner, *Polymer*, **16**, 497 (1975).
144. Dröscher, M., K. Herturg, H. Reimann and G. Wegner, *Makromol. Chem.*, **177**, 1695 (1976).
145. Dröscher, M., K. Herturg, H. Reimann and G. Wegner, *Makromol. Chem.*, **177**, 2793 (1976).
146. Michner, B. and R. Mateva, *Makromol. Chem.*, **187**, 223 (1986).
147. Mateva, R. and G. Sirashké, *Eur. Polym. J.*, **33**, 553 (1997).
148. Raine, H. C., R. B. Richards and H. Ryder, *Trans. Faraday Soc.*, **41**, 56 (1945).
149. Dole, M., W. P. Hellingen, N. R. Larson and J. A. Wethington, *J. Chem. Phys.*, **20**, 781 (1952).
150. Hunter, E. and W. G. Oakes, *Trans. Faraday Soc.*, **41**, 49 (1945).
151. Mandelkern, L., M. Hellman, D. W. Brown, D. E. Roberts and F. A. Quinn, Jr., *J. Am. Chem. Soc.*, **75**, 4093 (1953).
152. Manaresi, P., A. Munari, F. Pilati, G. C. Alfonso, S. Russo and M. L. Sartirana, *Polymer*, **27**, 955 (1986).
153. Lopez, L. C., G. L. Wilkes and J. F. Geibel, *Polymer*, **30**, 147 (1989).
154. Risch, B. G., G. L. Wilkes and I. M. Warakomski, *Polymer*, **34**, 2330 (1993).
155. Garbuglio, C., M. Ragazzini, O. Pilat, D. Carcono and Gb. Cevidalli, *Eur. Polym. J.*, **3**, 137 (1967).
156. Sibilia, J. P., L. G. Roldan and L. S. Chandrasekanan, *J. Polym. Sci. Pt. A-2*, **10**, 549 (1972).
157. Brubaker, M. M., D. D. Coffman and H. H. Hoehn, *J. Am. Chem. Soc.*, **74**, 1509 (1952).
158. Colombo, P., L. E. Kuckaka, J. Fontana, R. N. Chapman and M. Steinbeg, *J. Polym. Sci. Pt. A-1*, **4**, 29 (1966).

159. DeVito, S., F. Ciardelli, E. Benedette and E. Bramanti, *Polym. Adv. Technol.*, **8**, 53 (1997).
160. Sonmazzi, A. and F. Garbassi, *Prog. Polym. Sci.*, **22**, 1547 (1997).
161. Drent, E., S. A. M. Brookhaven and M. I. Doyle, *J. Organomet. Chem.*, **417**, 235 (1991).
162. Lommerts, B. J., E. A. Klop and J. Aerts, *J. Polym. Sci.: Pt. B: Polym. Phys.*, **31**, 1319 (1993).
163. Starkweather, H. W., Jr., *J. Polym. Sci.: Polym. Phys. Ed.*, **15**, 247 (1977).
164. Klop, E. A., B. J. Lommerts, J. Veurink, J. Aerts and R. R. Van Puijenbroek, *J. Polym. Sci.: Pt. B: Polym. Phys.*, **33**, 3151 (1995).
165. Wittner, H., P. Pino and U. W. Suter, *Macromolecules*, **21**, 1262 (1988).
166. Machado, J. M. and J. E. Flood, *Polym. Prepr., Polym. Chem. Div. Am. Chem. Soc.*, **36**, 291 (1995).
167. Chien, C. W. J., A. Z. Zhao and F. Xu, *Polym. Bull.*, **28**, 315 (1992).
168. Brückner, S., C. DeRosa, P. Corradini, W. Porzio and A. Musco, *Macromolecules*, **29**, 1535 (1996).
169. Chien, H., D. McIntyre, J. Cheng and M. Fone, *Polymer*, **36**, 2559 (1995).
170. Waymouth, R. W. and M. K. Leclerc, *Angew. Chem.*, **37**, 922 (1998).
171. Uozumi, T., K. Miyazana, T. Sono and K. Soga, *Macromol. Rapid Comm.*, **18**, 833 (1997).
172. Modena, M., C. Garbuglio and M. Ragozzini, *J. Polym. Sci.: Polym. Lett.*, **10B**, 153 (1972).
173. Garbuglio, C., M. Modena, M. Valera and M. Ragozzini, *Eur. Polym. J.*, **10**, 91 (1974).
174. Brown, D. W., R. E. Lowry and L. A. Wall, *J. Polym. Sci.: Polym. Phys. Ed.*, **12**, 1302 (1974).
175. Wilson, F. C. and H. W. Starkweather, Jr., *J. Polym. Sci.: Polym. Phys. Ed.*, **11**, 919 (1973).
176. Meier, D. J., *J. Polym. Sci.*, **26C**, 81 (1969).
177. Helfand, E. and Z. R. Wasserman, *Macromolecules*, **5**, 879 (1972); *ibid.*, **11**, 960 (1978); *ibid.*, **13**, 994 (1980).
178. Noolandi, J. and K. M. Hong, *Ferroelectrics*, **30**, 117 (1980).
179. Hong, K. M. and J. Noolandi, *Macromolecules*, **14**, 727 (1981).
180. Whitmore, M. D. and J. Noolandi, *J. Chem. Phys.*, **93**, 2946 (1990).
181. Matsen, M. W. and M. Schick, *Phys. Rev. Lett.*, **72**, 2660 (1994).
182. Matsen, M. W. and M. Schick, *Macromolecules*, **27**, 6761 (1994).
183. Brown, R. A., A. J. Masters, C. Price and X. F. Yuan, *Comprehensive Polymer Science*, vol. 2, Pergamon Press (1989) p. 155.
184. Bates. F. S. and G. H. Fredrickson, *Physics Today*, p. 37, February 1999.
185. Molau, G. E. and H. Kesskula, *J. Polym. Sci., Pt. A-1*, **4**, 1595 (1966).
186. Bates, F. Ş., M. F. Schulz, A. K. Khandpres, S. Förster, J. H. Rosedale, K. Almdal and K. Mortensen, *Faraday Discuss Chem. Soc.*, **98**, 7 (1994).
187. Schulz, M. F. and F. S. Bates, in *Physical Properties of Polymers Handbook*, J. E. Mark, ed., American Institute of Physics (1996) pp. 427ff.
188. Seow, P. K., Y. Gallot and A. Skoulios, *Makromol. Chem.*, **177**, 177 (1976).
189. Kenney, J. F., *Polym. Eng. Sci.*, **8**, 216 (1968).
190. Iwakura, Y., Y. Taneda and S. Vehida, *J. Appl. Polym. Sci.*, **13**, 108 (1961).
191. Weimann, P. A., D. A. Hajduk, C. Chu, K. A. Chaffin, J. C. Brodel and F. S. Bates, *J. Polym. Sci.: Pt. B: Polym. Phys.*, **37**, 2068 (1999).
192. Séquéla, R. and J. Prud'homme, *Polymer*, **30**, 1446 (1989).

193. Hamley, I. W., *The Physics of Block Copolymers*, Oxford University Press (1998). p. 281.
194. Quiram, D. J., R. A. Register and G. R. Marchand, *Macromolecules*, **30**, 4551 (1997).
195. Heuschen, J., R. Jérome and Ph. Teyssié, *J. Polym. Sci.: Pt. B: Polym. Phys.*, **27**, 523 (1989).
196. Ashman, P. C. and C. Booth, *Polymer*, **16**, 889 (1975).
197. Booth, C. and D. V. Dodgson, *J. Polym. Sci.: Polym. Phys. Ed.*, **11**, 265 (1973).
198. Ashman, P. C., C. Booth, D. R. Cooper and C. Price, *Polymer*, **16**, 897 (1975).
199. Booth, C. and C. J. Pickles, *J. Polym. Sci.: Polym. Phys. Ed.*, **11**, 249 (1973).
200. Ashman, P. C. and C. Booth, *Polymer*, **17**, 105 (1976).
201. Takahashi, A. and Y. Yamashuto, in *Advances in Chemistry Series No. 142, Copolymers, Polyblends and Composites*, N. A. I. Platzer, ed., American Chemical Society (1975) p. 267.
202. Hirata, E., T. Ijitsu, T. Hashimoto and H. Kawai, *Polymer*, **16**, 249 (1975).
203. Mohajer, Y., G. L. Wilkes, J. C. Wang and J. E. McGrath, *Polymer*, **23**, 1523 (1982).
204. Balsamo, V., F. von Gyldenfeldt and R. Stadler, *Macromol. Chem. Phys.*, **197**, 3317 (1996).
205. Shimura, Y. and T. Hatakeyamer, *J. Polym. Sci.: Polym. Phys. Ed.*, **13**, 653 (1975).
206. Quiram, D. J., R. A. Register, G. R. Marchand and A. J. Ryan, *Macromolecules*, **30**, 8338 (1997).
207. Perret, R. and A. Skoulios, *Makromol. Chem.*, **156**, 143 (1972).
208. Perret, R. and A. Skoulios, *Makromol. Chem.*, **162**, 147 (1972).
209. Perret, R. and A. Skoulios, *Makromol. Chem.*, **162**, 163 (1972).
210. Skoulios, A. E., in *Block and Graft Copolymers*, J. J. Burke and V. Weiss, eds., Syracuse University Press (1973) p. 121.
211. Nojima, S., M. Ohno and T. Ashida, *Polym. J.*, **24**, 1271 (1992).
212. Misra, A. and S. N. Garg, *J. Polym. Sci.: Pt. B: Polym. Phys.*, **24**, 983 (1986); *ibid.*, **24**, 999 (1986).
213. Coffey, D. H. and T. J. Meyrick, *Proc. Rubber Technol. Conf.* (1954) p. 170.
214. O'Malley, J. J., *J. Polym. Sci.: Polym. Phys. Ed.*, **13**, 1353 (1975).
215. O'Malley, J. J., T. J. Pacansky and W. J. Stanffor, *Macromolecules*, **10**, 1197 (1977).
216. Theil, M. H. and L. Mandelkern, *J. Polym. Sci., Pt. A-2*, **8**, 957 (1970).
217. Galen, J. C., P. Spegt, S. Suzuki and A. Skoulious, *Makromol. Chem.*, **175**, 991 (1974).
218. Onder, K., R. H. Peters and L. C. Spark, *Polymer*, **13**, 133 (1972).
219. Harrell, L. L., Jr., *Macromolecules*, **2**, 607 (1969).
220. Cella, R. J., *Encyclopedia of Polymer Science and Technology Supplement No. 2*, John Wiley and Sons (1977) p. 487.
221. Wolfe, J. R., Jr., *Rubber Chem. Tech.*, **50**, 688 (1977).
222. Valance, M. A. and S. L. Cooper, *Macromolecules*, **17**, 1208 (1984).
223. Lilaonitkul, A., J. G. West and S. L. Cooper, *J. Macromol. Sci., Phys.*, **12**, 563 (1976).
224. Stevenson, J. C. and S. L. Cooper, *Macromolecules*, **21**, 1309 (1988).
225. Tyagi, D., J. L. Hedrich, D. C. Webster, J. E. McGrath and G. L. Wilkes, *Polymer*, **29**, 833 (1988).
226. Merker, R. L., M. J. Scott and G. G. Haberland, *J. Polym. Sci., Part A*, **2**, 31 (1964).
227. Okui, N., H. M. Li and J. H. Magill, *Polymer*, **19**, 411 (1978).
228. Warwicher, J. O., *J. Mol. Biol.*, **2**, 350 (1960).

229. Takahashi, Y., M. Gehoh and K. Yuzuriha, *J. Polym. Sci.: Pt. B: Polym. Phys.*, **29**, 889 (1991).
230. Lucas, F., J. T. B. Shaw and S. G. Smith, *Adv. Protein Chem.*, **13**, 107 (1958).
231. Lucas, F., J. T. B. Shaw and S. G. Smith, *Nature*, **178**, 861 (1956).
232. Meyer, K. H. and H. Mark, *Chem. Ber.*, **61**, 1932 (1928).
233. Fraser, R. D. B., T. P. MacKae and F. H. C. Stewart, *J. Mol. Biol.*, **19**, 580 (1966).
234. Kaplan, D. L., S. Fossey, C. M. Mella, S. Arcidiacono, K. Senecal, W. Muller, S. Stockwell, R. Beckwitt, C. Viney and K. Kerkam, *Mater. Res. Bull.*, October 1992, p. 41.
235. Termonia, Y., *Macromolecules*, **27**, 7378 (1994).
236. Lucas, F., J. T. B. Shaw and S. G. Smith, *J. Textile Inst.*, **46**, T-440 (1955).
237. Astbury, W. T., *Trans. Faraday Soc.*, **29**, 193 (1933).
238. Coyne, K. J., X. X. Qin and J. H. Waite, *Science*, **277**, 1830 (1991).
239. Nakajima, A., T. Hayashi, K. Kugo and K. Shinoda, *Macromolecules*, **12**, 840 (1979).
240. Nakajima, A., K. Kugo and T. Hayashi, *Macromolecules*, **12**, 849 (1979).
241. Hayashi, T., A. G. Walton and J. M. Anderson, *Macromolecules*, **10**, 346 (1977).
242. DiMarzio, E. A., C. M. Guttman and J. D. Hoffman, *Macromolecules*, **13**, 1194 (1980).
243. Whitmore, M. D. and J. Noolandi, *Macromolecules*, **21**, 1482 (1988).
244. Vilgis, T. and A. Halperin, *Macromolecules*, **24**, 2090 (1991).
245. Beech, D. R., C. Booth, V. E. Dodgson, R. E. Sharp and J. R. S. Waring, *Polymer*, **13**, 73 (1972).
246. Mansfield, M. L., *Macromolecules*, **16**, 914 (1983).
247. Flory, P. J., D. Y. Yoon and K. A. Dill, *Macromolecules*, **17**, 862 (1984); *ibid.*, **17**, 868 (1984).
248. Marqusee, J. A. and K. A. Dill, *Macromolecules*, **19**, 2420 (1986).
249. Kumar, S. K. and D. Y. Yoon, *Macromolecules*, **22**, 3458 (1989).
250. Zuniga, I., K. Rodrigues and W. L. Mattice, *Macromolecules*, **23**, 4108 (1990).
251. Hamley, I. W., M. L. Mallwork, D. A. Smith, J. P. A. Fairlough, A. J. Ryan, S. M. Mai, Y. W. Yang and C. Booth, *Polymer*, **39**, 3321 (1998).
252. L. Mandelkern, *Chemtracts-Macromolecular Chemistry*, **3**, 347 (1992).
253. Douzinas, K. C., R. E. Cohen and A. Halasa, *Macromolecules*, **24**, 4457 (1991).
254. Nojima, S., S. Yamamoto and T. Ashida, *Polym. J.*, **27**, 673 (1995).
255. Rangarajan, R., A. Register and L. Fetters, *Macromolecules*, **26**, 4640 (1993).
256. Unger, R., D. Beyer and E. Donth, *Polymer*, **32**, 3305 (1991).
257. Kilb, R. W. and A. M. Bueche, *J. Polym. Sci.*, **28**, 285 (1958).
258. Stockmayer, W. H., L. D. Moore, Jr., M. Fixman and B. N. Epstein, *J. Polym. Sci.*, **16**, 517 (1955).
259. Krause, S., *J. Phys. Chem.*, **65**, 1618 (1961).
260. Beuche, A. M., *J. Am. Chem. Soc.*, **74**, 65 (1952).
261. Krigbaum, W. R. and N. Tokita, *J. Polym. Sci.*, **43**, 467 (1960).
262. Danusso, F., *Atti. Accad. Naz. Lincei*, **25**, 520 (1958).
263. Krigbaum, W. R. and I. Uematsu, *J. Polym. Sci.: Polym. Chem. Ed.*, **3**, 767 (1965).
264. Lemstra, P. J., T. Kooistra and G. Challa, *J. Polym. Sci.: Polym. Phys. Ed.*, **10**, 823 (1972).
265. Flory, P. J., *Principles of Polymer Chemistry*, Cornell University Press (1953) pp. 495ff.
266. Inoue, T., T. Soen, T. Hashimoto and H. Kawai, *J. Polym. Sci.*, A2, **7**, 1283 (1969).

267. Inoue, T., H. Kawai, M. Fukatsu and M. Kurata, *J. Polym. Sci.: Polym. Lett.*, **6B**, 75 (1968).
268. Dubin, P., *Microdomains in Polymer Solutions*, Plenum Press (1985).
269. Guenet, J. M., *Thermoreversible Gelations of Polymers and Biopolymers*, Academic Press (1992).
270. Mandelkern, L., C. O. Edwards, R. C. Domszy and M. W. Davidson, in *Microdomains in Polymer Solutions*, P. Dubin ed., Plenum Press (1985) pp. 121ff.
271. Chan, E. K. M. and L. Mandelkern, *Macromolecules*, **25**, 5659 (1992).
272. Newman, S., W. R. Krigbaum and D. K. Carpenter, *J. Phys. Chem.*, **60**, 648 (1956).
273. Heyman, E., *Trans. Faraday Soc.*, **31**, 846 (1935).
274. Ichise, N., Y. Yang, Z. Li, Q. Yuan, J. Mark, E. K. M. Chan, R. G. Alamo and L. Mandelkern, in *Synthesis, Characterization, and Theory of Polymeric Networks and Gels*, S. M. Aharoni ed., Plenum Press (1992) p. 217.
275. Prasad, A. and L. Mandelkern, *Macromolecules*, **23**, 5041 (1990).
276. Kobayashi, M., T. Nakaoki and N. Ishihara, *Macromolecules*, **23**, 78 (1990).
277. Kobayashi, M., T. Yoshioka, T. Kozasa, K. Tashiro, J. Suzuki, S. Funahashi, Y. Izumi, *Macromolecules*, **27**, 1349 (1994).
278. Xue, Gi., J. Zhang, J. Chen, Y. Li, I. Ma, G. Wang and P. Seen, *Macromolecules*, **33**, 229 (2000).
279. Wellinghoff, S. J., J. Shaw and E. Baer, *Macromolecules*, **12**, 932 (1979).
280. Berghmans, H., F. Gouaerts and N. Overbergh, *J. Polym. Sci.: Polym. Phys. Ed.*, **17**, 1251 (1979).
281. Garrett, R. R. and P. J. Flory, *Nature*, **177**, 176 (1956); P. J. Flory and R. R. Garrett, *J. Am. Chem. Soc.*, **80**, 4836 (1958).
282. Nakaoki, T., K. Tashiro and M. Kobayashi, *Macromolecules*, **33**, 4299 (2000).
283. Edwards, C. O. and L. Mandelkern, *J. Polym. Sci.: Polym. Lett. Ed.*, **20**, 355 (1982).
284. Price, C., in *Developments in Block Copolymers*, I. Goodman ed., Applied Science Publishers (1982) p. 39.

6

Thermodynamic quantities

6.1 Introduction

Since the fusion of polymers is classified as a first-order phase transition, the equilibrium melting temperature, T_m^0, of a homopolymer is well defined and an important theoretical quantity. This temperature represents the disappearance of the most perfect crystals made up of chains of infinite length. As was discussed in Chapter 2, this quantity is very difficult, if not impossible, to obtain by direct experiment. The melting temperatures that are determined by the usual conventional methods do not satisfy the equilibrium condition. Recourse can be made to theory to establish this temperature.(1) In many instances extrapolation procedures have been employed.(2) These methods take advantage of certain features of the crystallization process and the resulting morphology. The underlying basis for the extrapolation methods and their experimental validity will be discussed in Volume 3. In the present discussion strong efforts have been made to assign values as true as possible to T_m^0. It is a challenge for the future to develop both theoretical and experimental methods to reliably determine T_m^0.

It is a matter of interest to assess how T_m^0 of a homopolymer depends on the chemical nature and structure of its chain repeating unit. The melting temperature is uniquely described by the ratio of the heat of fusion to entropy of fusion, per repeating unit. Therefore, attention should be focused on how these two independent quantities depend on structure. The enthalpies of fusion per chain repeating unit are experimentally accessible for many polymers. From these data, and T_m^0, it is possible to develop an understanding of the molecular and structural basis of the thermodynamic quantities that govern fusion.

6.2 Melting temperatures, heats and entropies of fusion

A key quantity necessary to carry out the thermodynamic analysis is the enthalpy of fusion per repeating unit. This quantity is an inherent property of a polymer chain.

236

It is independent of the level of crystallinity and other morphological features of the crystalline state. From it, and the equilibrium melting temperature, one can obtain the entropy of fusion per repeating unit. There are two direct methods, based on straightforward thermodynamic principles, that are available to determine ΔH_u. There are also several indirect methods. One of the direct methods for determining ΔH_u has been discussed in detail in Chapter 3. The value of ΔH_u can be obtained from the depression of the melting temperature by low molecular weight diluents. According to Eq. (3.9)

$$\frac{1}{T_m} - \frac{1}{T_m^0} = \left(\frac{R}{\Delta H_u}\right)\left(\frac{V_u}{V_1}\right)\left[-\frac{\ln v_2}{x} + \left(1 - \frac{1}{x}\right)(1 - v_2) - \chi_1(1 - v_2)^2\right]$$

$$(3.9)$$

The validity of this equation for polymer–diluent mixtures has been amply demonstrated. The only parameters that are needed to analyze the experimental data are the respective molar volumes of the diluent and polymer repeating unit.

The other direct thermodynamic method that leads to reliable values for ΔH_u is the application of the Clapeyron equation to the change in the equilibrium melting temperature with applied hydrostatic pressure p. Accordingly

$$\frac{dT_m^0}{dp} = T_m^0 \frac{\Delta V_u}{\Delta H_u}$$

$$(6.1)$$

In order to apply Eq. (6.1) the volume of the repeating unit for the liquid and crystal (unit cell) needs to be known as a function of pressure at the melting temperature. From the experimentally determined T_m^0 and ΔV_u (the latent volume change per unit on melting) as a function of applied pressure, ΔH_u can be obtained by extrapolation to atmospheric pressure. It is important that the volume of both the liquid and the crystal be determined as a function of pressure and temperature if erroneous results are to be avoided.(3). In the following discussion we shall list the results obtained by the two direct methods in separate tables. In several cases ΔH_u values have been obtained by both methods. The results can then be compared to assess the consistency of the methods.

There are several indirect methods that also yield values of ΔH_u. They all require the determination of the enthalpy of fusion as well as the degree of crystallinity of the system. The degree of crystallinity can be obtained by different experimental techniques such as infra-red, wide-angle x-ray diffraction and density measurement among others. Quite often the enthalpy of fusion is measured as a function of density and the data extrapolated to the value of the unit cell to yield ΔH_u. The directly measured enthalpy of fusion, as well as the methods used to determine the crystallinity level, are dependent on morphological and structural detail. Moreover, all of the methods usually have different sensitivities to the phase structures. In our discussion, we shall only use the data obtained by the indirect methods either when they are the sole values available for a given polymer or if they can help resolve any discrepancies. Most of the data that will be discussed will include the direct determination of ΔH_u. We shall, however, have occasion to compare the different methods.

The values of ΔH_u determined from the melting temperature depression by diluent are given in Table 6.1 along with related quantities of interest. The T_m^0 values listed have been selected as representing the best for the given polymer, after carefully examining all the available data. To allow for different sizes the ΔH_u value has been divided by M_0, the molecular weight of the chemical repeating unit. Thus, the heat of fusion per gram of crystalline polymer is given in the fourth column of the table. Dividing ΔH_u by the absolute equilibrium melting temperature yields the entropy of fusion per repeating unit, ΔS_u. For polymers that are polymorphic, i.e. those that can crystallize in more than one ordered structure, the appropriate form is indicated. When pertinent, the specific stereo structure involved is also noted. The values for ΔH_u, and the related thermodynamic properties that are listed in the table encompass a large number of polymers that represent virtually all chemical and structural types of repeating units. These results will be discussed in detail after an examination of the thermodynamic properties determined by other methods.

The values of ΔH_u, and related thermodynamic quantities, that were obtained by the application of the Clapeyron equation are listed in Table 6.2. Here, although the number of polymers studied by this method is not as numerous as obtained by the diluent method, many different polymer types are represented. Some of the ΔH_u values in this table, as poly(tetrafluoroethylene), poly(hexamethylene adipamide), poly(aryl ether ether ketone) and some of the aliphatic poly(esters) are unique to this method. The ΔH_u values for the other polymers listed were also obtained by the diluent method. In most cases good agreement is obtained between the two methods. For example, there is almost exact agreement for poly(4-methyl pentene-1) between the two methods. However, another study with the same polymer gives a factor of two less for ΔH_u.(4) This difference can be attributed to the fact that too small a value was used for the crystal specific volume. This discrepancy points out the need for accurate values of the parameters involved, in addition to the melting point measurements themselves. With one exception, the other results obtained by the two direct methods agree with one another to about 10%, or better. The exception is the value of ΔH_u for poly(methylene oxide). The two values differ by almost a factor of two. If the higher value was accepted then the degree of crystallinity of solution formed crystallites deduced from enthalpy of fusion measurement would only be about 0.5. This value is too low to be acceptable. Hence the lower value for ΔH_u of poly(methylene oxide) is taken to be the more reliable.

Table 6.1. Thermodynamic quantities determined by use of diluent equation (Eq. (3.9))[1]

Polymer		T_m^0 (K)	ΔH_u (J mol⁻¹)	$\Delta H_u/M_0$ (J g⁻¹)	ΔS_u (J K⁻¹ mol⁻¹)	References
ethylene $+\text{CH}_2+_n$		418.7	4 142	295.8	9.9	a,b,c
iso.-propylene	α	465.2[2]	8 786	208.8	18.1	d,e,f,g,h,i,j
$+\text{CH}_2-\text{CH}+_n$, CH_3	β		8 201	194.9	17.6	
iso.-butene-1	(I)	408.7	6 318	112.5	15.5	k,l
$+\text{CH}_2-\text{CH}+_n$, CH_2-CH_3	(II)	397.2	6 276	111.9	15.8	
	(III)	379.7	6 485	115.6	17.1	
4-methyl pentene-1 $+\text{CH}-\text{CH}_2+_n$, $\text{CH}_2-\text{CH}(\text{CH}_3)-\text{CH}_3$		523.2	5 297	63.7	10.1	m

Table 6.1. (*cont.*)

Polymer	T_m^0 (K)	ΔH_u (J mol⁻¹)	$\Delta H_u/M_0$ (J g⁻¹)	ΔS_u (J K⁻¹ mol⁻¹)	References
1-methyl octamer $\left[CH(CH_3)-(CH_2)_6-CH_2 \right]_n$	268.2	10857	86.2	40.5	n
iso.-styrene $\left[CH_2-CH(C_6H_5) \right]_n$	>516.2[3]	8682[4]	83.4	16.8	o,p,q
syn.-styrene $\left[CH_2-CH(C_6H_5) \right]_n$	>560.5[4a]	8577	82.4	15.3	r,s
vinyl alcohol $\left[CH_2-CH(OH) \right]_n$	523.2	6862	156.1	13.1	t,u,v

acrylonitrile	$\left[CH_2 - CH \right]_n$ $\quad C \equiv N$	593.2	5021	94.7	8.5	w,x
iso.-isopropyl acrylate	$\left[CH_2 - CH \right]_n$ ... C, O, $O - CH$, CH_3, CH_3	450.2	5857	51.4	13.0	y
trans-1,4-butadiene (I)	$\left[CH_2 - C = C - CH_2 \right]_n$ $\quad H \quad H$	369.2	13 807	255.7	37.4	z
(II)		421.2	4 602	85.2	10.9	
cis-1,4-butadiene	$\left[CH_2 - C = C - CH_2 \right]_n$ $\quad H \quad H$	273.2	9 205	170.4	33.7	aa

Table 6.1. (cont.)

Polymer		T_m^0 (K)	ΔH_u (J mol^{-1})	$\Delta H_u/M_0$ (J g^{-1})	ΔS_u (J K^{-1} mol^{-1})	References
trans-1,4-isoprene	(α)	360.2	12 719	187.0	35.3	bb,cc,dd
	(β)	354.2	10 544	155.1	29.8	
$\left[\mathrm{CH_2-C(CH_3)=C(H)-CH_2}\right]_n$						
cis-1,4-isoprene		308.7	4 393	64.6	14.2	ee,ff
$\left[\mathrm{CH_2-C(H_3C)=C(H)-CH_2}\right]_n$						
trans-1,4-chloroprene		380.2	8 368.5	94.6	22.0	gg,hh,ii
$\left[\mathrm{CH_2-C(Cl)=C(H)-CH_2}\right]_n$						
trans-pentenamer		307.2	12 008	176.3	39.1	jj,kk
$\left[\mathrm{CH=CH-CH_2-CH_2-CH_2}\right]_n$						

trans-octenamer[5]	$+CH{=}CH{-}(CH_2)_5{-}CH_2 +_n$	350.2	23 765	215.7	67.9	ll
cis-octenamer	$+CH{=}CH{-}(CH_2)_5{-}CH_2 +_n$	311.2	21 000	190.9	67.5	mm
trans-decenamer[5]	$+CH{=}CH{-}(CH_2)_7{-}CH_2 +_n$	353.2	32 844	237.6	92.9	ll
trans-dodecenamer	$+CH{=}CH{-}(CH_2)_9{-}CH_2 +_n$	357.2	41 171	247.6	115.3	ll
methylene oxide	$+CH_2{-}O +_n$	479.2	7 012	233.7	14.6	nn,oo,pp,qq
ethylene oxide	$+CH_2{-}CH_2{-}O +_n$	353.2	8 703	197.8	24.6	rr,ss,tt,uu
iso.-propylene oxide	$+CH_2{-}\underset{\underset{CH_3}{\mid}}{CH}{-}O +_n$	355.2	7 531 8 368[6]	129.8	21.2	vv,ww
trimethylene oxide	$+(CH_2)_3{-}O +_n$	323.2	8 786	151.5	27.2	xx

Table 6.1. (*cont.*)

Polymer	T_m^0 (K)	ΔH_u (J mol^{-1})	$\Delta H_u/M_0$ (J g^{-1})	ΔS_u (J K^{-1} mol^{-1})	References
tetramethylene oxide $-\!\!+\!(CH_2)_4\!-\!O\!+_n$	330.2	15 899	220.8	48.2	yy
hexamethylene oxide $+\!(CH_2)_6\!-\!O\!+_n$	346.7	23 640	236.4	68.2	zz
1,3-dioxolane $+\!O\!-\!CH_2\!-\!O\!-\!(CH_2)_2\!+_n$	366.2	15 481	209.2	42.3	aaa
1,3-dioxocane $+\!O\!-\!CH_2\!-\!O\!-\!(CH_2)_5\!+_n$	319.2	7 740	75.9	24.3	bbb
3,3-dimethyl oxetane II	349.2	9 205	107.0	26.3	ccc
III	329.2	7 448	86.6	22.6	

$$+\!O\!-\!CH_2\!-\!\underset{\underset{CH_3}{|}}{\overset{\overset{CH_3}{|}}{C}}\!-\!CH_2\!+_n$$

Compound	Structure					
3-ethyl 3-methyl oxetane	$+\!\!O-CH_2-C(C_2H_5)(CH_3)-CH_2+_n$	334.2	6276	62.8	18.8	ddd
3,3-diethyl oxetane —monoclinic —orthorhombic	$+\!\!O-CH_2-C(C_2H_5)(C_2H_5)-CH_2+_n$	373.2 / 353.2	10460 / 10042	91.8 / 88.1	28.0 / 28.4	eee
3,3-bis-ethoxy methyl oxetane	$+\!\!O-CH_2-C(CH_2-O-CH_2-CH_3)(CH_2-O-CH_2-CH_3)-CH_2+_n$	398.2	9414	54.1	23.6	fff
3,3-bis-azido methyl oxetane	$+\!\!O-CH_2-C(CH_2N_3)(CH_2N_3)-CH_2+_n$	401.2	53555	318.8	133.5	fff

Table 6.1. (*cont.*)

Polymer	T_m^0 (K)	ΔH_u (J mol^{-1})	$\Delta H_u/M_0$ (J g^{-1})	ΔS_u (J K^{-1} mol^{-1})	References
2,6-dimethyl 1,4-phenylene oxide	548.2	5 230	43.6	9.5	ggg,hhh
2,6 dimethoxy 1,4-phenylene oxide	560.2	3 184	20.9	5.7	iii
trimethylene sulfide	363.2	10 460	141.4	28.8	jjj
ethylene azelate	338.2	43 095	138.1	127.6	kkk

Name	Structure				
decamethylene adipate	$+O-C(=O)-(CH_2)_4-C(=O)-O-(CH_2)_{10}+_n$	352.7	42 677	150.3	121.0	lll
decamethylene azelate	$+O-C(=O)-(CH_2)_7-C(=O)-O-(CH_2)_{10}+_n$	342.2	41 840	129.7	121.3	kkk
decamethylene sebacate	$+O-C(=O)-(CH_2)_8-C(=O)-O-(CH_2)_{10}+_n$	353.2	50 208	147.7	142.2	mmm
ethylene terephthalate	$+O-C(=O)-C_6H_4-C(=O)-O(CH_2)_2+_n$	613.2	23 430	122.0	38.2	nnn,ooo,ppp, qqq
tetramethylene terephthalate	$+O-C(=O)-C_6H_4-C(=O)-O(CH_2)_4+_n$	503.2	31 798	144.5	63.2	rrr

Table 6.1. (cont.)

Polymer	T_m^0 (K)	ΔH_u (J mol^{-1})	$\Delta H_u/M_0$ (J g^{-1})	ΔS_u (J K^{-1} mol^{-1})	References
hexamethylene terephthalate	433.7	35 564	143.4	82.0	kkk
decamethylene terephthalate	411.2	46 024	151.4	111.9	kkk
tetramethylene isophthalate	425.7 438.2	42 258	192.1	99.3	rrr,sss
diethylene glycol terephthalate	373.2	39 748	168.4	106.5	ttt

	Structure					
β-propiolactone	$+\!\!-\!C\!-\!O\!-\!(CH_2)_2\!-\!\!+_n$, O=C	357	8577	119.1	24.0	uuu
ε-caprolactone	$+\!\!-\!C\!-\!O\!-\!(CH_2)_5\!-\!\!+_n$, O=C	337	16 297	142.9	48.3	uuu
α,α′-dimethyl propiolactone	$+\!\!-\!C\!-\!O\!-\!CH_2\!-\!C\!-\!\!+_n$, CH_3, CH_3, O=C	542.2	14 853	148.5	27.4	vvv,www, xxx
α,α′-diethyl propiolactone	$+\!\!-\!C\!-\!O\!-\!CH_2\!-\!C\!-\!\!+_n$, H_5C_2, C_2H_5, O=C	531.2	20 920	163.4	39.4	yyy

Table 6.1. (*cont.*)

Polymer	T_m^0 (K)	ΔH_u (J mol^{-1})	$\Delta H_u/M_0$ (J g^{-1})	ΔS_u (J K^{-1} mol^{-1})	References
α-methyl, α-N-propyl, β-propiolactone	425.2	14 602	114.1	34.3	zzz
decamethylene azelamide	487.2	36 819	112.9	75.3	kkk
decamethylene sebacamide	489.2	34 727	102.7	71.0	kkk
N,N'-sebacoyl piperazine	453.2	25 941	102.9	57.2	aaaa

	502.2	17 949	158.8	35.7	bbbb
caprolactam γ^7					
ester amide[8] 6-6	526	92 885	188.3	176.6	cccc
ester amide[8] 12-2	517	10 258	194.6	198.3	cccc
ester amide[8] 12-6	487	116 315	200.8	238.8	cccc
ester amide[8] 12-12	470	140 164	211.3	298.2	cccc
urethane[9] $n = 2$	440.2	44 267	192.5	100.6	dddd
$n = 5$	428.2	45 522	167.4	106.3	
$n = 10$	427.2	61 505	179.8	144.0	
urethane[10] $n = 5$	462.2	54 810	154.8	118.6	dddd
$n = 10$	465.2	70 961	167.4	152.5	
vinyl fluoride	470.2	7 531	163.7	16.0	eeee
vinylidene fluoride (α)	532.2	6 694	104.6	12.6	ffff,gggg, hhhh

Table 6.1. (cont.)

Polymer	T_m^0 (K)	ΔH_u (J mol^{-1})	$\Delta H_u/M_0$ (J g^{-1})	ΔS_u (J K^{-1} mol^{-1})	References
chloro trifluoro ethylene	483.2	5021	43.1	10.4	iiii
2,2'-bis 4,4'(oxyphenyl) propane carbonate	590.2	34 008	133.9	57.6	jjjj,kkkk
dimethyl siloxane[11]	233	2767	36.7	11.9	llll

tetramethyl-p-silphenylene siloxane	433.2	11 340	54.4	26.2	mmmm

cellulose tributyrate[12] $X = -O-C(=O)-(CH_2)_2CH_3$	480.2	12 552	33.7	26.1	nnnn
cellulose trinitrate[12] $X = -O-NO_2$	>973	3 765–6 276	12.6–21.1	3.9–6.4	oooo
cellulose (2.44) nitrate[12] cellulose $(O-NO_2)_{2.44}$	890.2	5 648	21.5	6.3	pppp
cellulose tricaprylate[12] $X = -O-C(=O)-(CH_2)_6CH_3$	389.2	12 970	24.0	33.3	qqqq
collagen	418.2[13]	9 414	100.4	22.5	rrrr

[1] Adapted with permission from L. Mandelkern and R. G. Alamo, in *American Institute of Physics Handbook of Polymer Properties*, J. E. Mark ed., American Institute of Physics Press (1996) p. 119.

[2] The reported equilibrium melting temperature of iso. poly(propylene) has ranged from 458 K to 493 K. (Phillips, R. A. and M. D. Wolsowicz, in *Polypropylene Handbook*, E. P. Moore, Jr. ed., Hansen Publications, Inc.)

Notes to Table 6.1. (*cont.*)

[3] 516.2 K is the highest T_m observed (q). Therefore, T_m^0 should be greater.

[4] Average value of references cited.

[4a] Extrapolated equilibrium melting temperatures of the α and β forms are very close to one another. Depending on the method used they are close to 545 K or 573 K. (Ho, R. M., C. P. Lin, H. Y. Tsai and E. M. Wo, *Macromolecules*, **33**, 6517 (2000).)

[5] Extrapolated to all trans.

[6] Obtained by direct determination of activity coefficients in polymer–diluent mixtures.

[7] That these data belong to the γ form is deduced from the reported specific volumes (V_c^γ) and the heat of fusion data of Fig. 13 in K. Illers and K. H. Haberkorn, *Makromol. Chem.*, **142**, 31 (1971).

[8] Ester-amide (*n-m*)

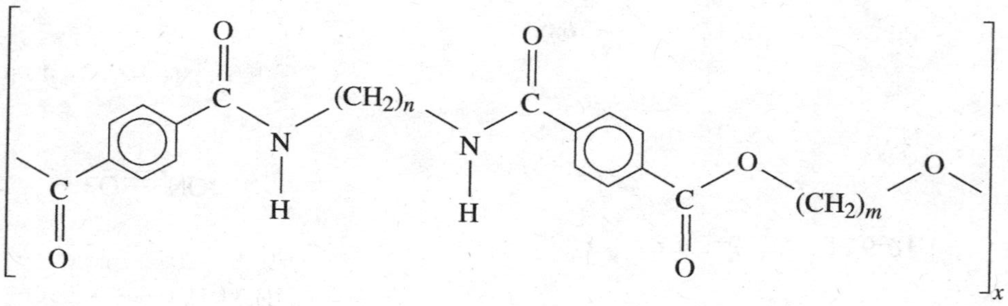

9

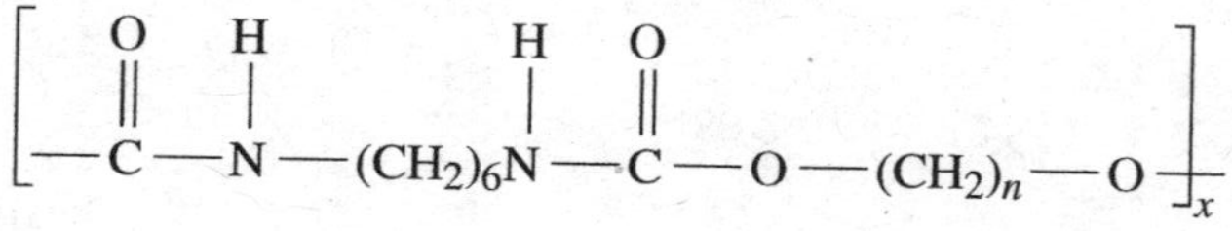

10

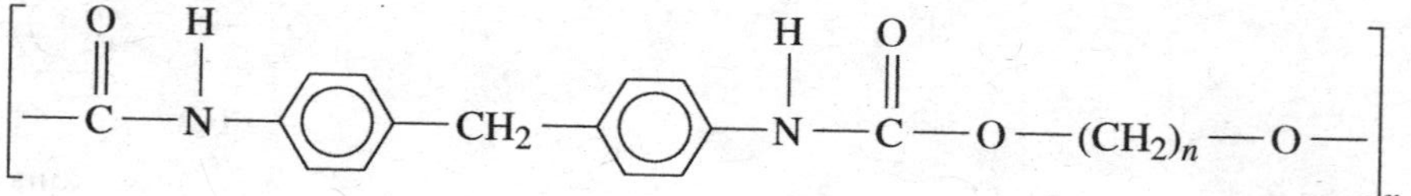

[11] The value of ΔH_u was determined with only one diluent. It was originally reported in terms of calories per mole of chain atoms and misinterpreted. A clarification in terms of joules per gram was subsequently given. (Arangurem, M. L., *Polymer*, **39**, 4897 (1998).)

[12]

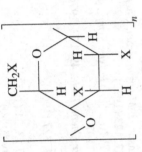

[13] Extrapolated from the melting point of glycol mixtures.

References

a. Flory, P. J. and A. Vrij, *J. Am. Chem. Soc.*, **85**, 3548 (1963).
b. Quinn Jr., F. A. and L. Mandelkern, *J. Am. Chem. Soc.*, **80**, 3178 (1958); L. Mandelkern, *Rubber Chem. Tech.*, **32**, 1392 (1959).
c. Nakajima, N. and F. Hamada, *Kolloid Z. Z. Polym.*, **205**, 55 (1965).
d. Mucha, M., *J. Polym. Sci: Polym. Symp.*, **69C**, 79 (1981).
e. Fatou, J. G., *Eur. Polym. J.*, **7**, 1057 (1971).
f. Monasse, B. and J. M. Haudin, *Coll. Polym. Sci.*, **263**, 822 (1985).
g. Fujiwara, Y., *Coll. Polym. Sci.*, **265**, 1027 (1987).
h. Krigbaum, W. R. and I. Uematsu, *J. Polym. Sci: Polym. Chem. Ed.*, **3**, 767 (1965).
i. Danusso, F. and G. Gianotti, *Eur. Polym. J.*, **4**, 165 (1968).
j. Shi, G., B. Huang and J. Zhang, *Makromol. Chem. Rapid Comm.*, **5C**, 573 (1984).
k. Danusso, F. and G. Gianotti, *Makromol. Chem.*, **61**, 139 (1963).
l. Wilski, H. and T. Grewer, *J. Polym. Sci.: Polym. Symp.*, **6**, 33 (1964).
m. Charlet, G. and G. Delmas, *J. Polym. Sci.: Polym. Phys. Ed.*, **26**, 1111 (1988).
n. Gianotti, G., G. Dall'Asta, A. Valvassori and V. Zamboni, *Makromol. Chem.*, , **149**, 117 (1971).
o. Dedeurwaerder, R. and J. F. M. Oth, *J. Chim. Phys.*, **56**, 940 (1959).
p. Danusso, F. and G. Moraglio, *Rend. Accad. Naz., Lincei*, **27**, 381 (1959).

Notes to Table 6.1. (*cont.*)

q. Lemstra, P. J., T. Kooistra and G. Challa, *J. Polym. Sci.: Polym. Phys. Ed.*, **10**, 823 (1972).
r. Gianotti, G. and A. Valvassori, *Polymer*, **31**, 473 (1990).
s. Gvozdic, N. V. and D. J. Meier, *Polym. Comm.*, **32**, 183 (1991).
t. Ohgi, H. and T. Sato, *Macromolecules*, **26**, 559 (1993).
u. Fujii, F., *J. Polym. Sci.: Macromol. Rev.*, **5**, 431 (1971).
v. Tubbs, R. K., *J. Polym. Sci.: Polym. Phys. Ed.*, **3**, 4181 (1965).
w. Krigbaum, W. R. and N. Takita, *J. Polym. Sci.*, **43**, 467 (1960).
x. Hinrichsen, V. G., *Angew. Makromol. Chem.*, **121** (1971).
y. Wessling, R. A., J. E. Mark and R. E. Hughes, *J. Phys. Chem.*, **70**, 1909 (1966).
z. Natta, G. and G. Moraglio, *Rubber Plastics Age*, **44**, 42 (1963).
aa. Natta, G. and G. Moraglio, *Makromol. Chem.*, **66**, 218 (1963).
bb. Mandelkern, L., F. A. Quinn, Jr., and D. E. Roberts, *J. Am. Chem. Soc.*, **78**, 926 (1956).
cc. Lovering, E. G. and D. C. Wooden, *J. Polym. Sci.: Polym. Phys. Ed.*, **9**, 175 (1971).
dd. Flanagan, R. D. and A. M. Rijke, *J. Polym. Sci.: Polym. Phys. Ed.*, **10**, 1207 (1972).
ee. Roberts, D. E. and L. Mandelkern, *J. Am. Chem. Soc.*, **77**, 781 (1955).
ff. Dalal, E. N., K. D. Taylor, and P. J. Phillips, *Polymer*, **24**, 1623 (1983).
gg. Mochel, W. E. and J. T. Maynard, *J. Polym. Sci.*, **13**, 235 (1954).
hh. Krigbaum, W. R. and J. H. O'Mara, *J. Polym. Sci.: Polym. Phys. Ed.*, **8**, 1011 (1970).
ii. Garrett, R. R., C. A. Hargreaves II and D. N. Robinson, *J. Macromol. Sci. Chem.*, **A4**, 1679 (1970).
jj. Capizzi, A. and G. Gianotti, *Makromol. Chem.*, **157**, 123 (1972).
kk. Wilkes, C. E., M. J. P. Pekló and R. J. Minchak, *J. Polym. Sci.: Polym. Symp.*, **43C**, 97 (1973).
ll. Gianotti, G. and A. Capizzi, *Eur. Polym. J.*, **6**, 743 (1970).
mm. Gianotti, G., A. Capizzi and L. DelGiudice, *Rubber Chem. Tech.*, **49**, 170 (1976).
nn. Inoue, M., *J. Polym. Sci.*, **51**, 518 (1961).
oo. Wissbrun, K. F., *J. Polym. Sci., Polym. Phys. Ed.*, **4**, 827 (1966).
pp. Majer, T., *Kunststoffe*, **52**, 535 (1963).
qq. Korenga, T., F. Hamada and A. Nakajima, *Polym. J.*, **3**, 21 (1972).
rr. Allen, R. S., Master Thesis, Florida State University (1980).
ss. Alfonso, G. C. and T. P. Russell, *Macromolecules*, **19**, 1143 (1986).
tt. Mandelkern, L., *J. Appl. Phys.*, **26**, 443 (1955).

uu. Rijke, A. M. and S. McCoy, *J. Polym. Sci.: Pt. A-2*, **10**, 1845 (1972).

vv. Booth, C., C. J. Devoy and G. Gee, *Polymer*, **12**, 327 (1971).

ww. Booth, C., C. J. Devoy, D. V. Dodgson and I. H. Hillier, *J. Polym. Sci.: Polym. Phys. Ed.*, **8**, 519 (1970).

xx. Perez, E., J. G. Fatou and A. Bello, *Eur. Polym. J.*, **23**, 469 (1987).

yy. Takahashi, T. and Y. Yamishita, in *Copolymers, Blends and Composites, Advances in Chemistry*, Series 142, N. A. J. Plazec ed., American Chemical Society (1975) p. 207.

zz. Marco, C., A. Bello and J. G. Fatou, *Makromol. Chem.*, **179**, 1333 (1978).

aaa. Alamo, R., Doctoral Thesis, University of Madrid (1981).

bbb. Alamo, R., A. Bello and J. G. Fatou, *J. Polym. Sci.: Polym. Phys. Ed.*, **28**, 907 (1990).

ccc. Perez, E., J. G. Fatou and A. Bello, *Eur. Polym. J.*, **23**, 469 (1987).

ddd. Bello, A., E. Perez and J. G. Fatou, *Macromolecules*, **19**, 2497 (1986).

eee. Cited in M. A. Gomez, J. G. Fatou and A. Bello, *Eur. Polym. J.*, **22**, 43 (1986).

fff. Hardenstine, K. E., G. V. S. Henderson, Jr., L. H. Sperling, C. J. Murphy and G. E. Mauser, *J. Polym. Sci.: Polym. Phys. Ed.*, **23**, 1597 (1985).

ggg. Shultz, A. R. and C. R. McCullough, *J. Polym. Sci: Polym. Phys. Ed.*, **10**, 307 (1972).

hhh. Janeczelk, H., H. Turska, T. Szeholy, M. Lengyel and F. Till, *Polymer*, **19**, 85 (1975).

iii. Savolainen, A., *Eur. Polym. J.*, **10**, 9 (1974).

jjj. Sanchez, A., C. Marco, J. G. Fatou and A. Bello, *Eur. Polym. J.*, **24**, 355 (1988).

kkk. Flory, P. J., H. D. Bedon and E. H. Keefer, *J. Polym. Sci.*, **28**, 151 (1958).

lll. Mandelkern, L., R. R. Garrett and P. J. Flory, *J. Am. Chem. Soc.*, **74**, 3939 (1952).

mmm. Evans, R. D., H. R. Mighton and P. J. Flory, *J. Am. Chem. Soc.*, **72**, 2018 (1950).

nnn. Blundell, D. J. and B. N. Osborn, *Polymer*, **24**, 953 (1983).

ooo. Roberts, R. C., *Polymer*, **10**, 113 (1969).

ppp. Wlochowicz, A. and W. Przygock, *J. Appl. Polym. Sci.*, **17**, 1197 (1973).

qqq. Slade, P. E. and T. A. Orofino, *Anal. Calorimetry*, **1**, 63 (1968).

rrr. Conix, A. and R. Van Kerpel, *J. Polym. Sci.*, **40**, 521 (1959).

sss. Phillips, R. A., J. M. McKenna and S. L. Cooper, *J. Polym. Sci.: Pt. B: Polym. Phys.*, **32**, 791 (1994).

ttt. Guzman, J. and J. G. Fatou, *Eur. Polym. J.*, **14**, 943 (1978).

uuu. Crescenzi, V., G. Manzini, G. Calzalari and C. Borri, *Eur. Polym. J.*, **8**, 449 (1972).

vvv. Borri, C., S. Brückner, V. Crescenzi, G. Della Fortuna, A. Mariano and P. Scarazzato, *Eur. Polym. J.*, **7**, 1515 (1971).

www. Marand, H. and J. D. Hoffman, *Macromolecules*, **23**, 3682 (1990).

Notes to Table 6.1. (*cont.*)

xxx. Noah, J. and R. E. Prud'Homme, *Eur. Polym. J.*, **17**, 353 (1981).

yyy. Normand, Y., M. Aubin and R. E. Prud'Homme, *Makromol. Chem.*, **180**, 769 (1979).

zzz. Grenier, D., A. Leborgne, N. Spassky and R. E. Prud'Homme, *J. Polym. Sci.: Polym. Phys. Ed.*, **19**, 33 (1981).

aaaa. Flory, P. J., L. Mandelkern and H. K. Hall, *J. Am. Chem. Soc.*, **73**, 2532 (1951).

bbbb. Gechele, G. B. and L. Crescentini, *J. Appl. Polym. Sci.*, **7**, 1349 (1963).

cccc. Manzini, G., V. Crescenzi, A. Ciana, L. Ciceri, G. Della Fortuna and L. Zotteri, *Eur. Polym. J.*, **9**, 941 (1973).

dddd. Kajiyama, T. and W. J. Macknight, *Polymer J.*, **1**, 548 (1970).

eeee. Sapper, D. I., *J. Polym. Sci.*, **43**, 383 (1960).

ffff. Nandi, A. K. and L. Mandelkern, *J. Polym. Sci.: Polym. Phys. Ed.*, **29**, 1287 (1991).

gggg. Welch, G. J. and R. L. Miller, *J. Polym. Sci.: Polym. Phys. Ed.*, **14**, 1683 (1976).

hhhh. Nakagawa, K. and Y. Ishida, *J. Polym. Sci.: Polym. Phys. Ed.*, **11**, 2153 (1973).

iiii. Bueche, A. M., *J. Am. Chem. Soc.*, **74**, 65 (1952).

jjjj. Legras, R. and J. P. Mercier, *J. Polym. Sci.: Polym. Phys. Ed.*, **15**, 1283 (1977).

kkkk. Wineman, P. L. quoted by L. D. Jones and F. E. Karasz, *J. Polym. Sci.: Polym. Lett.*, **4B**, 803 (1966).

llll. Lee, C. L., O. K. Johannson, O. L. Flaningan and P. Hahn, *Polym. Prepr.*, **10**, 1311 (1969).

mmmm. Okui, N., H. M. Li and J. H. Magill, *Polymer* **19**, 411 (1978).

nnnn. Mandelkern, L. and P. J. Flory, *J. Am. Chem. Soc.*, **73**, 3026 (1951).

oooo. Flory, P. J., R. R. Garrett, S. Newman and L. Mandelkern, *J. Polym. Sci.*, **12**, 97 (1954).

pppp. Newman, S., *J. Polym. Sci.*, **13**, 179 (1954).

qqqq. Goodman, P., *J. Polym. Sci.*, **24**, 307 (1957).

rrrr. Flory, P. J. and R. R. Garrett, *J. Am. Chem. Soc.*, **80**, 4836 (1958).

Table 6.2. *Thermodynamic quantities determined by the use of Clapeyron equation (Eq. (6.1))*[1]

Polymer	T_m^{02} (K)	ΔH_u (J mol^{-1})	$\Delta H_u/M_0$ (J g^{-1})	ΔS_u (J K^{-1} mol^{-1})	References
ethylene $\left[\text{CH}_2\right]_n$	414.6	4 059	289.9	9.8	a
iso.-butene-1 (I)	406.2	7 782[3]	138.7	19.2	b,c
(II)	392.9	7 531[3]	134.2	19.2	
$\left[\text{CH}_2\text{—CH}\right]_n$ $\;\;\;\;\;$ CH$_2$—CH$_3$					
4-methyl pentene-1 $\left[\text{CH—CH}_2\right]_n$ CH$_2$—CH(CH$_3$)CH$_3$	506.2	5 205	61.9	10.3	d

Table 6.2. (*cont.*)

Polymer	T_m^{02} (K)	ΔH_u (J mol^{-1})	$\Delta H_u/M_0$ (J g^{-1})	ΔS_u (J K^{-1} mol^{-1})	References
methylene oxide $+CH_2-O+_n$	456.2	11 673	389.1	25.6	e
ethylene oxide $+CH_2-CH_2-O+_n$	339.2	9037	205.4	26.6	f
tetramethylene oxide $+(CH_2)_4-O+_n$	315.9	14 728	204.6	46.6	f
ethylene adipate $+O-C(=O)-(CH_2)_4-C(=O)-O-CH_2+_n$	326.2	20 150	127.5	61.8	g
ethylene pimelate $+O-C(=O)-(CH_2)_5-C(=O)-O-CH_2+_n$	309.2	27 489	159.8	88.9	g

ethylene suberate	$\left[\!-\!O\!-\!C(=O)\!-\!(CH_2)_6\!-\!C(=O)\!-\!O\!-\!CH_2\!-\!\right]_n$	336.2	24 451	131.4	72.8	g
ethylene azelate	$\left[\!-\!O\!-\!C(=O)\!-\!(CH_2)_7\!-\!C(=O)\!-\!O\!-\!CH_2\!-\!\right]_n$	320.2	40 488	202.4	126.4	g
ethylene terephthalate	$\left[\!-\!O\!-\!C(=O)\!-\!C_6H_4\!-\!C(=O)\!-\!O(CH_2)_2\!-\!\right]_n$	535.2	26 150	136.2	48.9	h
hexamethylene adipamide α_2	$\left[\!-\!N(H)\!-\!(CH_2)_6\!-\!N(H)\!-\!C(=O)\!-\!(CH_2)_4\!-\!C(=O)\!-\!\right]_n$	542.2	43 367	191.9	79.9	i

Table 6.2. (*cont.*)

Polymer	T_m^{02} (K)	ΔH_u (J mol^{-1})	$\Delta H_u/M_0$ (J g^{-1})	ΔS_u (J K^{-1} mol^{-1})	References
tetrafluoro ethylene (virgin)	619.2	5 105	102.1	8.2	j
tetrafluoro ethylene (melt cured)	601.2	4 632	92.6	7.7	
aryl-ether-ether-ketone	611.2	47 359	164.4	77.5	k
2,2'-bis(4,4'-oxyphenyl) propane carbonate	506.2	39 497	155.5	78.0	l

[1] Adapted with permission from L. Mandelkern and R. G. Alamo, in *American Institute of Physics Handbook of Polymer Properties*, J. E. Mark ed., American Institute of Physics Press (1996) p. 119.

[2] Melting temperature actually used in calculation

[3] Average values

References

a. Davidson, T. and B. Wunderlich, *J. Polym. Sci.: Polym. Phys. Ed.*, **7**, 377 (1969).

b. Starkweather, H. W., Jr., G. A. Jones, *J. Polym. Sci.: Polym. Phys. Ed.*, **24**, 1509 (1986).

c. Leute, U. and W. Dollhopt, *Coll. Polym. Sci.*, **261**, 299 (1983).

d. Zoller, P., H. W. Starkweather and G. A. Jones, *J. Polym. Sci.: Polym. Phys. Ed.*, **24**, 1451 (1986).

e. Starkweather, H. W., Jr., G. A. Jones and P. Zoller, *J. Polym. Sci.: Polym. Phys. Ed.*, **26**, 257 (1988).

f. Tsujita, Y., T. Nose and T. Hata, *Polym. J.*, **6**, 51 (1974).

g. Ueberreiter, K., W. H. Karl and A. Altmeyer, *Eur. Polym. J.*, **14**, 1045 (1978).

h. Starkweather, H. W., Jr., P. Zoller and G. A. Jones, *J. Polym. Sci.: Polym. Phys. Ed.*, **21**, 295 (1983).

i. Starkweather, H. W., Jr., P. Zoller and G. A. Jones, *J. Polym. Sci.: Polym. Phys. Ed.*, **22**, 1615 (1984).

j. Starkweather, H. W., Jr., P. Zoller, G. A. Jones, and A. D. Vega, *J. Polym. Sci.: Polym. Phys. Ed.*, **20**, 751 (1982).

k. Zoller, P., T. A. Kehl, H. W. Starkweather, Jr. and G. A. Jones, *J. Polym. Sci.: Polym. Phys. Ed.*, **27**, 993 (1989).

l. Jones, L. D. and F. E. Karasz, *J. Polym. Sci.: Polym. Lett.*, **4B**, 803 (1966).

Table 6.3 lists a set of polymers whose thermodynamic parameters have only been obtained by an indirect method. The 'method' column gives the experimental method that complemented the calorimetric measurement. These data represent some important polymers and follow the general expectation from the results obtained by the direct methods. For many polymers a comparison can be made between the direct and indirect methods. In some cases the agreement is very good while in others there is wide disagreement. For example, very good agreement is found by use of the Clapeyron equation and ΔH–density measurement for two aliphatic polyesters, poly(ethylene adipate) and poly(ethylene suberate).(5) A similar agreement is found for poly(tetra methyl-p-silphenylene) between the diluent method and two indirect methods, ΔH–density and the extrapolation of ΔH–thickness relations to infinite thickness.(6,7) In contrast, for isotactic poly(propylene) the extrapolation of ΔH–density measurements to the density of the unit cell leads to a much lower value than that obtained by the diluent method.(8,9) For the polycarbonate, poly-(4,4'-dioxydiphenyl-2-2 propane carbonate), the x-ray method gives a value that is in agreement with the direct methods. However the ΔH–density method gives consistently lower values.(10,11) There is a serious discrepancy in ΔH_u obtained by the Clapeyron method and by calorimetry combined with density of poly(ether ether ketone).(12) There is a difference in sensitivity to the phase structure by the different physical measurements. For many others of the polymers that have been studied reasonable agreement has been obtained with one or the other of the absolute methods. Although in principle, and in practice, reliable values for ΔH_u can be obtained by indirect methods, care must be exercised in using such data since some serious discrepancies have been observed. It should be recalled that while an analysis of the melting of copolymers can yield values for ΔH_u, such data have not been used in the compilations because of complications that were discussed in Chapter 5.

When examining the data in these tables it is apparent that no simple or obvious correlation exists between the melting temperatures and enthalpies of fusion. The heats of fusion for the different polymers listed fall mainly into two general classes. In one, the values of ΔH_u are of the order of a few thousand calories per mole of repeating unit. In the other category they are about $10\,000\ \mathrm{cal\ mol^{-1}}$. The values for a few polymers lie in between. Many of the high melting polymers are characterized by lower heats of fusion; conversely, a large number of the low melting polymers possess relatively large heats of fusion. It should be emphasized that ΔH_u and ΔS_u values represent the difference in the enthalpy and entropy, respectively, between the liquid and crystalline states. Therefore, it is the changes that occur in these quantities on fusion that are important. Consequently, proper attention must be given to these properties in both states.

Table 6.3. Unique values of thermodynamic parameters determined by indirect methods[1]

Polymer	T_m^0 (K)	ΔH_u (J mol⁻¹)	$\Delta H_u/M_0$ (J g⁻¹)	ΔS_u (J K⁻¹ mol⁻¹)	Method	References
syn.-propylene $\left[\!\!\left[CH_2\!-\!CH \right]\!\!\right]_n$ with CH_3	455.2[2]	8 400	200	18.8	DSC–density	a,b,c
iso.-methyl methacrylate $\left[\!\!\left[CH_2\!-\!C \right]\!\!\right]_n$ with CH_3, $C\!=\!O$, CH_3	493.2[3]	5 021	50.2	12.2	DSC–x-ray	d,e
syn.-vinyl chloride $\left[\!\!\left[CH_2\!-\!CH \right]\!\!\right]_n$ with Cl	538.2[3] 658.2[4]	4 937 6 694	79.0 107.1	9.2 10.2	DSC–x-ray DSC–x-ray	f
octamethylene oxide $\left[\!\!\left[(CH_2)_8\!-\!O \right]\!\!\right]_n$	356.2	32 401	253.1	91.0	DSC–density	g

Table 6.3. (cont.)

Polymer	T_m^0 (K)	ΔH_u (J mol^{-1})	$\Delta H_u/M_0$ (J g^{-1})	ΔS_u (J K^{-1} mol^{-1})	Method	References
3-tert-butyl oxetane 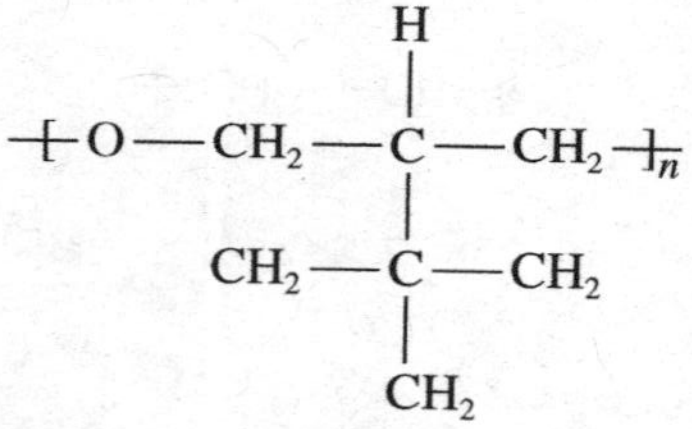	350.2	5 021	45.2	14.3	DSC–x-ray	h
1,4-phenylene ether 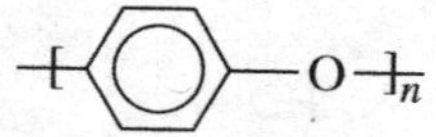	535.2 563.2	7 824	85.0	14.6	DSC–x-ray	i,j
2,6-diphenyl 1,4-phenylene ether 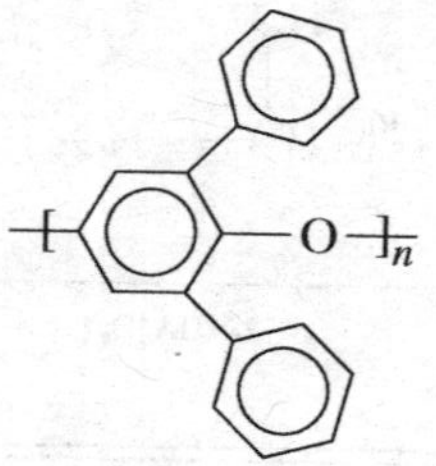	757.2	12 201	50.0	16.1	DSC–x-ray	k
ethylene sulfide	489.2	14 226	237.1	29.1	Tm–mol. wt.[5]	l

3-tert-butyl oxetane

$$\left[\!\!\left[O-CH_2-\underset{\underset{\displaystyle CH_2}{|}}{\overset{\overset{\displaystyle H}{|}}{C}}-\underset{\underset{\displaystyle CH_2}{|}}{C}-CH_2 \right]\!\!\right]_n$$

1,4-phenylene ether

2,6-diphenyl 1,4-phenylene ether

ethylene sulfide

$$\left[\!\!\left[CH_2-CH_2-S \right]\!\!\right]_n$$

Polymer	Structure					Method	Ref.
3,3'-dimethyl thietane	$\left[\,S-CH_2-\underset{CH_3}{\overset{CH_3}{C}}-CH_2\,\right]_n$	286.2	5 442	56.7	19.02	DSC–x-ray	m
p-phenylene sulfide	$\left[\!-\!\bigcirc\!-\!S\,\right]_n$	621.7	12 092	112.0	19.4	DSC–x-ray	l,n,o
ethylene sebacate	$\left[\,O-\overset{O}{\overset{\|}{C}}-(CH_2)_8-\overset{O}{\overset{\|}{C}}-O-(CH_2)_2\,\right]_n$	356.2	36 765	161.2	103.2	DSC–density	p
3-hydroxy butyrate	$\left[\,\underset{CH_3}{CH}-CH_2-\overset{O}{\overset{\|}{C}}-O\,\right]_n$	476.2[6]	13 286[7]	154.5	27.9	DSC–x-ray–density	q,r,s
tetrachloro bis phenol-adipate	(structure)	556.2	33 890	55.2	60.9	DSC–x-ray	t

Table 6.3. (*cont.*)

Polymer		T_m^0 (K)	ΔH_u (J mol^{-1})	$\Delta H_u/M_0$ (J g^{-1})	ΔS_u (J K^{-1} mol^{-1})	Method	References
ethylene 2,6-naphthalene dicarboxylate		610.2	24 987	93.9	40.9	DSC–x-ray–density	u
caprolactam	α	533.2	30 271[8]	267.9	56.8	DSC–specific volume	v
hexamethylene sebacamide		498.2	50 626	179.5	101.6	DSC–density	w
tetramethylene adipamide		623.2	41 618	210.2	66.8	DSC–x-ray	x

Structure					
undecane amide $+\!\!C(=\!O)\!-\!(CH_2)_{10}\!-\!N(H)\!-\!\}_n$	514.2	35 982	196.6	70.0	DSC–density y,z
imide[9] $n = 1$	613.2	72 467	143.8	118.2	DSC–x-ray aa
$n = 2$	577.2	80 165	146.3	138.9	DSC–x-ray
$n = 3$	541.2	87 956	148.6	162.5	DSC–x-ray
imide (LARC-DPI) 3,3′-bischloro methyl oxacyclobutane	663.2	98 060	124.5	147.0	DSC–x-ray bb
	476.2	19 456	126.3	40.9	DSC–density cc
L-lactic acid	480.2	5 858	81.4	12.2	DSC–x-ray dd,ee

3,3′-bischloro methyl oxacyclobutane:

$+\!O\!-\!CH_2\!-\!C(CH_2Cl)(CH_2Cl)\!-\!CH_2\!\}_n$

L-lactic acid:

$+\!C(CH_3)(H)\!-\!C(=\!O)\!-\!O\!\}_n$

Table 6.3. (cont.)

Polymer	T_m^0 (K)	ΔH_u (J mol^{-1})	$\Delta H_u/M_0$ (J g^{-1})	ΔS_u (J K^{-1} mol^{-1})	Method	References
dichlorophosphazene	306.2	8380	71.0	27.4	DSC–1 – λ	ff
urethane[10]						
n = 2	440.2	41 547	180.6	94.4	DSC–x-ray	gg
n = 3	434.2	41 840	171.5	96.4	DSC–x-ray	
n = 4	453.2	48 534	188.1	107.1	DSC–x-ray	
n = 5	428.2	41 463	152.4	96.8	DSC–x-ray	
n = 6	438.2	51 882	181.4	118.4	DSC–x-ray	
n = 7	419.2	58 534	161.8	115.8	DSC–x-ray	
n = 8	430.2	55 229	175.9	128.4	DSC–x-ray	
n = 9	420.2	52 300	159.5	124.5	DSC–x-ray	
n = 10	427.2	56 484	165.2	132.2	DSC–x-ray	
urethane[11]						
n = 2	510.2	48 534	155.6	95.1	DSC–x-ray	gg
n = 3	500.2	47 279	145.0	94.5	DSC–x-ray	
n = 4	505.2	52 718	155.1	104.4	DSC–x-ray	
n = 5	462.2	51 045	144.2	110.4	DSC–x-ray	
n = 6	470.2	51 882	141.0	110.3	DSC–x-ray	
n = 7	464.2	50 626	132.5	109.1	DSC–x-ray	
n = 8	469.2	59 413	150.0	126.6	DSC–x-ray	
n = 9	463.2	58 576	142.9	126.4	DSC–x-ray	
n = 10	465.2	69 036	162.8	148.4	DSC–x-ray	

| new-TPI[12] | >679.2 | 63 800 | 116 | 93.9 | DSC–density | hh |
| ethyl-aryl-ether-ether ketone | >513.2 | 22 348 | 74.0 | <43.5 | DSC–x-ray | ii |

[1] Adapted with permission from L. Mandelkern and R. G. Alamo, *American Institute of Physics Handbook of Polymer Properties*, J. E. Mark ed., American Institute of Physics Press (1996) p. 119.

[2] For a sample 94% syndiotactic content (diads based analysis by ^{13}C NMR), $T_m = 160\,°C$, $\Delta H_u = 1920\,cal\,mol^{-1}$. Data from S. Z. D. Cheng *et al.*, *Polymer*, **35**, 1884 (1994).

[3] For 64% syndiotactic polymer.

[4] Calculated in Ref. (107) for 100% syndiotactic material by modifying the data of D. Kockott, *Kolloid Z. Z. Polym.*, **198**, 17 (1964).

[5] Samples do not have most probable molecular weight distribution.

[6] Cited in R. P. Pearce and R. H. Marchessault, *Macromolecules*, **27**, 3869 (1994).

[7] Average of literature values.

[8] Taking $V_c^\alpha = 0.814\,cm^3\,g^{-1}$ from D. R. Holmes, C. W. Bunn and D. J. Smith, *J. Polym. Sci.*, **17**, 159 (1955). If all literature values for V_c^α are considered, ΔH_u ranges from 6215 cal mol^{-1} to 7250 cal mol^{-1} (26 000 J mol^{-1} to 30 334 J mol^{-1}).

[9]

Notes to Table 6.3. (cont.)

10

11

12

References

a. Galambos, A., M. Wolkowicz, R. Zeigler and M. Galimberti, *ACS Preprints, PMSE Div.* April (1991).
b. Haftka, S. and K. Könnecke, *J. Macromol. Sci. Phys.*, **B30**, 319 (1991).
c. Rosa, C. D., F. Auricemma, V. Vinte and M. Galimberto, *Macromolecules*, **31**, 6206 (1998).
d. Kusy, R. P., *J. Polym. Sci.: Polym. Chem. Ed.*, **14**, 1527 (1976).
e. deBoer, A., G. O. A. vanEkenstein and G. Challa, *Polymer*, **16**, 930 (1975).
f. Guinlock, E. V., *J. Polym. Sci.: Polym. Phys. Ed.*, **13**, 1533 (1975).
g. Marco, C., J. G. Fatou, A. Bello and A. Blanco, *Makromol. Chem.*, **181**, 1357 (1980).
h. Bello, A., E. Perez and J. G. Fatou, *Makromol. Chem., Macromol. Symp.*, **20/21**, 159 (1988).

i. Wrasidlo, W., *J. Polym. Sci.: Polym. Phys. Ed.*, **10**, 1719 (1972).

j. Boon, J. and E. P. Magré, *Makromol. Chem.*, **126**, 130 (1969).

k. Wrasidlo, W., *Macromolecules*, **4**, 642 (1971).

l. Nicco, A., J. P. Machon, H. Fremaux, J. Ph. Pied, B. Zindy and M. Thiery, *Eur. Polym. J.*, **6**, 1427 (1970).

m. Bello, A., S. Lazcano, C. Marco and J. G. Fatou, *J. Polym. Sci.: Polym. Chem. Ed.*, **22**, 1197 (1984).

n. Cheng, S. Z. D., Z. Q. Wu and B. Wunderlich, *Macromolecules*, **20**, 2802 (1987).

o. Huo, P. and P. Cebe, *Coll. Polym. Sci.*, **270**, 840 (1992).

p. Hobbs, S. W. and F. W. Billmeyer, *J. Polym. Sci.: Polym. Phys. Ed.*, **8**, 1387 (1970).

q. Organ, S. J. and P. J. Barham, *Polymer*, **34**, 2169 (1993).

r. Barham, P. J., A. Keller, E. L. Otun and P. A. Holmes, *J. Mater. Sci.*, **19**, 2781 (1984).

s. Bloembergen, S., D. A. Holden, T. L. Bluhm, G. K. Hamer and R. H. Marchessault, *Macromolecules*, **22**, 1656 (1989).

t. Lanza, E., H. Berghmann and G. Smets, *J. Polym. Sci.: Polym. Phys. Ed.*, **11**, 75 (1973).

u. Cheng, S. Z. D. and B. Wunderlich, *Macromolecules*, **21**, 789 (1988).

v. Illers, K. H. and H. Haberkorn, *Makromol. Chem.*, **142**, 31 (1971).

w. Inoue, M., *J. Polym. Sci.: Polym. Chem. Ed.*, **1**, 2697 (1963).

x. Gaymans, R. J., D. K. Doeksen and S. Harkema, in *Integration of Fundamental Polymer Science and Technology*, L. A. Kleintjens and P. J. Lemstra eds., Elsevier (1986) p. 573.

y. Fakirov, S., N. Avramova, P. Tidick and H. G. Zachmann, *Polym. Comm.*, **26**, 26 (1985).

z. Gogolewski, S., *Coll. Polym. Sci.*, **257**, 811 (1979).

aa. Cheng, S. Z. D., D. P. Heberer, H. S. Lien and F. W. Harris, *J. Polym. Sci.: Polym. Phys. Ed.*, **28**, 655 (1990).

bb. Muellerlede, J. T., B. G. Risch, D. E. Rodrigues and G. L. Wilkes, *Polymer*, **34**, 789 (1993).

cc. Wiemers, N. and G. Wegner, *Makromol. Chem.*, **175**, 2743 (1974).

dd. Vasanthakumari, R. and A. J. Pennings, *Polymer*, **24**, 175 (1983).

ee. Fischer, E. W., H. J. Sterzel and G. Wegner, *Kolloid Z. Z. Polym.*, **251**, 980 (1973).

ff. Allcock, H. R. and R. A. Arcus, *Macromolecules*, **12**, 1130 (1979).

gg. Kajiyama, R. and W. J. Macknight, *Polym. J.*, **1**, 548 (1970).

hh. Hsiao, B. S., B. B. Sauer and A. Biswas, *J. Polym. Sci.: Polym. Phys. Ed.*, **32**, 737 (1994).

ii. Handa, Y. P., J. Roovers and F. Wang, *Macromolecules*, **27**, 551 (1994).

It is instructive to examine the T_m^0 values of polymers in terms of their structure and the thermodynamic quantities governing fusion. However, it is informative to first examine trends in monomeric systems. The melting temperatures of rigid monomeric molecules depend primarily on energetic interactions and the overall molecular shape.(13) As would be expected, flexible, monomeric molecules melt lower than comparable rigid molecules. A striking example of the influence of molecular flexibility in the melt is found in the melting temperatures of the polyphenyls. When the rings are linked para they are colinear, even though individual rings can rotate. The melting temperatures of these components rise rapidly with the number of rings in the molecule. However, when the rings are meta linked, so that rotation about the joining bonds causes large conformational changes, the melting points only rise slowly with the number of rings. For example, p-pentaphenyl melts at 395 °C; the corresponding meta compound melts at 112 °C. The seven ring para compound melts at 545 °C. In the meta series the compound with sixteen rings only melts at 321 °C. The flexibility in the melt makes an important contribution to the entropy of the melt and thus to the entropy of fusion. A similar influence of flexibility would be expected to be carried over to long chain polymers.

The Flory and Vrij analysis, based on the melting temperatures and enthalpies of fusion of the n-paraffins, led to the equilibrium melting temperature of linear polyethylene.(1) This value is listed in Table 6.1. The enthalpy of fusion of this polymer was obtained from several studies that used the diluent method. The theoretical equilibrium melting temperature has been confirmed by many studies involving extrapolation methods.(14–21) Based on the Flory–Vrij analysis, the melting temperature of a given type chain is expected to be the limiting value of the corresponding series of shorter chain oligomers. This expectation is fulfilled by the data sets that are available. As a corollary, the relation between the melting points of different polymers corresponds to that of their respective monomeric analogues. In the following, the relation between the melting temperature and chain constitution for different classes of polymers will be examined. The main purpose will be to seek general trends and principles.

A compilation of the melting temperatures of the homologous series of isotactic poly(1-alkenes) is given in Fig. 6.1.(22–24) Here the melting temperature is plotted against the number of carbon atoms in the side-group.[1] The results among the different studies are quite good. Since the melting temperature of isotactic poly(propylene) is uncertain, a value of 200 °C has been arbitrarily taken for this polymer.(25,26) The melting temperature of isotactic poly(propylene) is at least 50 °C greater than that of polyethylene. The data in Table 6.1 indicate that this

---

[1] When the polymer has more than one crystal structure the highest melting polymorph is plotted.

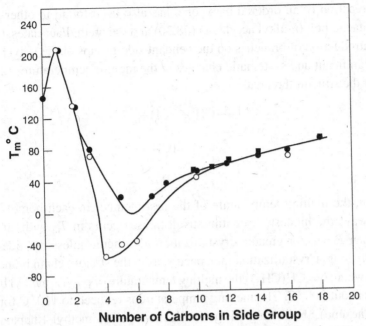

Fig. 6.1 Plot of observed melting temperature against number of carbon atoms in side-group for the isotactic poly(1-alkenes). (○ from Reding (22); ● from Turner-Jones (23); ■ from Trafara *et al.* (24))

increase in melting temperature is caused by both enthalpic and entropic factors. As the length of the side-group increases there is a rapid decrease in T_m. A minimum is reached when there are four to five carbons in the side-chain. Up to this point the backbone maintains the helical crystalline structure. The large decrease in T_m between poly(propylene) and poly(butene-1) is apparently due in part to the decrease in ΔH_u. There does not appear to be any influence of ΔS_u on this large change in T_m, although intuitively it might be expected that the flexibility of the side-group in the melt would have some effect. The minimum in the melting temperature corresponds to a change in the crystal structure. The ordering of the side-groups now plays a predominant role with a new ordered backbone structure.(23,24) The side-chains become fully extended and are packed parallel to one another on either side of the main chain axis in a paraffin-like manner. A steady increase in T_m then ensues as the number of carbons in the side-group increases. The melting temperatures are close to and parallel the melting temperatures of the corresponding *n*-alkanes.(23) For stereoregular polymers the participation of the chain backbone in the crystallization process has been demonstrated.(23,24) In stereo-irregular chains, such as atactic poly(1-octadecene), side-chain crystallization without backbone participation has also been shown.(27) However, a significant reduction in T_m relative to that of the stereoregular polymer is observed. Side-chain crystallization, without the participation of an ordered backbone, has also been found in other polymers such as atactic poly(*n*-alkyl acrylates) (28,30) and the methyl/acrylates.(29,31,32)

The introduction of branches on the pendant side-groups of the poly(1-alkenes) causes significant and systematic changes in the melting temperatures.(22,33–37) Consider the structural repeat

$$\left[\!-CH_2\!-\!CH\!-\right]_n$$
$$\underset{\displaystyle R}{\overset{\displaystyle |}{(CH_2)_y}}\qquad\qquad I$$

In general, the melting temperature of the first member in each homologous series is always the highest. A continuous decrease occurs in T_m with an increase in R. When $R = 3$, or greater, crystallization is difficult unless the side-chain is sufficiently long. Crystallization then occurs, as in the straight chain branches. For example, when $R = CH(CH_3)_2$ the melting temperature of poly(3-methyl butene-1) ($y = 0$) is about 300 °C. The melting temperature is reduced to 110 °C for poly(5-methyl-1 hexene).[2] Following this generalization, poly(6-methyl-1 heptene) is not crystalline. The melting temperature of poly(4-methyl-1 pentene 1), is 250 °C. From Table 6.1 we note that the ΔH_u values for these polymers are low relative to polyethylene and poly(propylene), while ΔS_u also decreases. The entropy change on fusion is indicated as the cause for the elevation of the melting temperature. The closer the branches (or the isopropyl group) are to the backbone the greater the steric crowding. This in turn influences both the crystal structure and the chain conformation in the melt. This steric effect is quite severe for poly(3-methyl-1 butene), as can be seen from models.

The type and placement of the branch also has a strong influence on T_m. The melting temperature of poly(3-methyl-1 pentene) is more than 100 °C greater than its isomer poly(4-methyl-1 pentene). Similarly, T_m of poly(4-methyl-1 hexene) is about 100 °C greater than that of poly(5-methyl-1 hexene). Replacing the methyl substituent by an ethyl group substantially increases T_m. For example, T_m of poly(3-ethyl-1 pentene) is greater than 425 °C as compared to 362 °C for poly (3-methyl-1 pentene). Similarly the melting temperature of poly(4-ethyl-1 pentene) is about 50 °C greater than poly(4-methyl-1 hexene). In all of these cases the influence of the substituent is tempered by the length of the side group.

When the substituent in structure I is $C(CH_3)_3$ a substantial increase in T_m results. The melting temperatures of both poly(4,4-dimethyl-1 pentene) and poly(4,4-

---

[2] The melting temperatures given in this section are taken from the sources cited and do not represent equilibrium melting temperatures. However, despite the uncertainty in the equilibrium melting temperatures, the discussion still reflects the change in melting temperature with chain structure. When polymorphism exists the highest melting crystalline structure is again used.

dimethyl-1 hexene) are reported to be greater than 380 °C. Steric effects, influencing the chain conformation, would be expected to be the major influence in elevating the melting temperatures.

Polymers with cyclic substituents show similar trends in melting temperatures. For example, when R is a phenyl group T_m varies from about 250 °C for polystyrene ($y=0$) to 160 °C for poly(4-phenyl-1 butene) ($y=2$). Poly(5-phenyl-1 hexene), $y=3$, has not been crystallized. Following the established pattern, T_m of poly(3-phenyl-1 butene) is about 200 °C greater than that of poly(4-phenyl-1 butene). When R is a cyclohexyl group in structure I, T_m decreases continuously from 372 °C for $y=0$, poly(vinyl cyclohexane), to 170 °C for poly(4-cyclohexyl-1 butene), $y=2$. When R is a cyclopentyl group, T_m of poly(vinyl cyclopentane) is 270 °C. The T_m's of poly(allyl cyclopentanes) and poly(allyl cyclohexanes) have similar melting temperatures in the range 225–230 °C. The melting temperature of poly(vinyl cyclohexane), 372 °C, seems to be abnormally high when compared to the next member of the series, poly(allyl cyclohexane) as well as to poly(styrene). The reason for those differences can be explained by the crystal–crystal transition that is observed in this polymer. This transition increases the entropy of the crystal and hence reduces the entropy of fusion. In summary, a wide range in the melting temperatures of hydrocarbon polymers can be achieved by altering the side-group structure. The major reason is the steric effect in influencing the chain conformation and the resultant change in the entropy of fusion.

Hydrocarbon polymers with aromatic rings in the backbone have high melting temperatures due to the extended chain conformations that approach those of liquid crystals. For example, poly(p-xylene) based on the repeat unit

$$-\!\!\!+CH_2-\!\!\!\left\langle\!\!\bigcirc\!\!\right\rangle\!\!-CH_2\!+\!\!\!_n \qquad \text{II}$$

displays several low temperature polymorphic transitions. Melting to the isotropic state takes place in the vicinity of 430 °C.(38–41a) This very high melting temperature is a dramatic example of the influence of chain rigidity. As the number of methylene groups increases the melting temperature is reduced. The melting temperature of poly(p-phenylene butylene) is reported to be in the range of 200–215 °C.(41b) The introduction of ring substituents can cause large variations in the melting temperature.(41a) The placement of two methyl groups in the meta positions of poly(p-xylene) reduces the melting temperature to about 355 °C.(39) The insertion of a chlorine atom into the ring reduces the melting temperature to 284 °C.(40) However, introducing ring substituents into meta poly(xylene) results in a polymer that melts at 135 °C. The high melting temperatures of the para substituted polymers are most likely caused by relatively low entropies of fusion, based on the ΔH_u values generally observed for hydrocarbons. The naphthalene analogue of II melts at 310 °C.(41a)

Vinyl polymers have high melting temperatures accompanied by low enthalpies of fusion. A case in point is isotactic poly(styrene) that has a relatively high melting temperature with a low heat of fusion. On a weight basis its heat of fusion is comparable to the much lower melting poly(cis-isoprene). Poly(acrylonitrile) and poly(vinyl alcohol) are examples of polar, high melting polymers whose heats of fusion are low when compared to those of hydrocarbon polymers. The corresponding low entropies of fusion are characteristic of these polymers. They are the basis for the high melting temperatures and indicate significant restraints in conformational freedom in transferring a structural unit from the crystalline to the liquid states. Crystallizable polymers prepared from esters of acrylic and methacrylic acid are also high melting with low heats of fusion.

It is of interest to compare the melting temperatures and thermodynamic parameters that govern fusion of isotactic and syndiotactic polymers that have the same chemical repeating unit. However, in order to make a meaningful comparison between a given pair, it is necessary that the polymers have the same level of structural irregularity and be crystallized under similar conditions. The reported melting temperatures for stereo-irregular poly(methyl methacrylates), each polymerized in its own way, illustrates this problem. In one case the melting temperature of the isotactic polymer is reported as 160 °C, while that of the syndiotactic polymer is given as greater than 200 °C.(42) In another report the melting temperature of the syndiotactic polymer is given as 190 °C while that of the isotactic one is 160 °C.(43) It is tempting to conclude that the melting temperature of the syndiotactic polymer is greater than the isotactic one. However, since these studies were done prior to the availability of NMR analysis the chain microstructures are not known. In an isotactic polymer, containing 94% triads, the melting temperatures range from 140 to 160 °C after isothermal crystallization from the melt.(44) Extrapolation of this data leads to an equilibrium melting temperature of 220 °C. Until comparable studies are carried out with the syndiotactic polymer, a rational comparison cannot be made between the poly(methyl methacrylates). The available data for poly(isopropyl acrylate) indicate that the melting temperature of the isotactic form is about 60 °C greater than the syndiotactic counterpart.(45)

According to the compilation given in Table 6.1 the melting temperature of syndiotactic poly(styrene) is much greater than that of the isotactic polymer. The situation for poly(propylene) is not as clear. As is indicated in the table, the reported equilibrium melting temperatures of isotactic poly(propylene) range from 185 to 220 °C. Possible reasons for these large differences have been given.(26) The equilibrium melting temperature of syndiotactic poly(propylene) is reported to lie between 140 and 170 °C. It is complicated by polymorphism and low levels of structural regularity.(46) However, a sample of relatively high stereoregularity, 96% pentads, gives an extrapolated equilibrium melting temperature of 172 °C. Extrapolation of melting point data of polymers with 92–76% pentads leads to an equilibrium melting temperature of 182 °C for completely syndiotactic poly(propylene).

The reported melting temperature of 120 °C for isotactic 1,2 poly(butadiene)(47) can be contrasted with 154 °C for the syndiotactic polymer.(48) It should be noted that these melting temperatures are for polymers, without any determination of the stereo structure. The equilibrium melting temperature of poly(4-methyl-1 pentene) is given as 250 °C in Table 6.1. The melting temperature of an as prepared syndiotactic polymer is reported to be 220 °C.(49) This fragmentary data suggests that the equilibrium melting temperatures of the two stereo isomers are comparable to one another. The equilibrium melting temperature of the polymorphs of isotactic poly(1-butene) range from 106 °C to 135 °C. In contrast the observed melting temperature of a syndiotactic sample of the polymer (93% pentads) is 50 °C.(50)

Poly(vinyl alcohol) is another example where relating the melting temperature and stereoisomerism is difficult.(51,52) High degrees of structural regularity have not been achieved with either the isotactic or syndiotactic polymer. The determinations of ΔH_u and ΔS_u suffer from the same problem.(51,52,52a) The level of stereo regularity, based on triads, is in the order of 75%. The available evidence indicates that the melting temperatures of the isotactic and syndiotactic polymers are comparable to one another.

The data summarized above does not allow for any conclusions to be reached with respect to the influence of stereoisomerism on the melting temperature. The available evidence is conflicting and does not allow for the assessment of any general trends. Clearly, melting temperatures for polymers having the same chain defect concentration and crystallized under comparable conditions are needed.

The melting temperatures of the diene type polymers follow the same general trends that are found in their monomeric analogues. It is well known that the cis isomers of hydrocarbon derivatives are lower melting than the corresponding trans ones. It is not surprising, therefore, that poly(1,4-trans-isoprene) is higher melting than the 1,4 cis polymer. Similarly poly(1,4-cis-butadiene) has a melting temperature of +1 °; that of the corresponding trans polymer is 148 °C. The isomers of poly(1,4 2,3-dimethyl butadiene) show a difference of about 70 °C between the isomers. This difference in melting temperatures is also observed between the cis and trans poly(pentenamers) and poly(octenamers).(53–55) The cis–trans effect is quite general. It is also found among poly(esters) and poly(urethanes) containing the CH=CH group in the main chain.(56) The melting temperature of the cis isomer of the poly(urethane) prepared from 1,4-cyclohexanediol and methylene bis(4-phenyl isocyanate) is substantially less than the corresponding trans isomer. However, when 1,4-cyclohexane dimethanol is used as the glycol, comparable melting temperatures are observed for the two isomers.(57)

When examining the thermodynamic parameters that govern the fusion of the poly(1,4-isoprenes) it can be noted that both the enthalpy and entropy of fusion of the higher melting polymorph of the trans form are greater than that of the all cis chain. In contrast, although the melting temperature of the high melting trans form of poly(butadiene) is about 150 °C greater than the cis structure, it is accompanied by a significant decrease in both the enthalpy and entropy of fusion. Based on the rotational isomeric state theory, there is no reason for any unusual chain conformations in the melt to be the source of the low entropy of fusion. Therefore, the origin of this low entropy of fusion must reside in the crystalline state. This possibility was pointed out by Natta and Corradini (58) who demonstrated that there is a certain element of disorder in the chain conformation within the crystal of the high temperature form of poly(1,4-trans-butadiene). This partial disorder of the chain in the crystalline state is thus the basis for the low entropy of fusion. We should also note from Table 6.1 that the entropy of fusion of poly(1,4-cis-polyisoprene) is substantially less than that of poly(1,4-cis-polybutadiene). It has been suggested that this difference can be attributed in part to a disorder in chain packing. However, in analyzing the entropy of fusion account must also be taken of the volume change on melting. (cf. seq.)

The melting temperatures of the trans poly(alkenamers) increase with the number of carbon atoms in the repeating unit for the same crystal structure. In the limit of an infinite size repeat the melting temperatures approach that of linear polyethylene.(59) Differences in melting temperatures are found between polymers having either an even or odd number of carbons in the repeating unit, reflecting differences in the crystal structure.(59,60) The thermodynamic quantities in Table 6.1 indicate that there is a steady increase in the enthalpy of fusion, on either a mole or weight basis, with an increase in the size of the repeating unit. Although the entropy of fusion per repeating unit also increases, the value per single bond remains effectively constant and is similar to that for polyethylene.

There are several groupings of polymers that are of particular interest. Polymers commonly considered to be elastomers at ambient temperature must have low glass temperatures and be either noncrystalline or, if crystallizable, have low melting temperatures. Some typical polymers in this category are poly(1,4-cis-isoprene), natural rubber, poly(isobutylene), poly(dimethyl siloxane), and poly(1,4-cis-butadiene). Their melting temperatures are 35 °C, 5 °C,(61) −38 °C and 0 °C respectively. Poly(1,4-cis-isoprene), poly(1,4-cis-polybutadiene) and poly(dimethyl siloxane) are all characterized by low values of ΔH_u. We can assume that poly(isobutylene) will follow the same pattern. Poly(dichlorophosphazene) is also an elastomer at room temperature. T_g for this polymer is $-65\,°C$. Although its melting point, $33\,°C$, is above room temperature it only crystallizes at temperatures well below ambient.(62) This behavior is reminiscent of poly(1,4-cis-isoprene) where equilibrium melting temperature is $35.5\,°C$ but it supercools quite easily. It is estimated that for the phosphazene polymer $\Delta S_u = 6.5$ e.u. mol^{-1}. Therefore, based on the low melting temperatures of these polymers, their classification as elastomers resides in the large entropic contribution to the fusion process.

In contrast to the elastomers, polymers with high glass temperatures and high melting temperatures have been termed engineering, or high performance, plastics. Included among this group of polymers, along with the respective melting temperatures, are poly(tetrafluoroethylene) $346\,°C$; poly(2,6-dimethyl 1,4-phenylene oxide) $275\,°C$; poly(2,6-dimethyoxy 1,4-phenylene oxide) $287\,°C$; the poly(carbonate), poly(4,4'-dioxydiphenyl 2,2-propane carbonate) $317\,°C$; poly(ethylene 2,6-naphthalene dicarboxylate) $337\,°C$ and poly(phenylene sulfide) $348\,°C$, to cite just a few examples. None of these high melting polymers have large values of ΔH_u. In fact, in examining the data in the tables one finds that associated with these polymers are some extremely low values of ΔH_u. For example, ΔH_u for poly(tetrafluroethylene) is about 1200 cal mol^{-1}, which is about the same as for poly(2,6-dimethyl 1,4-phenylene oxide). The ΔH_u value for poly(2,6-dimethoxy 1,4-phenylene oxide) is only 760 cal mol^{-1}. Poly(2,6-diphenyl 1,4-phenylene ether), with a melting temperature of $484\,°C$ has a ΔH_u of only about 3000 cal mol^{-1}. It is clear that the high melting temperatures cannot be attributed to enthalpic effects. Rather they are caused by low values of ΔS_u. Polymers with similar properties, whose fusion parameters are not available, include the crystalline poly(sulfones) $[(CH_2)-SO_2]$ and poly(phenylene sulfide). The melting temperatures of the alkyl type poly(sulfones) are in the range $220-271\,°C$. The smaller the number of methylene groups between the sulfones, the higher the melting temperature.(63) The melting temperature of poly(phenylene sulfide) is $285\,°C$.(64)

Poly(esters) have many different chemical repeating units and structures. These in turn are reflected in the observed and equilibrium melting temperatures. The aliphatic poly(esters) melt at temperatures lower than polyethylene. This finding is consistent with the melting of the corresponding monomeric system. Monomeric chain esters melt at lower temperatures than do the *n*-alkanes of the same chain length. This observation, and the fact that usually the greater the number of ester groups the lower the melting point, appear surprising.[3](13) A plot of the melting temperatures against the number of chain atoms in the structural repeating unit

---

[3] The two poly(esters) containing the largest concentration of ester groups in the chain are anomalous with respect to this generalization. Poly(ethylene succinate) melts at $108\,°C$ while poly(ethylene malonate) is a liquid at room temperature.(65,66)

of the aliphatic polyesters indicates that, except for the perturbation of the "odd–even" effect, there is an apparent approach to the melting temperature of polyethylene.(65,66) There is, however, about a 70 °C difference in melting temperatures between the highest recorded values of an aliphatic poly(ester) and of polyethylene. Thus, whether the trend observed represents a true asymptotic approach to the melting temperature of linear polyethylene still remains to be established. The fact that the aliphatic poly(esters) melt lower than polyethylene, and are lower the greater the number of ester groups, argues against the polarity of the chain having a major influence on the melting temperature. These results are contrary to what would be expected if the determining factor were an increase in intermolecular interactions due to the polar ester groups. However, it is the difference between interactions in the crystalline and liquid state as well as entropic effects, that are important. The thermodynamic parameters found in the tables do not give any direct clues as to the reason for this behavior. It has been suggested (65) that bonds in the region of the ester group are more flexible in the melt than the CH_2–CH_2 bond, and this is the reason for the observed melting behavior.

The data in the tables indicate that the melting temperatures of the aliphatic poly(esters) with an odd number of CH_2 units are lower than those containing an even number. The alternation of melting temperatures in this particular series is found throughout the organic chemistry of low molecular weight substances, including the *n*-alkanes. It is also observed in other homologous polymer series such as poly(amides), poly(urethanes), poly(ureas), poly(ethers) and aromatic poly(esters) containing an aliphatic chain portion.(66) An example of this effect is given in Fig. 6.2 for several different polymer series. Curve (a) is the plot for the series of aliphatic polyesters based on decamethylene glycol. The observed melting temperatures are plotted against the number of carbons in the dibasic acid. The well-known zigzag line is observed. As the number of carbon atoms increases the difference in melting temperatures between successive odd–even polymers decreases. There is indication that for a sufficiently high number of carbon atoms this effect will be lost.

The reason for the odd–even effect on the melting temperatures has usually been attributed to different positions of the ordered planar zigzag conformation resulting in different alignments of the carbonyl group.(65–68) It has been pointed out (70), as is evident from the table, that ΔH_u alternates with carbon number in an opposite manner to the melting temperature. Consequently enthalpy differences cannot be the cause of the alternation. The alternation must result from differences in the entropies of fusion.

The introduction of ring structures directly into the chain backbone results in substantially higher melting temperatures compared to the corresponding aliphatic polymers.(70) This behavior has already been described for hydrocarbon

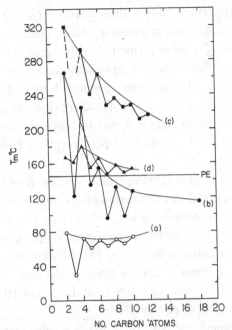

Fig. 6.2 Composite plot of different homologous series illustrating "odd–even" effect. Plot of melting temperature against number of carbon atoms. (a) Aliphatic poly(esters) based on decamethylene glycol; number of carbon atoms in dibasic acid. (b) Aromatic poly(esters) based on terephthalic acid; number of carbon atoms in diol. (c) Aliphatic poly(amides) based on hexamethylene diamine; number of carbon atoms in diacid. (d) Poly(urethanes).

polymers. Similar effects are found in other polymers such as the poly(amides), poly(anhydrides) and poly(urethanes). Although the aromatic poly(esters) have higher melting points than their aliphatic counterparts the ΔH_u values are comparable to one another. The significantly higher melting temperatures of the aromatic polymers must, therefore, result from a lower entropy of fusion per repeating unit as is indicated in the tables. The melting temperature of poly(ethylene naphthalene 2,6-dicarboxylate) is about 70 °C greater than that of poly(ethylene terephthalate), despite the fact that the enthalpy of fusion of the latter poly(ester) is slightly the higher of the two. Consequently, and not unexpectedly, the role of the naphthalene group is to reduce the entropy of fusion, probably by stiffening the chain in the melt.

The importance of the *p*-phenylene linkage is illustrated by the melting temperature of the poly(esters) based on terephthalic acid. As the length of the diol increases the melting temperature decreases.(71–74) For example, the melting temperatures of the polymers based on ethylene, trimethylene, tetramethyl, hexamethylene and decamethylene terephthalate gradually decrease from about 300 °C to 130 °C as the distance between the ester groups increase. This decrease in melting temperature is accompanied by an increase in the values of ΔH_u. The importance of ΔS_u in establishing the melting temperature is illustrated once again. The poly(alkylene 2,6-naphthalene dicarboxylates) follow a similar pattern. There is a progressive decrease in T_m with an increasing number of methylene groups.(74) In contrast, the melting temperature of poly(ethylene 1,4-diphenoxy-oxybutane, p,p'-dicarboxylate) remains high at 252 °C, although the spacing of the ester groups relative to poly(ethylene terephthalate) is doubled. The disposition of the p-phenylene linkage is the same and thus the melting temperature remains essentially unaltered. The melting temperature of poly(2-methyl-1,3-propane glycol terephthalate) is reduced to the range of 73–82 °C.(75)

The odd–even effect in melting temperatures is also found in the aliphatic portion of the aromatic poly(esters). An example is also illustrated in Fig. 6.2 for poly(esters) based on terephthalic acid.(71–73) Here, the difference in the odd–even melting temperatures is much greater than for the corresponding aliphatic polymers indicating the entropic influence. The differences in the melting temperatures decrease as the number of carbon atoms in the diol increases. The melting temperatures appear to be approaching those of the completely aliphatic polymers at a temperature below that of polyethylene.

In analogy to monomers, the position of the ring substituent dramatically influences the location of T_m. For example, T_m of poly(tetramethylene isophthalate) is about 80 °C less than the corresponding polyester based on terephthalic acid. This decrease in T_m is accompanied by a 50% increase in ΔH_u. The difference in melting points is due to an increase in ΔS_u and reflects the difference in shape and conformational versatility of the two isomers. This difference in melting temperature is also found in other aromatic poly(esters).(73) Poly(decamethylene terephthalate) melts 100 °C higher than poly(decamethylene isophthalate); the difference between poly(ethylene terephthalate) and poly(ethylene isophthalate) is about 130 °C.

The introduction of branches, such as methyl groups, into the hydrocarbon portion of either aliphatic or aromatic poly(esters) reduces the melting temperature.(66,74,76) In some cases the polymers have not as yet been crystallized. Poly(esters) based on terephthalic acid are particularly interesting in this regard. The introduction of a single methyl group into either poly(ethylene terephthalate) or poly(triethylene terephthalate) results in a reduction of the melting points by about 135–140 °C. However, the introduction of two symmetrically arranged methyl groups into poly(triethylene terephthalate) results in a much smaller reduction. Another example of disruption of the aliphatic portion of the chain is the introduction of ether linkages. When ether linkages are inserted into the aliphatic portion of the chain the melting temperature is reduced considerably. For initially low melting aliphatic poly(esters) crystallinity can be eliminated. This effect is also observed with aromatic type poly(esters). Polymers with the repeat $-[OCC_6H_4-COOCH_2CH_2OCH_2CH_2-O]-$ cannot be crystallized from the melt. However, when admixed with diluent crystallization ensues.(77) Extrapolations from polymer–diluent mixtures gives a melting temperature somewhat less than 180 °C. This represents a significant decrease from that of the pure aliphatic sequence.

The analysis of the melting temperature of poly(esters) has led to the recognition of certain behavior patterns. These are similar, in many respects, to those involved in the melting of similarly constituted monomeric systems. There is the "odd–even" effect well known in *n*-alkanes and other low molecular weight substances. The introduction of ring structures into the chain backbone results in a significant elevation of the melting temperature. Isomerization of the ring alters the melting temperature. As the subsequent discussion will indicate these patterns are also found in other type polymers. There is, however, a unique feature to polymer crystallization. Quantitative analysis of the thermodynamic parameters that govern fusion has indicated the importance of ΔS_u in determining the melting temperature. Many examples have been given in the foregoing discussion of the poly(esters). The influence of ΔS_u on the melting temperatures of polymers is a feature not usually found in monomeric systems.(65)

The poly(amides) also offer a rich diversity of repeating units and melting temperatures with which to assess the principles that have evolved so far. The aliphatic poly(amides) melt much higher than the corresponding poly(esters) and polyethylene. A wide range of melting temperatures are observed that depend on the number and arrangement of the carbon atoms in the repeating unit. The odd–even effect is again observed as is illustrated in Fig. 6.2 for poly(amides) based on hexamethylene diamine. It is apparent that the same pattern that was observed with the poly(esters) is being followed. A set of melting temperature data is compiled in Fig. 6.3 for poly(amides) prepared from diamines and diacids that each contain an even number of carbon atoms.(66,78–81) The melting temperatures are plotted against the total number of carbon atoms in the repeating unit. The polymers in this and similar figures are designated by the conventional notation. The first digit represents the number of carbons in the diamine, the second that in the diacid. It is evident that a smooth curve results.[4]

Within experimental error, the melting temperatures given in Fig. 6.3 only depend on the total number of carbons in the repeat, irrespective of their distribution between the diamine and diacid. For example, the melting temperatures of the polymers 10,10; 8,12; and 6,14 are very close to one another. Other combinations give very similar results. The melting temperatures of 12,10 and 10,12 are also close to each other.(82) The melting temperatures decrease very rapidly with increasing size of

---

[4] The general trends found here and in Fig. 6.4 are clear. However, most of the melting temperatures have been determined by rapid heating and are thus subject to error. The T_m value for a given polymer does not always agree among different reports. When discrepancies exist, the highest T_m value was selected.

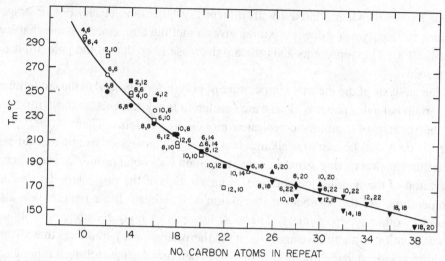

Fig. 6.3 Plot of observed melting temperatures against the total number of carbon atoms in the structural repeating unit of poly(amides). Both the diamines and diacids contain an even number of carbon atoms.

the repeating unit, reflecting the dilution of the amide group. For example, the melting temperature of the 4,6 poly(amide) is 295 °C; in the other extreme, T_m of the 18,20 poly(amide) is reduced to 146 °C. The melting temperatures of 6,24 and 6,34 follow the pattern established by the plot in Fig. 6.3.(82a) Thus, in contrast to the poly(esters) there is an increase in melting temperature as the proportion of polar groups in the chain increases. This observation has been explained by the hydrogen bonding capacity of the amide groups. As the number of carbon atoms in the repeat increases, the melting temperatures can be thought of as approaching that of linear polyethylene.

The melting temperatures of the poly(amides) where either one or both of the diamines or diacids contains an odd number of carbons, are plotted in Fig. 6.4. The dashed line is a replot of the curve in Fig. 6.3. The data are based on the series pimelic, azelaic and undecanoic acids. The odd–even effect is made clear in this figure. For a given diacid, the polymers with an even number of carbons in the diamine melt higher than the corresponding odd-numbered component. All the hydrogen bonds are still formed in the crystals of the poly(amides) that have an odd number of CH_2 groups. The crystal structures are different, however, from the polymers that have an even number. In this series, polymers with a total even number of carbon atoms melt significantly lower than the corresponding polymer with even carbon numbers in both species. However, with just one exception, the polymers with an even number of carbon atoms in the diamine have melting temperatures close to the dashed line in the figure. Thus, if T_m of these poly(amides) were plotted in Fig. 6.3 (all even number diamines) they would fall on the solid line in the figure.

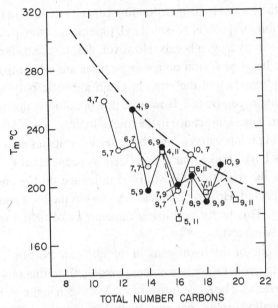

Fig. 6.4 Plot of melting temperatures against the total number of carbons of poly(amides) where either one or both of the diamine and diacid contain an odd number of carbons.

A close examination of the data in Fig. 6.4 indicates that for most, but not all, of the polymers, the melting temperatures depend only on the total number of carbons. As examples, compare the pairs 6,9 and 4,11; 8,9, 6,11 and 10,7 and 9,9 and 7,11. Deviations from this generalization are observed at the lower carbon numbers. The poly(amides) 2,5 and 4,3 have been reported to have the same melting temperatures. However, 1,6 decomposes before melting, at a much higher temperature.(83) This difference in melting temperatures has been attributed to differences in crystal structure. The dilution effect is also observed when either the diamine or diacid has an odd number of carbons. For the series n,3 the melting temperature decreases from 276 °C to 217 °C as n values increase from 4 to 12.(84) A smaller, but similar, effect has been observed in 1,n poly(amides).(85)

Poly(amides) based on ω-amino acids also display a wide range in melting temperatures. The T_m values generally decrease with the number of carbon atoms in the amino acids.(80,86) For example, the melting temperature of poly(glycine) (nylon-2) is reported to be 365 °C, while that for poly(laurolactam) (nylon-12) is about 185 °C. There is also a distinct alternating effect in T_m. The melting temperatures of the poly(amides) formed from ω-amino acids containing an odd number of carbon atoms are greater than those formed from even numbered ones.

A comparison of the data in Table 6.1 for the aliphatic poly(amides) with the corresponding poly(esters) indicates that, despite the much higher melting temperatures of the former, their heats of fusion are substantially less. This is true, notwithstanding the greater hydrogen bonding capacity of most of the poly(amides). With the exception of poly(N,N'-sebacoyl piperazine) the poly(amides) listed in the tables all form hydrogen bonds. However, despite the differences in melting temperatures, the heats of fusion on a weight basis are all comparable to one another. It is quite obvious that the straight chain aliphatic poly(amides) can, and in fact do, form hydrogen bonds. However, this bonding is not manifested in the enthalpy of fusion. Any significance in the role of hydrogen bond formation in causing the high melting temperatures must be reflected in its influence on the entropy of fusion. Studies have shown that a significant concentration of hydrogen bonds are maintained in the melt.(87–90) Minimal influence on the enthalpy of fusion would then be expected. However, a local ordering in the melt can be attributed to hydrogen bonding. This factor will lower the entropy of fusion relative to that of corresponding poly(esters).

Substitution of one of the hydrogens in the aliphatic portion of a poly(amide) reduces the melting temperature in a manner analogous to that observed in the corresponding poly(esters). The substitution of the hydrogen in the —NHCO— group by methyl, or other alkyl groups, has an even more pronounced effect.(67,91,92) For example, the polymer of N-methyl undecanoic acid ($N{=}CH_3{-}(CH_2)_{10}CO$) melts at 60 °C, while the unsubstituted polymer melts at 188 °C. Although poly(N-methyl laurolactam) is partially crystalline, the homopolymers poly(N-ethyl laurolactam) and poly(N-benzyl laurolactam) are completely amorphous.(93) The reductions in T_m have been ascribed to the decreased capacity for hydrogen bond formation when the H atom is replaced by an alkyl group.

The melting temperatures of poly(amides) of the type

$$-CO-CH-CO-NH-(CH_2)_6-N-H$$
$$\mathord{\mid}$$
$$(CH_2)_n$$
$$\mathord{\mid}$$
$$CH_3$$

III

have been determined for odd values of n that range from 3 to 17.(94) Initially, there is a large decrease in T_m with increasing n. A minimum in T_m is observed for $n = 7$–9 followed by a continuous increase with n. This result is similar to what was already noted for the poly(1-alkenes). It is a general phenomenon for polymers with side-chain branches.

Poly(thioamides) have much lower melting points than the corresponding poly(amides).(95) For example, the difference in melting points is about 100 °C for polymers that have 12 CH_2 groups in the repeating unit. This difference becomes progressively smaller as the number of methylene groups is increased. This difference is only about 25 °C for 22 CH_2 groups.

The introduction of ring structures into the chain backbone results in a substantial increase in the melting temperature. Although this is a general phenomenon for all polymers it manifests itself in extremely high melting temperatures for poly(amides).(67,96,97) This behavior is similar to that of monomeric diamides.(98) The introduction of an aromatic group, or groups, can be accomplished in several different ways. The poly(amides) can be synthesized from either aliphatic diamines and diaromatic acids, or phenylenediamines and aliphatic diacids (or diacid chlorides). Polymers in the latter category are geometrically asymmetric and form liquid crystals. For the poly(amides) prepared with either aromatic acids or aromatic diamines melting temperatures increase in going from the ortho to meta to para derivatives.(97) The melting temperature of poly(amides) derived from phenylene diamine and sebacic acid increases from 135 °C to 332 °C with the change in isomers. For the corresponding polymers based on phthalamides, T_m increases from 115 °C to 316 °C. The melting temperatures also increase as the chain length of the aliphatic sequence decreases. Suggestions of the odd–even effect are also found.

The introduction of cyclohexane rings into the chain results in a number of structural differences. In addition to position isomers and geometric cis and trans isomers, chair, boat and twist conformations are also possible. There are large numbers of possibilities for copolymeric behavior from the point of view of crystallization. The crystallization behavior of this group follows the principles elucidated in Chapter 5.(99,100) The discussion here will be restricted to homopolymers and consideration of the influence of position and geometric isomers on the melting temperature.

Studies of poly(amides) prepared from cyclohexane bis(alkyl amines) have focused on the role of 1,4, 1,3, and 1,2 linkages and the type and length of the dicarboxylic acid. Comparison of properties with polymers based on *p*-xylene-α, α′ diamine and *p*-phenylene bis(ethyl amine) have also been reported.(99,101) For each series the expected alternation in T_m with the number of carbon atoms in the dicarboxylic acid is observed.(99,101) In a series based on 1,4-cyclohexane bis(methyl amine), 1,4-CBMA, and diacids of varying length the melting temperature of the trans isomer is about 100 °C greater than the corresponding cis polymer.(99) This difference has been attributed to the greater symmetry and rigidity of the trans form. The melting temperatures of poly(amides) based on trans 1,4-cyclohexane bis(ethyl amine), 1,4-CBEA, are about 16 °C higher than those of the corresponding trans CBEA polymers. The nature of the linkage has a profound effect on the relation of T_m to the geometric isomers. The cis isomers of the 1,3 disubstituted cyclohexanes are generally more thermodynamically stable than the trans ones. In turn, this fact is reflected in the melting temperatures. As examples, the melting temperature of the trans isomer of the poly(amide) formed with adipic acid and 1,4-cyclohexane bis(methyl amine) (1,4-CBMA-6) is 340 °C, while that of the corresponding cis isomer is reduced to 240 °C. However, for the 1,3-CBMA-6 the situation is reversed. The cis isomer melts at 240 °C, while the trans conformer melts at 150–160 °C. In the 1,4-CBMA-6 polymer the two bonds connecting the substituent to the ring are parallel to one another in the trans conformation. The situation is reversed for 1,3-CBMA-6 and the two bonds are parallel in the cis structure. When the size of the alkyl group is increased to ethyl as in 1,4-CBEA-6, T_m of trans is 340 °C and that of cis 120 °C.

Poly(amides) based on 1,2-cyclohexyl rings should have larger melting point depressions relative to the 1,4 structure than is shown by the 1,3 ring. For 1,2 rings the more stable isomer is the trans configuration. Therefore, T_m of the trans polymer should be higher than the corresponding cis structure. Although the pure isomers having the 1,2 linkage have not been synthesized, only short extrapolations are needed on available data for geometric copolymers based on 1,2-CBEA-6 for a reliable estimate of T_m of the pure species.(100) These expectations are in fact found. For the pure polymers T_m of trans is 140–150 °C while that of the cis is 120 °C.

A comparison of melting temperatures can also be made between 1,4-CBMA, poly(amides) based on p-xylene-α,α' diamine and on p-phenylene bis(ethyl amine) with a series of dicarboxylic acids. The melting points of the poly(amides) based on p-xylene-α,α' diamine and on p-phenylene bis(ethyl amine) differ by only about 20–30 °C for polymers whose dicarboxylic acids have the same number of methylene units. The melting temperatures of the trans CBMA's are about 10–15 °C higher than the p-xylene-α,α' diamine poly(amides).

The poly(urethanes) are similar to the poly(amides) in that hydrogen bond formation is possible between the functional group —(O—CO—NH)—. However, the O—CH$_2$ bond typical of polyesters is also present. It is, therefore, not surprising that the melting temperatures of the poly(urethanes) fall between those of the poly(esters) and poly(amides). The generalizations established for the poly(esters) and poly(amides) are also applicable to the melting temperatures of the poly(urethanes). Melting temperatures progressively decrease with increasing length of the hydrocarbon portion.(65,102,103) The poly(urethanes) with the lowest concentration of O—CO—NH groups have melting temperatures somewhat less than that of polyethylene.(65) The odd–even effect is also observed, as is illustrated in Fig. 6.2. Introduction of ring structures, either aromatic or cycloaliphatic, substantially raises the melting temperature.(66,102–104)

Specific details that illustrate these generalizations are found in Tables 6.1 and 6.3. The data are based on two series of polyurethanes, hexamethylene diisocyanate and 4,4'-diphenyl methane diisocyanate each with a variety of diols. The number of methylene sequences in the hydrocarbon portion of these polymers varies from 2 to 10.(102,103) The enthalpies of fusion per mole are qualitatively similar to those of the aliphatic polyesters. The ring containing polymers in the two series have higher melting temperatures. When compared with poly(amides) based on hexamethylene diamine, the values of ΔH_u are within 10% of one another. The entropies of fusion per rotatable bond are in the range 1.5–1.9 e.u. so that there is no major difference between the two classes of polymers. A similar situation was observed for the aliphatic and aromatic poly(esters) and indicates that the simple comparison is not adequate when a ring is present in the backbone.

An interesting comparison in the fusion parameters can be made between the aliphatic poly(urethanes) and the corresponding poly(amides) based on either 1,6 hexamethylene diamine, or hexamethylene diisocyanate each with similar dicarboxylic acids.(102,103) The difference is in the replacement of the $\mathrm{O} - \overset{\overset{\text{O}}{\|}}{\underset{\underset{\text{N}}{|}}{\text{C}}} - \overset{\text{H}}{|}$ group by $- \overset{\overset{\text{O}}{\|}}{\underset{\underset{\text{N}}{|}}{\text{C}}} - \overset{\text{H}}{|} -$. The melting points of the poly(amides) are about 70–90 °C greater than the corresponding polyurethanes. Where quantitative comparisons can be made, the ΔH_u values are virtually identical. The melting point differences can again be attributed to the entropy of fusion. The slightly enhanced entropy of fusion of the poly(urethanes) is due to the oxygen in the chain backbone.

The thermodynamic properties governing fusion for a series of poly(esteramides) are given in Table 6.1.(105) As would be expected from the structure of the repeating unit, all of these polymers are high melting. It can also be anticipated that the melting temperatures decrease monotonically with an increase in the number of methylene units in the repeat. Concomitantly, ΔH_u increases by a factor of about one and a half. There is then a corresponding increase in ΔS_u.

The relation between the melting temperature and structure of the poly(anhydrides) essentially follows the same principles found with other polymer types. The melting temperatures of the aliphatic poly(anhydrides) are lower than that of polyethylene.(106) For example, poly(sebacic anhydride) melts at 83 °C. The melting temperatures of rings containing poly(anhydrides) can be varied over wide limits by changing the ring type, the phenylene linkage and the length and type of the intervening groups.(93,107,108) Polymers containing aromatic rings in the chain backbone melt substantially higher than the aliphatic poly(anhydrides). Poly(terephthalate anhydride), for example, melts at 410 °C, while poly(isophthalate anhydride) melts at 259 °C. Changing the nature of the intervening group can cause a large variations in T_m. The introduction of methylene units into the chain can reduce melting temperatures to as low as 92 °C. Replacing the aromatic rings by heterocyclic ones, such as furan, tetrahydrofuran and thiophene causes a substantial decrease in T_m. For example, with furan in the chain T_m is reduced from 410 °C to 67 °C.(107)

The crystalline, aromatic poly(imides) are high melting polymers. Even within this class of polymers significant changes in T_m occur with different repeating units.

Some examples are given in the following. A class of poly(imides) was synthesized by the reaction of aromatic dianhydrides with diamines and contained carbonyl and ether connecting groups between the aromatic rings.(109) One of the repeating units has the structural formula

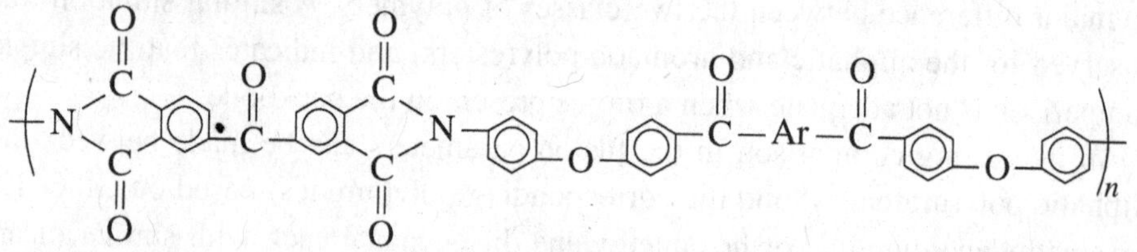

IV

The melting temperature of this polymer is 350 °C when Ar, an aromatic ring, is inserted in the meta position; T_m increases to 427 °C when the para isomer is formed. A dependence of T_m on the ring isomer is usually observed.(109,110) Melting temperatures can also be altered by introducing flexible species into the chain backbone. A case in point is found in poly(imides) of the type described in Table 6.3.(111) In these polymers, ethylene glycol sequences have been incorporated into the chain. Polymers with $n = 1, 2$ and 3 have been studied. The thermodynamic parameters governing fusion are listed in the table. As the number of glycol groups increases, T_m^0 decreases as would be expected. Concomitantly, ΔS_u increases. Again, it is the increase in ΔS_u that governs the decrease in T_m^0. Similar behavior has been found in other polymers. The increase in ΔS_u can be attributed to the enhanced molecular flexibility of the chain backbone in the melt as the concentration of —O—CH_2—CH_2—O— groups is increased.

The melting temperature of the LARC-CPI poly(imide) listed in the table is about 50 °C higher than the polymers just described, with $n = 1$.(112) Both ΔH_u and ΔS_u are greater for the higher LARC-CPI polymer. Similar influences of structure are found in alkyl–aromatic poly(imides)based on

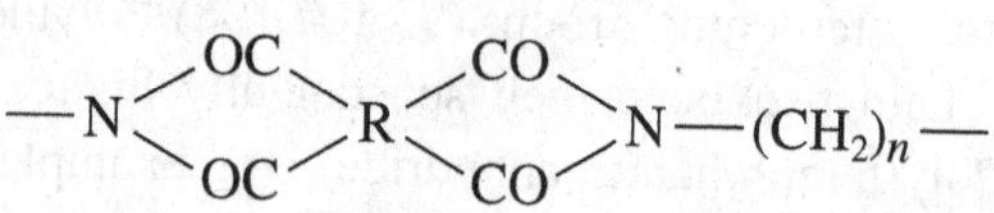

V

as the repeating unit.(113) The R group is one of the following:

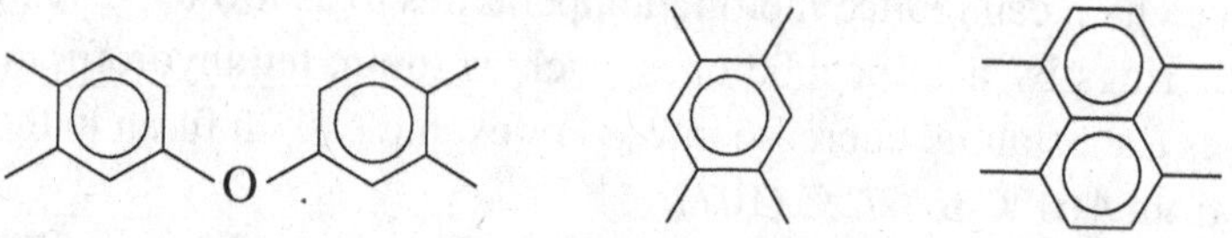

The melting temperatures of the polymers based on 3,3', 4,4'-diphenyl oxide tetracarboxylic acid decrease from 352 °C for $n = 4$ to 202 °C for $n = 8$; a substantial change in melting temperature. Polymers with $n = 9$ to 12 have low levels of crystallinity and are difficult to crystallize, indicating a further reduction in T_m.

The aliphatic poly(lactones) are low melting and have relatively low values of ΔH_u (see Table 6.1). However, by inserting a dimethyl substituent, as in poly(α,α'-dimethyl propiolactone), T_m is raised to 269 °C. This increase in melting temperature, above that of poly(β-propiolactone), is accompanied by about a 50% increase in ΔH_u and a modest increase in ΔS_u. One might have expected that the steric hindrance due to the two methyl groups on the same carbon would restrict the chain conformation in the melt and be the cause of the melting point elevation. The melting temperature of poly(α,α'-diethyl propiolactone) is 258 °C with, however, slightly higher values of ΔH_u and ΔS_u than the dimethyl polymer.

In examining the fusion properties of the poly(ethers), it is found that the equilibrium melting temperature of poly(methylene oxide), 206 °C, is much greater than that of polyethylene and the other members of the poly(oxyalkane) series. There is a difference of more than 100 °C in melting temperatures between poly(methylene oxide) and poly(ethylene oxide). Figure 6.5 illustrates that as the length of the aliphatic group increases a minimum in T_m is reached in the vicinity of the three and four methylenes. The melting temperatures then increase with the number of carbons in the alkane group. However, the values of the hexa, octa and decamethylene polymers are very close to one another. The fusion parameters characteristic of poly(methylene oxide) do not directly explain its high melting temperature. Its heat of fusion is less than that of poly(ethylene oxide) on a mole basis, but greater on a weight basis. The entropy of fusion per repeating unit is less than that of

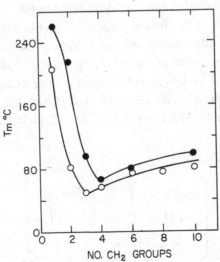

Fig. 6.5 Plot of melting temperatures, T_m, against number of CH_2 groups in poly(ethers) O and poly(thioethers) ●.

poly(ethylene oxide). The situation is reversed, however, when compared on the basis of rotatable single bonds. Bearing on this matter is the fact that the characteristic ratio of poly(methylene oxide) is about two and half times greater than that of poly(ethylene oxide). The larger spatial expanse of the poly(methylene oxide) chain should affect the entropy of fusion, and appears to play an important role in causing its high melting temperature. The enthalpies of fusion of poly(ethylene oxide) and poly(trimethylene oxide) are very close to one another. The decrease in melting point that occurs here can be attributed to an enhanced entropy of fusion. As the series progresses to poly(tetramethylene oxide) and poly(hexamethylene oxide) a large increase occurs in ΔH_u that is accompanied by a modest increase in the melting temperature. What might be expected to be a substantial increase in the melting temperature, based on ΔH_u values, is tempered by the concomitant increase in ΔS_u.

The melting temperatures, as well as the entropies and enthalpies of fusion are virtually identical for poly(ethylene oxide) and isotactic poly(propylene oxide). Bulkier side groups do, however, have a significant influence on the melting temperature. For example, the melting temperature of poly(t-butyl ethylene oxide) is 149 °C;(114) that of poly(isobutylene oxide) is 177 °C;(115) and that of poly-(styrene oxide) is 162 °C.(114) A wide range in melting temperatures can be achieved by varying the nature of the side group.(116) It can be surmised that changes in ΔS_u are involved.

Not surprisingly, the introduction of aromatic rings into the chain backbone causes a substantial increase in T_m (see Tables 6.1 and 6.3). The melting temperature of poly(1,4-phenylene ether) is about 290 °C. The melting temperature increases to 484 °C when two phenyl groups are substituted in the 2,6 position of the ring. Indirect measurements of the fusion parameters (Table 6.3) indicate that a large increase in ΔH_u, but only a slight increase in ΔS_u accompany this large increase in T_m. However, when direct fusion parameters are available, as with poly(2,6-dimethyl 1,4-phenylene oxide), T_m is virtually the same as that of the unsubstituted polyether. In this case there is a decrease in both ΔH_u and ΔS_u. In contrast, poly(2,6-dimethoxy 1,4-phenylene oxide), with a T_m^0 of 287 °C, is characterized by low values of ΔH_u and ΔS_u.

Polyethers of the type

$$\left[-CH_2 - \underset{\underset{R_2}{|}}{\overset{\overset{R_1}{|}}{C}} - CH_2 - O - \right]_n \qquad\qquad VI$$

show some interesting melting features.(117) There is only a small difference in melting temperatures between poly(3,3-dimethyl oxetane) and the corresponding

3,3-diethyl polymer. The enthalpies and entropies of fusion per mole increase slightly for this pair. When R_1 and R_2 differ, as in poly(3-ethyl 3-methyl oxetane) the melting temperature is reduced relative to the symmetrically substituted polymer, and is accompanied by a large decrease in ΔH_u. This difference may reflect packing difficulties in the crystalline state. The decrease in ΔH_u is sufficient to offset the decrease in ΔS_u. When the substituents are more complex, such as in poly(3,3-bis-ethoxy methyl oxetane) and poly(3,3-bis-azido methyl oxetane), the melting temperatures are essentially the same for both polymers. However, there is an almost sixfold change in ΔH_u. This pair of polymers represents a classical example of compensation between ΔH_u and ΔS_u to yield a constant set of melting temperatures.

A similar pattern in melting temperatures is found in the series poly(3,3-bis-hydroxy methyl oxetane), poly(3-methyl-3-hydroxy methyl oxetane) and poly(3-ethyl-3 hydroxy methyl oxetane).(118) The melting temperature of the first polymer is 303 °C; that of the methyl substituted one is reduced to 152 °C, while that of the ethyl substituted polymer increases slightly to 163 °C.

The melting temperatures of the poly(formals), $-[CH_2-O-(CH_2)_x-O-]_n$, initially decrease with n, reach a minimum value for poly(1,3-dioxepane), $n = 3$, and then increase.(119) The thermodynamic parameters that are given in Table 6.1, for poly(1,3-dioxolane) and poly(1,2-dioxocane) indicate a two-fold decrease in ΔH_u as the melting temperature decreases from 93 °C to 46 °C. At the same time there is a comparable decrease in ΔS_u, giving another example of compensation in the two quantities.

The melting temperatures of the poly(alkyl thioethers) are greater than the corresponding oxygen containing ones.(13) These differences are also illustrated in Fig. 6.5, where the melting temperatures for both types of poly(ethers) are compared.(120) As with the oxyethers, the thioethers show a minimum in T_m in the vicinity of three to four methylene units. For the lower carbon number polymers there is a large difference, about 140 °C, in the melting temperatures between the two types. However, the melting temperatures of the poly(ethers) containing four or more methylene units are close to one another. These observations indicate the importance of the proportions of ether linkage per repeating unit in determining the melting temperature. The substantial difference in ΔH_u between poly(ethylene oxide) and poly(ethylene sulfide) can account for the increased melting temperature of the latter. Steric effects are also manifest by the fact that the T_m for poly(isobutylene sulfide) is 201.5 °C (121) as compared to 67 °C for poly(butylene sulfide).(122) The melting temperature of poly(3,3-diethyl thiotane) is comparable to that of the corresponding oxygen poly(ether).(122) However, in an anomalous situation, T_m of poly(3,3-dimethyl thiotane) is 13 °C as compared to 76 °C for the corresponding oxygen containing polymer.(123).

The poly(aryl ether ketones) are engineering, or high performance, polymers that have high glass and melting temperatures. For example, poly(aryl ether ether ketone) has an observed melting temperature of about 340 °C and an extrapolated equilibrium melting temperature of 395 °C.(124) The observed melting temperatures that have been reported for a given polymer in this group vary. This is probably due in part to differences in molecular weight and crystallization conditions. For analysis purposes the highest melting temperature reported will be used. The inherent assumption has been that only para linkages are involved. Thus, from a crystallization point of view we are dealing with a homopolymer. However, this important structural feature has not been established in many cases.(125) The incorporation of isomers into the chain can reduce the melting temperature and can affect the melting temperature of the all para polymer.(126,126a)

The phenyl rings and the ether and ketone moeites can be arranged in different ways within a repeating unit. The poly(arylates) of interest have similar crystal structures. They crystallize in an orthorhombic lattice with the chains aligned parallel to the *c*-axis of the unit cell. There is a definite change in the dimensions of the unit cell with the composition of the repeating unit.(127) As the concentration of ketone groups increases the length of the *c*-axis also increases. Changes also take place in the basal plane. The length of the *a*-axis decreases, while the length of the *b*-axis increases. The net result is an increase in the volume of the unit cell with increasing ketone content. The x-ray diffraction patterns suggest that the melting temperatures be examined in terms of the composition of the repeating unit. This suggestion is carried out in the plots shown in Fig. 6.6. Here, both the observed and extrapolated equilibrium melting temperatures are plotted against the mole fraction of ketone linkages.(127) For either set of melting temperatures, a straight line

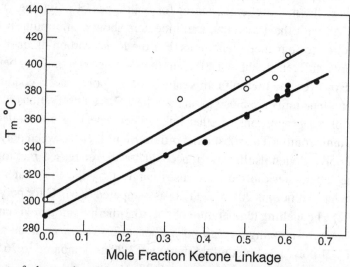

Fig. 6.6 Plot of observed melting temperatures ● and extrapolated equilibrium melting temperatures ○ of poly(aryl ether ketones) as a function of mole fraction of ketone linkage.

results that extrapolates to the respective melting temperatures of poly(phenylene ether), i.e. the poly(aryl ether) devoid of ketone groups. As orginally noted by Harris and Robeson (128) the melting temperature depends only on the concentration of ketone groups, irrespective of how they are arranged in the repeating sequence. For example, the melting temperatures of poly(ether ketone) and poly(ether ether ketone ketone) are the same. Similarly poly(ether ketone ether ketone ketone) and poly(ether ketone ketone ether ketone) have the same melting temperatures.

It has been surmised that the increase in melting temperature with ketone content is due to a decrease in the entropy of fusion because of an enhanced rigidity of the chain.(129) It is further argued that the isomorphism in the lattice of the diphenyl ether and diphenyl ketone groups results in an essentially constant enthalpy of fusion over the composition range of a repeating unit. Although these arguments may be plausible, the determination of the full set of enthalpies and entropies of fusion is necessary to explain the variation in melting temperatures.

The halocarbon polymers are good examples of chain molecules that have relatively high melting temperatures and low heats of fusion. The high melting temperatures are thus caused by low entropies of fusion and are exemplified by the fluorine substituted polymers. Taking polyethylene as a reference, the melting temperature smoothly increases from 145.5 °C to 346 °C with the systematic replacement of a hydrogen atom by a fluorine until poly(tetrafluoroethylene) is reached. This change represents a rather substantial increase in the melting temperature. At the same time the enthalpy of fusion decreases from about 1960 cal mol^{-1} of ethylene to 1219 cal mol^{-1} of the completely fluorinated repeat. Obviously the entropy of fusion is the major factor in determining the melting temperature. Replacing one of the fluorine atoms by a chlorine, in the repeat of poly(tetrafluoroethylene), reduces T_m by more than 100 °C. The enthalpy of fusion remains unaltered, so again there is a significant decrease in ΔS_u.

The fusion parameters that govern poly(vinyl chloride) are difficult to analyze. The polymers that have been studied, in this regard, have a low level of stereoregularity, small crystallites and low levels of crystallinity.(130,131) An extrapolated value of 385 °C has been given to the melting temperature of the completely stereoregular syndiotactic poly(vinyl chloride).(132) This value is to be compared with 157 °C that is given for poly(vinyl fluoride) in Table 6.1. In contrast, the reported melting temperature of 195 °C for poly(vinylidene chloride)(133) appears low relative to that listed for poly(vinylidene fluoride) and the extrapolated value for poly(vinyl chloride). Since values of ΔH_u and ΔS_u are not available for most of these polymers it is difficult to give a rational interpretation of their melting temperatures. Stereoregularity, and regio defects, can make an important difference when comparing the melting temperatures of this class of polymers.

Cellulose, and its derivatives, are polar polymers that are characterized by very high melting temperatures. However, these polymers possess extremely low heats of fusion. For example, cellulose trinitrate and cellulose 2,4 nitrate have the highest reported melting temperatures of any polymer. Their enthalpies of fusion are only in the range of about 700 to 1400 cal mol^{-1}. The entropy of fusion, reflecting the chain structure, governs the high T_m values. The characteristic low values of ΔH_u of cellulose derivatives and their low levels of crystallinity have made it difficult to experimentally determine latent enthalpy effects. Therefore, it is not surprising that at one time these polymers were not considered to be crystalline.(134–137) The melting temperature of cellulose tributyrate is about 120 °C less than that of cellulose triacetate.(138) There is a further decrease in melting temperature between cellulose tributyrate and cellulose tricaprylate from 207 °C to 116 °C, with essentially no change in ΔH_u. This change is clearly due to concomitant increase in ΔS_u. This increase in ΔS_u can be attributed to an additional gain in the configurational entropy of the ester side groups. As the length of the side group of the tri-substituted derivatives increases, the melting point initially decreases, reaches a minimum value, and then increases.(139) This behavior, due to the side-chain crystallization, is similar to what has already been noted for other homologous series. Qualitatively similar results have been reported with hydroxy propyl cellulose with appended alkyl substituted branches.(140)

The symmetrically substituted polysiloxanes, $\left(- \underset{\underset{R}{|}}{\overset{\overset{R}{|}}{Si}} - O - \right)_n$, display some different and interesting fusion characteristics.(141) Although poly(dimethyl siloxane) displays classical behavior, the polymers containing longer alkyl substituents are different. Poly(dimethyl siloxane) has a glass temperature in the vicinity of −120 °C and, depending on molecular weight and crystallization conditions, an observed melting temperature of about −40 °C.(142–147) It is not surprising that with these characteristics poly(dimethyl siloxane) behaves as an elastomer over a wide temperature range. The thermodynamic parameters governing the fusion of this polymer are given in Table 6.1.[5] The relatively small ΔH_u value can be attributed in part to the cohesive energy density that is characteristic of the silicone polymers. Thus the low enthalpy of fusion and somewhat higher entropy of fusion lead to the low melting temperature.

When the two methyl groups are replaced by longer alkyls such as ethyl, propyl and butyl the fusion process becomes more complex. An example of the thermogram, obtained by differential scanning calorimetry, is given in Fig. 6.7 for the fusion of poly(dipropyl siloxane).(148) Two of the endotherms, labeled A and B

---

[5] The value of ΔH_u was originally reported in terms of calories per mole of chain atoms and was misinterpreted. It was subsequently amended so as to be expressed in terms of repeating units.(145)

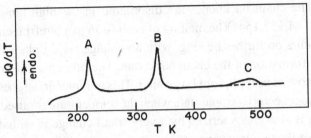

Fig. 6.7 Thermogram from differential scanning calorimetry of poly(dipropyl siloxane). (From Godovsky *et al.*(148))

respectively, are sharp. The other, a broader endotherm, is labeled C. The identification of these transitions can be made by wide-angle x-ray diffraction. The transition at A (218 K) corresponds to a crystal–crystal transition of two well-defined forms. The number of Bragg reflections is reduced after this transition indicating the development of a more symmetrical crystalline system. On heating to point B the crystal transforms to an inhomogeneous phase. All the Bragg crystalline reflections disappear, with the exception of the 100 reflection. The intensity and half-width of this reflection remain unchanged. Thus, only order in the lateral direction of the chain persists through this transition. This phase is heterogeneous, birefringent and hence anisotropic. It is said to be mesomorphic and displays liquid-crystal behavior. In the temperature region between B and C, the chain backbone is partially disordered throughout its length. Conformational disorder along the chain, following a crystal–crystal transformation, is not uncommon.(149,150) Such behavior was evidenced quite early in poly(1,4-trans-butadiene)(151) as well as in other polymers. The poly(dipropyl siloxane) sample becomes homogeneous and isotropic after passing through point C at about 480 K. At this point the Bragg reflection disappears and a typical amorphous x-ray diffraction pattern results. The temperature interval for the mesophase is about 100 K in this case. The enthalpy change characteristic of the anisotropic–isotropic transition is about 10% of that at the melting transition at B, and is typical of all polymers of this type.[6](144)

The fusion of poly(diethyl siloxane) follows a path similar to that of poly(dipropyl siloxane). There are, however, some differences in the specifics.(146,152–155) The main difference is that poly(diethyl siloxane) exists in two different low temperature modifications each having similar chain structures but differing in packing characteristics. Each of the polymorphs undergoes a crystal–crystal transition at similar temperatures, 260 K and 214 K respectively (corresponding to point A

[6] The conventional melting point–composition relation, Eq. (3.9), cannot be used for the higher dialkyl substituted poly(siloxanes) because the equilibrium is between a crystalline phase and an anisotropic melt. Thus, the Flory–Huggins relation is not applicable.

in Fig. 6.7). The chain backbones are disordered in the high temperature crystalline forms.(144,152,154) The melting of each form to a birefringent, anisotropic
mesophase occurs on further heating, with no change taking place in the average
disordered conformation of the chain backbone. The properties of this mesophase
are similar to those of poly(dipropyl siloxane). This polymer undergoes a broad transition to a homogeneous isotropic state when the temperature is raised above 50 °C.
This transition is also characterized by a very small change in enthalpy.(141,153)
Thus, except for the two low temperature crystalline modifications the diethyl and
dipropyl substituted siloxanes behave similarly.

Poly(di-*n*-butyl siloxane) shows one sharp endothermic peak at 217 K representing the transformation from a crystalline to birefringent anisotropic state.(156)
Calorimetric measurements of this polymer indicate that the loss of anisotropy,
and the formation of the isotropic state, take place over a very broad temperature
range of approximately 190 °C. Only one crystalline form is reported by differential
calorimetry for this polymer.[7] Other symmetrically substituted poly(dialkyl siloxanes), such as pentyl, hexyl and decyl side-groups, show two endothermic peaks.
The systems eventually become isotropic as they melt into the liquid or amorphous state.(157) The temperatures of the two endothermic peaks increase with
the number of carbon atoms in the substituent. The temperature for isotropy also
increases. There is, however, evidence from birefingence measurements that the
temperature for isotropy levels off, although this temperature has not been reported
for poly(dodecyl siloxane). Poly(methyl alkyl siloxane) with long alkyl groups
(8–20 carbon number) show side-chain crystallization in a standard manner.(158)
An extrapolation of the available data for the symmetrically substituted poly(dialkyl
siloxanes) indicates that for poly(dimethyl siloxanes) the transition temperature
from the anisotropic phase to the isotropic one would be considerably below its
melting temperature.(141) In a formal sense this is the reason why the anisotropic
phase is not observed with this polymer.

The poly(diphenyl siloxane) chain is a rigid molecule because of the interaction of the phenyl substituents. Only a single melting temperature in the range
of 247 °C to 260 °C has been reported for the homopolymer.(141,159,160) Other
diaryl siloxane and aryl methyl siloxane polymers also have relatively high melting
temperatures.(161)

As has been indicated, the type of anisotropic mesomorphic phase that is observed in the poly(dialky siloxanes) is also observed in other polymers. In particular
this type of phase structure is generally observed in poly(dialkyl silylenes)(162)
and poly(dialkyl oxy phosphazenes).(141) The classification of this type of phase

---

[7] There are conflicting reports as to how many crystalline endothermic peaks are observed for poly(di-*n*-butyl
siloxane).(151,152)

structures and their relation to conventional liquid-crystal phases has been discussed in detail.(163).

The fusion of the poly(siloxanes) presents some unusual problems. The isolated chains are relatively flexible as manifested by their low glass and melting temperatures. Calculations of the characteristic ratios differ only slightly with the nature of the side-groups and confirm the flexibility.(164) Yet, with no traditional mesogens in their chain structure, these flexible chains are able to form thermodynamically stable anisotropic phases in the solid state.(165) Polarizing microscopy and small-angle neutron scattering have indicated that in the anisotropic state the chains are in a highly extended rodlike conformation.(141,166–168) A minimum molecular weight needs to be exceeded in order for the anisotropic phase to be formed.(160,167,168) The common feature of the group of polymers that display this behavior is their inorganic backbone and predominantly organic side-groups. In a condensed system one can expect that the interactions between these two different moieties will favor a very strong preference for one another. The numbers of contacts between like groups will be enhanced by elongated rodlike molecules rather than random coils. It can then be postulated that as the melt is cooled a temperature is reached where the enthalpic interactions are such that the decrease in entropy that occurs when a random coil spontaneously transforms to a rodlike extended structure is overcome. The interaction between like groups can be optimized with only lateral order between the chains. Upon further lowering of the temperature the full three-dimensional order will develop. We thus have the specific situation where a collection of flexible chain molecules can exhibit behavior that is similar to that of liquid crystals.

A different type of silicone polymer containing a ring in the chain, poly(tetra methyl-*p*-silphenylene siloxane), has a melting temperature of 160 °C, due in part to the aromatic ring in the backbone. There is, however, no indication that this polymer forms an anisotropic phase and the fusion parameters are normal ones.(169)

The backbones of the poly(silylenes) consist entirely of silicon atoms, to which are appended different substituents. The structural formula of the polymers of interest can be represented as

$$\left[\begin{array}{c} R_1 \\ | \\ -Si- \\ | \\ R_2 \end{array}\right]_n \qquad\qquad \text{VII}$$

We limit consideration to the case where R_1 and R_2 are alkyl groups. Hydrodynamic and thermodynamic studies in dilute solution have shown that isolated chains of these polymers are substantially more extended than those of carbon backbone polymers with similar degrees of polymerization. Their characteristic ratios are relatively high.(170,171) Because of the catenated silicon chain these polymers display interesting electronic, photochemical and spectroscopic properties.(162,172) The dialkane substituted poly(silylenes) are characterized by a first-order phase transition from a low temperature crystalline form to an anisotropic, birefringent mesomorphic form.(173–175) This transition strongly influences properties. The formation of a mesomorphic phase is typical of a chain with an inorganic backbone and organic side groups. Essentially, a conventional three-dimensional well-ordered phase is transformed to one with intermolecular disorder but with a high degree of intramolecular organization. The wide-angle x-ray patterns are similar to those found with the poly(dialkyl siloxanes).(173) It has been shown that poly(di-*n*-alkylsilanes) that contain up to 14 carbon atoms show this type of transition to a liquid-crystalline type phase.(174) Poly(dimethyl silylene) appears to be an exception, in that the chain conformation is maintained. However, the packing is transformed to hexagonal while the three-dimensional order is maintained.(176)

A plot of the temperature of transition from the crystal to anisotropic phase is given in Fig. 6.8 for a series of poly(dialkyl silanes).(174) The alkyl side-chains range from butyl to tetradecyl. Initially, as the length of the alkyl group increases there is a significant drop in the transition temperature. However, beyond eight carbons there is an increase in the transition temperatures with only a small change for the higher carbon numbers. This pattern is similar to that observed for other polymers that display side chain crystallization. On heating through the anisotropic

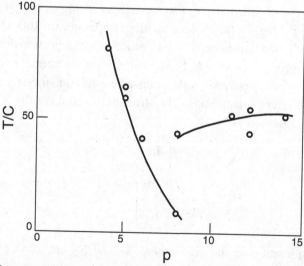

Fig. 6.8 Crystal to mesophase transition temperatures plotted as a function of the number of carbon atoms, *p*, in the alkyl group in poly(dialkyl silanes). (From Weber *et al.* (174))

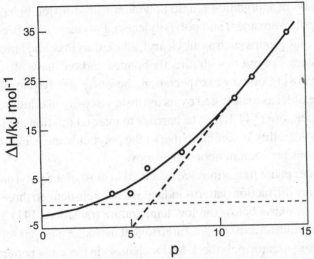

Fig. 6.9 The enthalpy change at the crystal to mesophase transition as a function of the number of carbon atoms in the alkyl group in poly(dialkyl siloxanes). (From Weber *et al.* (174))

phase, birefringence measurements indicate that the mesomorphic phase is transformed to an isotropic liquid. The temperatures for this transition to isotropy depend on the length of the side-groups. This temperature is 245 °C for poly(dipentyl silylene); 260 °C for poly(dihexyl silylene); and 210 °C for poly(ditetradecyl silylene).(175)

The enthalpy change involved in the transition to the mesomorphic phase is plotted, in Fig. 6.9, against the number of carbon atoms in the alkyl group.(174) The enthalpy change for the polymer with the butyl side-group is $2\,\text{kJ}\,\text{mol}^{-1}$. It increases very rapidly with increasing length of the alkyl group, reaching a value of $35\,\text{kJ}\,\text{mol}^{-1}$ for the polymer with the largest alkyl group. It is clear that the ordered structure of the alkyl groups plays a major role in the transformation to the mesomorphic phase.

Asymmetric substituted poly(silylenes), such as $R_1 = $ butyl and $R_2 = $ hexyl, show similar behavior.(177) In many cases the anisotropic mesophase is present at room temperature so that the typical first-order phase transition occurs on cooling below ambient temperature.(178)

The linear poly(phosphazenes), having the structure formula,

$$\left[\begin{array}{c} R_1 \\ | \\ P=N \\ | \\ R_2 \end{array} \right]_n$$

VIII

are another class of inorganic-organic polymers whose fusion behavior is similar to that of the poly(siloxanes) and poly(silylenes). Polymers have been synthesized with a variety of R groups such as alkyl, aryl, alkoxy, aryloxy and amino groups and amino acid esters. Polymers with directly bonded carbons have also been synthesized.(141,180,181) Contrary to expectation, the isolated poly(phosphazene) chain is relatively flexible, as manifested by its intrinsic viscosity in dilute solution and its low glass temperature.(144,181) The barriers to internal rotation are relatively low. The backbone flexibility is similar to that of the poly(siloxanes) and is governed by steric interactions between nonbonded groups.

Two first-order phase transitions are observed in most of the poly(phosphazenes). Wide-angle x-ray diffraction patterns indicate that a well-defined three-dimensional ordered structure exists below the low temperature transition.(141) The transition involves melting into an anisotropic, birefringent mesomorphic phase, as was found in other inorganic-organic polymers. Major changes in the x-ray pattern accompany this transition. There is a rich diffraction pattern at low temperatures that is typical of crystalline polymers. However, after the low temperature transformation to the anisotropic state the crystalline reflections disappear. They are replaced by a single sharp reflection at small angles and a diffuse halo at the larger ones. In the isotropic state, only the amorphous halo characteristic of the liquid state is observed.(182) These changes are similar to those that have been described for the poly(siloxanes) and poly(silylenes). The nature of the transitions is illustrated in Fig. 6.10 by

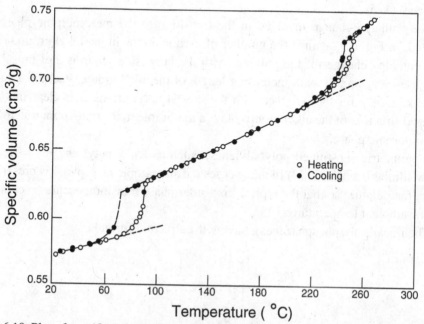

Fig. 6.10 Plot of specific volume against temperature for poly(bis trifluoroethoxy phosphazene). (From Masuko *et al.* (183))

Table 6.4. *Selected examples of transition temperatures and
enthalpy changes in the poly(phosphazenes)(185–189)*

Group	T_m	ΔH (cal g^{-1})	T_i (°C)
F			−68, −40
Cl			
Br			270
CF_3CH_2O	92, 83	8.6	240, 240
C_6H_4O	160	10.0	390
OP-FC_6H_4O	125	4.0	
m-FC_6H_4O	122	8.5	
p-FC_6H_4O	169	10.8	345
m-Cl C_6H_3O	90	5.8	370
p-Cl C_6H_4O	169, 165	6.6	365, 405
m-$CH_3 C_6H_4O$	90	8.3	348
p-$CH_3 C_6H_4O$	152	3.7	340
p-$CH_3 O C_6H_4O$	106	2.2	
p-$CH_3CH_2C_6H_4O$	43	1.1	
p-$C_2H_5(CH_3)CH C_6H_4O$	103	0.2	
p-$(CH_3)_3C C_6H_4O$	237		345
p-$C_6H_5CH_2C_6H_4O$	109	10.4	
p-$C_6H_5C_6H_4O$	160		398
3,4-$(CH_3)_2 C_6H_3O$	96	4.6	325
3,5-$(CH_3)_2C_6H_3O$	67	1.2	320
3-$CH_3ClC_6H_3O$	123	5.2	
5-$CH_3C_6H_4O$	142		
$N(CH_3)_2C_6H_4O$	203		

dilatometric studies of poly(bis trifluoroethoxy phosphazene).(183) The existence
of two first-order transitions in this polymer is quite clear. The onset and termination
of each of the transitions appear quite distinct by this experimental technique.

Table 6.4 is a compilation of the transition temperatures for selected poly-
(phosphazenes).(185–189) Here T_m is the transition temperature from the ordered
to the anisotropic phase; T_i is the transition temperature from the anisotropic to
the isotropic phase and ΔH is the enthalpy change between the isotropic and the
anisotropic phase. The exact value of these parameters will depend on the specific
crystallization conditions.(186) However, the general trends shown in the table are
still important. The two transitions are separated by more than 100 °C, depending
on the nature of the substituent. Their values depend specifically on the nature of
the substituent on the phosphorus atom. In many cases the transition between the
anisotropic and isotropic states is close to the decomposition temperature. The melting temperature, T_m, is sensitive to small changes in the composition and structure
of the substituent. The enthalpy change at T_m is small. The change at the high temperature transition is only about one-tenth that of the low temperature one.(186) An

estimate has been made of the fusion parameters that govern the low temperature transition in poly(dichlorophosphazene) (Table 6.3). ΔH_u has been estimated to be 2000 cal mol^{-1} corresponding to a value of ΔS_u of 6.5 e.u. mol^{-1}. This entropy change is consistent with the elastomeric properties of this polymer.

A series of poly(amino acid ester phosphazenes) have also been studied.[188] Many of these polymers show a single first-order transition. Heating just a few degrees above this transition temperature leads to decomposition. Hence, the possible existence of an anisotropic phase could not be established for this polymer.

It is well established that isotopic substitution alters the melting temperatures of monomeric substances. Polymers behave in a similar manner. This phenomenon is particularly evident when deuterium is substituted for hydrogen. The difference in melting temperatures between the two species, and the resulting phase diagram, are especially important in interpreting small-angle neutron scattering[190] and certain aspects of vibrational spectroscopy.[191,192] When interpreting experimental data type a crucial issue to be resolved is whether the two species are uniformly dispersed, or if there are concentration fluctuations. It is not appropriate at this point to discuss, or interpret, scattering and vibrational spectroscopic results. However, the melting temperatures play a pivotal role in analyzing these types of data and thus fit into our present discussion. The basic reason is that the phase diagram of a binary mixture of hydrogenated and deuterated components depends on the melting temperatures of the pure species, as does the crystallization kinetics from the melt. Both of these factors play important roles in establishing the homogeneity of the crystalline phase. It also should be noted that phase separation in the melt, or in solution, can occur between two isotopically different species.[193] For these reasons we examine the difference, if any, between hydrogenated and deuterated polymers.

The melting temperatures of hydrogenated and deuterated n-alkanes, and of linear polyethylene, have been studied in detail. A summary of the results is given in Fig. 6.11.[194] Here, the melting point difference between the hydrogenated and deuterated species is plotted against the melting temperature of the hydrogenated component, which increases with carbon number. The numbers of carbons in the n-alkanes are listed on the right side of the figure. The data in this plot obey a linear relation. The 5–6 °C difference in linear polyethylene is consistent with the results for the lower molecular weight analogues.[195,196] The molecular packing in a hydrogenated n-alkane and polyethylene crystal are essentially the same. The main difference is that the a and b dimensions of the unit cell of the deuterated species are slightly smaller.[194,197] The melting temperatures of hydrogenated and deuterated poly(butadiene) have also been reported.[198] These polymers are random ethylene copolymers with ethyl branches. Since the addition of either the hydrogen or deuterated atoms was made with the same precursor, poly(butadiene),

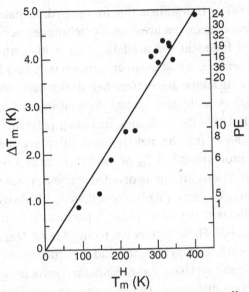

Fig. 6.11 Plot of isotope melting point difference, ΔT_m against T_m^H for some n-alkanes and linear polyethylene. The number of carbon atoms in the n-alkanes is indicated on the right side of the figure. (From Dorset, Strauss and Snyder (194))

the microstructure of both copolymers is the same. Hence, a rational comparison of melting temperatures can be made. In this case T_m of the hydrogenated species is about 2 °C greater than the deuterated polymer. This reduction, relative to the linear homopolymers, can be expected since the isotopic substitution is limited to 40% in the deuterated copolymer. The melting temperature of hydrogenated poly(ethylene oxide) depends on molecular weight and crystallization conditions. It is 2 °C to 5 °C higher than the corresponding deuterated polymer.(199)

In comparing melting temperatures of polymers that have stereo and or regio defects it is important that the level of structural regularity be the same in order for a meaningful comparison to be made. The melting temperatures of isotactic poly(hydrogenated propylene) and a companion deuterated polymer have been reported to be the same.(200) The level of isotacticity was in the 97–99% range. Possible differences in chain structure can result in melting temperature differences of the magnitude of interest here. Hydrogenated isotactic poly(styrene) has a 5.5 °C greater T_m than the deuterated counterpart, under the tacit assumption that the stereoregularity is the same for both polymers.

Deuterated poly(vinylidene fluoride) is reported to have a 6–8 °C greater melting temperature than the hydrogenated polymer.(201) It is difficult to prepare this polymer in a pure regiospecific form. The polymers cited are not regiospecific and each has a different content of reverse monomer addition. The reversion for the deuterated polymer is less (2.8%) than that for the hydrogenated one (3.5%). This difference is sufficient to cause an inversion in melting temperatures.(202)

The general trend found in the available data, with perhaps isotactic poly-(propylene) as an exception, is that the hydrogenated polymers have small, but significantly higher melting temperatures than their deuterated counterparts. A similar behavior is found in low molecular species. Several reasons have been proposed for this difference.(194) For the n-alkanes, including polyethylene, it is shown by corresponding state theory that the melting point differences between the isotopic species are directly proportional to T_m of the hydrogenated polymers. This is in agreement with observations. In a more detailed analysis, it has been theorized that for nonpolar polymers, T_m (deuterated) is less than T_m (hydrogenated) because of the same isotopic effects that lead to phase separation in liquid mixtures of the same two polymers.(203) These factors are reduced bond lengths and molar volume. In addition, the reduced bond lengths lead to reduced bond polarizability and thus molecular polarizability. Using these results, and invoking corresponding state theory, the melting point differences can be approximated. They are in reasonable agreement with experimental results.

When certain crystalline cyclic monomers are subject to high energy ionizing radiation in the solid state, they can be polymerized to high molecular weight crystalline polymers. A unique feature of polymers prepared in this manner is that they can be studied in their native crystalline state without being rendered molten and then crystallized. Thus, problems involving crystallization mechanism from the melt, and the attendant morphological features, are avoided. Such polymers have the outward appearance of large single crystals. It is not surprising, therefore, that polymers prepared in this manner have much higher melting temperatures and crystallinity levels than the corresponding polymers synthesized in the conventional manner and subsequently crystallized from the melt.(204–206)

The solid-state polymerization of trioxame to poly(methylene oxide) is a classical example of this procedure. The observed melting temperature of poly(methylene oxide) prepared by the solid-state polymerization is 198 °C. It is comparable to the extrapolated equilibrium melting temperature of 206 °C attributed to the conventionally synthesized and crystallized polymer.(204) It is significantly higher than the usual, directly measured value. Similarly, the melting temperature determined for poly(β-propiolactone), 122 °C, is much greater than the value of 84 °C that is directly observed for the melt crystallized polymer. Polymers prepared from 3,3-bis-chloromethyloxycyclobutane and diketene follow a similar pattern.(204)

The melting–crystallization of naturally occurring macromolecules of biological interest can also be analyzed within the framework that has been developed here. A good example is found in the behavior of the fibrous protein collagen. Present in major proportions of the repeating units in the collagen molecule are the amino acid glycine, and the imino acids proline and hydroxyproline. Although the concentration of amino and imino acid residues varies both in vertebrate and invertebrate species the glycine content remains essentially constant. It comprises about a third of the total amount. Hence, the chain repeat can be represented as the triplet (Gly–X–Y) where X and Y are residues other than glycine. Depending on the specific collagen, the sum of the proline and hydroxyproline residues ranges from 150 to 300 units per thousand.

Flory and Garrett have analyzed the fusion of a particular collagen, rat tail tendon, in detail.(207) The appropriate thermodynamic quantities involved in fusion are given in Table 6.1. These quantities, characteristic of this naturally occurring polymer, are similar to those characteristic of synthetic polymers. Both the heat and entropy of fusion appear to be normal. The heat of fusion, on a weight basis, is similar to that of the synthetic poly(amides). Any enhanced stability endowed to the crystalline state of the polymer by virtue of hydrogen bond formation is not in evidence, unless this contribution is much smaller than believed.

In order to analyze the dependence of the melting temperature on composition for the wide variety of collagens that exist, the specific triplets that participate in the ordered structure, as well as their sequence distribution, need to be specified. In addition, the role played by proline, or hydroxyproline, in positions X or Y, in the ordered state is important, as are the changes in melt structure with composition. In a formal manner the collagens exemplify a complex problem in copolymer melting. Despite the compositional, and presumably sequence variations, that occur among the collagens certain correlations exist between the melting temperature (determined at fixed total polymer concentration) and the imino acid content. However, it should be noted that exceptions have been found to all correlations that have been proposed, reflecting in part the complexity of the problem. Gustavson noted that the melting temperatures of the collagens increased with increasing concentration of hydroxyproline.(208) The increased stability was attributed to hydrogen bond formation involving the hydroxyl groups of hydroxyproline. Thus, the increase in T_m is presumed to be due to an increase in ΔH_u. Possible changes in ΔS_u were ignored. However, it has been subsequently pointed out that a correlation also exists between melting temperature and the total concentration of imino acid residues. Garrett has suggested that the reason for the increased melting temperature may be a decrease in ΔS_u which accompanies the increase in total proline and hydroxyproline content.(209) The increasing concentration of pyrrolidine rings in the chain backbone will suppress the conformational freedom of the chains in the molten state. A lower entropy of fusion would result even if the crystalline phase was unaffected. Consequently the melting temperature would increase. In principle, increased stability can be obtained, even in a fibrous protein, by suitably altering the conformational structure of the melt. The development of thermodynamically more stable structures does not necessarily require an increased concentration of hydrogen bonds in the crystalline state. A good correlation in melting temperatures has also been obtained with hydroxyproline in the Y position in the triplet. The melting temperature increases with hydroxyproline content. Theoretical calculations have shown that hydroxyproline in the Y position will enhance the stability of the triple stranded ordered collagen structure.(210) The enhanced stability, however, does not appear to come from hydrogen bonding involving water molecules but rather from the conformational properties of the hydroxyproline residues.(211)

The collagens as a class represent a classical example of the natural selection of species. The melting temperature of a particular collagen, as manifested in shrinkage temperatures (see Chapter 8) correlates with the environmental temperature of the particular collagen.(208,212,213) For example, cold-water fish collagens are low in pyrrolidine content and have low melting temperatures, in the range 10–20 °C. The melting temperature and pyrrolidine content of the collagens progressively increase with increase in environmental temperature. Collagens found in the cells of animals have the highest melting temperatures.

6.3 Entropy of fusion

The basis for the relation between the melting temperature and polymer structure as embodied in the quantities ΔH_u and ΔS_u can now be examined. The entropies and enthalpies of fusion in both the liquid and crystalline states need to be considered. Attempts to correlate the melting temperatures of polymers with intermolecular interactions, based on cohesive energy densities, have been unsuccessful.(13,65,214,215) In this kind of analysis attention is focused solely on ΔH_u. It is evident from the discussion of the thermodynamic parameters that there is no obvious, or simple, relation between T_m^0 and ΔH_u. Many polymers with low values of ΔH_u are high melting. Conversely many low melting polymers have relatively high values of ΔH_u. There are several homologous series where T_m^0 increases while ΔH_u decreases. From the survey of melting temperatures and thermodynamic quantities it is evident that the entropy of fusion is the major, but not necessarily the sole, factor in establishing the value of the melting temperature. A causal relation can be developed between ΔS_u and T_m^0. This relation is particularly striking for very high melting polymers where low values of ΔS_u are invariably observed.

The crystalline state is one of high three-dimensional order. Thus a low entropy is usually assigned to a repeating unit in this state. There are, however, exceptions to this generalization. As was pointed out earlier, there are classes of polymers that are conformationally disordered to some degree before the transformation to the completely isotropic, liquid state. Examples were found in the poly(siloxanes), poly(silylenes) and poly(phosphazenes) among others. The departure from structural regularity in these chains will increase the entropy of the crystalline state. With other factors being equal, this will result in decrease in the value of ΔS_u.

Poly(trans-1,4-butadiene) is polymorphic and undergoes a crystal–crystal phase transition prior to complete melting.(151) The low temperature crystalline form has a conventional, highly ordered crystalline structure. There is a regular repeat of the internal rotational angles between the carbon atoms adjacent to the double bonds. In contrast, in the high temperature crystalline form the degree of three-dimensional order is reduced due to disordering along the chain backbone. This disordering has been attributed to the random distribution of the rotational angles along the chain. Table 6.1 indicates that the structural difference between the two polymorphs is manifested in the thermodynamic parameters governing the fusion to the isotropic state. The higher melting polymorph has lower values of ΔH_u and ΔS_u relative to the lower melting form. Thus, the decrease in ΔS_u results from the higher entropy in the crystalline state prior to melting, because of the disorder.

Certain poly(amides) undergo a polymorphic transition from triclinic to hexagonal form at elevated temperatures.(216,217) Hexagonal packing allows for a greater amount of rotational freedom about the chain axis and thus an increase of the entropy in the crystalline state. The suggestion has been made that this phenomenon accounts in part for the higher melting temperatures of the aliphatic poly(amides) as compared with the corresponding poly(esters).(218) The poly(esters) do not exist in hexagonal form. The examples cited represent just a few cases where ΔS_u is reduced as a consequence of partial disorder in the crystalline state. It cannot be tacitly assumed that the crystalline state of a polymer necessarily represents one of perfect three-dimensional order.

The entropy in the liquid state, where the polymer molecules assume a multitudinous number of conformations, must also be taken into account. The conformations assumed depend on the specific nature of the chain repeating unit and their mutual interactions. The potentials that hinder the rotation of one chain unit relative to another are governed by steric repulsions and the interactions between neighboring chain substituents. The entropy of the liquid state depends on the conformation and relative extension of the individual polymer molecules. Depending on the polymer, conformations can vary from random or statistical coils to elongated rodlike molecules. The entropy of fusion reflects, in part, the conformational properties of the chain in the molten state. A large variation in the entropy of fusion among different classes of polymers can be expected based on their known conformational differences.

The conformational properties of a chain are reflected in dilute solution properties, and in particular its characteristic ratio, C_∞ which was defined in Chapter 1. The characteristic ratio can be determined experimentally by straightforward methods, or calculated theoretically using rotational isomeric state theory.(219) The characteristic ratios of a large number of polymers have been determined and it is instructive to ascertain what relation, if any, exists between the melting temperatures and characteristic ratios. A plot of the melting temperature against the characteristic ratio is given in Fig. 6.12 for a variety of polymers. Although some trends may be discerned in this plot, there is clearly no correlation between the two quantities. Most of the high melting polymers have high values of C_∞. Calculated values of C_∞ for p phenylene poly(amides) and poly(esters) are the order of 200.(220) These polymers are obviously very highly extended and show liquid-crystal behavior. A definite pattern is also found for the poly(alkane oxides),(221) (nos. 17–22 in the figure). However, in the main no pattern has emerged. For example, in the vicinity of C_∞ of 6–7 the melting temperatures of many polymers range from 270 °C to −40 °C. In another example, the characteristic ratios of poly(ethylene sulfide) (no. 26) and poly(ethylene oxide) (no. 18) are 4.5 and 6.2 respectively. Yet the melting temperature of the former is about 130 °C greater than the latter. This difference has been attributed to intermolecular interactions (222) and is supported by the differences in ΔH_u between the two polymers. Poly(acrylonitrile) and isotactic poly(styrene), (nos. 29 and 28) display similar behavior. Both have very similar characteristic ratios but melting temperatures that differ by 130 °C. An examination of the data for the poly(dienes), (nos. 10–13), shows that C_∞ for the cis isomers is less than that of the corresponding trans one. Yet T_m of the former is substantially less than that of the latter. The calculated characteristic ratios of corresponding aliphatic poly(amides) and poly(esters) are very similar to one another.(223) However, there is a substantial difference in their melting temperature. Poly(ethylene terephthalate) has a relatively low value for C_∞. Similarly poly(bisphenol-A carbonate) has a low value of C_∞ and a melting temperature of 300 °C.(224) Based on this extensive set of data, it can be concluded that, except for a few special cases, there is no correlation between T_m and C_∞. In retrospect this is not a surprising conclusion. Although there are good reasons to believe that the chain conformation plays an important role in determining the melting temperature, the conformational entropy is not the sole contribution to the entropy of fusion.

To proceed further with the analysis it is necessary to isolate the contribution of the conformational entropy change from the total entropy of fusion. A simplifying assumption that can be made is that ΔS_u is the sum of two parts: the entropy change due to the latent volume change in melting, ΔS_v, and entropy change that takes place at constant volume $(\Delta S_v)_v$.(225,226) Thus, in this approximation

$$\Delta S_u = \Delta S_v + (\Delta S_v)_v \qquad (6.2)$$

The first term on the right can be expressed formally by the Maxwell relation

$$\left(\frac{\partial S_v}{\partial V} \right)_T = \left(\frac{\partial P}{\partial T} \right)_V = -\frac{\alpha}{\beta} \qquad (6.3)$$

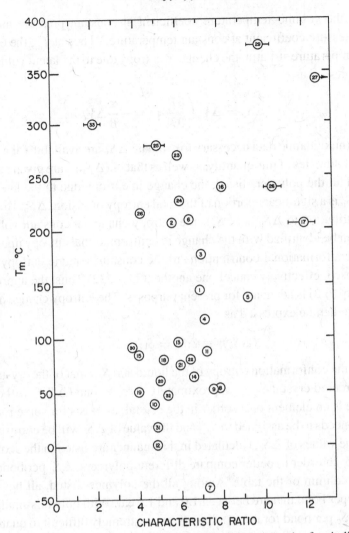

Fig. 6.12 Plot of melting temperature against characteristic ratio for indicated polymers. (1) Polyethylene; (2) i-poly(propylene); (3) i-poly(isopropyl acrylate); (4) s-poly(isopropyl acrylate); (5) i-poly(methyl methacrylate); (6) s-poly(methyl methacrylate); (7) poly(dimethyl siloxane); (8) poly(diethyl siloxane); (9) poly(dipropyl siloxane); (10) poly(cis-1,4-isoprene); (11) poly(trans-1,4-isoprene); (12) poly(cis-1,4-butadiene); (13) poly(trans-1,4-butadiene); (14) poly(caprolactone); (15) poly(propiolactone); (16) poly(pivalolactone); (17) poly(oxymethylene); (18) poly(ethylene oxide); (19) poly(trimethylene oxide); (20) poly(tetramethylene oxide); (21) poly(hexamethylene oxide); (22) poly(decamethylene oxide); (23) poly(hexamethylene adipamide); (24) poly(caprolactam); (25) poly(ethylene terephthalate); (26) poly(ethylene sulfide); (27) poly(tetrafluoroethylene); (28) i-poly(styrene); (29) poly(acrylonitrile); (30) poly(1,3-dioxolane); (31) poly(1,3-dioxopane); (32) poly(1,3-dioxocane); (33) bisphenol A-poly(carbonate).

Here α is the volume–temperature coefficient at constant pressure and β is the volume–pressure coefficient at constant temperature.[8] Thus at T_m^0, the equilibrium melting temperature at 1 atm, the change in entropy due to the latent volume change can be expressed as

$$\Delta S_v = \Delta V_v \left(\frac{\partial P}{\partial T} \right)_v = -\frac{\alpha}{\beta} \Delta V_v \qquad (6.4)$$

The thermodynamic data necessary to evaluate ΔS_v are available for a number of polymers. Estimates of this quantity, as well as that of $(\Delta S_v)_v$, are given in Table 6.5. For almost all the polymers listed, the change in entropy due to the latent volume change ΔS_v is a significant portion of the total entropy of fusion, ΔS_u. However, the major contribution to ΔS_u is $(\Delta S_v)_v$ the entropy change at constant volume. This quantity can be identified with the change in conformational entropy upon melting. Other nonconformational contributions to the constant volume entropy that have been proposed, effectively cancel one another.(227–232) Thus, the approximation made by Eq. (6.2) is adequate for present purposes. The entropy change at constant volume can then be expressed as

$$(\Delta S_v)_v \equiv \Delta S_c = S_{1,c} - S_{c,c} \qquad (6.5)$$

Here $S_{1,c}$ is the conformation entropy of the liquid and $S_{c,c}$ that of the crystal. For the perfectly ordered crystal $S_{c,c}$ is approximately zero, so that $(\Delta S_v)_v$ will equal $S_{1,c}$. When there is an element of disorder in the crystal, as exists for some polymers, a finite value needs to be assigned to $S_{c,c}$ and the value of ΔS_c will be correspondingly reduced. The values of ΔS_c, calculated in this manner are listed in the sixth column of Table 6.5. In order to better compare different polymers, ΔS_c per bond is given in the fifth column of the table. Among all the polymers listed, all but five have ΔS_c values per bond that are between 1.0 and 1.6 e.u. With only this small range in values of ΔS_c per bond for most polymers, it is extremely difficult to quantitatively relate the change in conformational entropy to the melting temperature.

The remaining five polymers have extremely low values of ΔS_c. These values can be related to some element of disorder within the crystal structure. The low value of ΔS_c per bond found for poly(tetrafluoroethylene) can be attributed to the room temperature polymorphic transition and the associated entropy change.(233,234) On the other hand, poly(cis-1,4-isoprene), natural rubber, is not known to undergo any polymorphic transitions at atmospheric pressure. Yet ΔS_c per bond is only 0.43 e.u. There is, however, some disagreement in interpreting the x-ray diffraction pattern of this polymer. Nyburg has concluded that the crystal structure is statistically

---

[8] Strictly speaking the pressure–temperature coefficient, $(\partial p/\partial T)$, should be integrated with respect to the volume, between the limits of the completely crystalline and liquid polymer.(227) Equation (6.3) however suffices for present purposes.

Table 6.5. *Contributions to entropy of fusion*[a]

Polymer	$\Delta S_u^{(1)}$	ΔS_v	$(\Delta S_v)_v^{(3)}$	ΔS_c/bond[4]	$\Delta S_c^{(5)}$ (calc)	References
Ethylene	2.29	0.52	1.77	1.77	1.76	a,b,c,d,e
	2.36	0.46	1.90	1.90	1.83	
	2.53	0.91	1.62	1.62		
1,4-cis-Isoprene (natural rubber)	3.46	1.80	1.70	0.43	5.41	f,g
1,4-trans-Isoprene (gutta percha)	8.75	3.7	5.1	1.28	5.47	g,h,l
	8.00	2.8	5.2	1.30		
i-Propylene	4.37	0.88–1.30	3.07–3.49	1.54–1.75	1.92	j,k,l
Methylene oxide	3.55	0.77	2.73	1.37	3.00, 2.80	b,d,g
Ethylene oxide	5.91	1.13	4.78	1.59	5.10, 4.28	c,d,g,m,n,
	6.12	1.68	4.44	1.48		
i-Styrene	4.00	1.30	2.70	1.35		o
Tetrafluoroethylene[2]	1.91	0.55	1.36	0.68	3.20	p
Ethylene adipate	14.77					q,r
	13.83	3.55	10.28	1.10	9.72	
Ethylene suberate	17.39					
	17.15	4.35	12.80	1.12	13.26	g,r
Ethylene sebacate	20.32	5.02	5.30	1.16	16.36	q,r
Ethylene terephthalate[4]	10.20	1.74	8.46	1.41	6.42	r,s,t,u
	11.67	5.02	6.65	1.10		
1,4-cis-Butadiene	8.03	1.72	6.31	1.58	5.52	g,t
4,4'-Dioxy diphenyl 2,2-propane carbonate	15.40	6.10	9.30	1.16	—	v
β-Propiolactone	5.72	2.45	3.27	0.93	3.36	w
ε-Caprolactone	11.55	3.50	7.98	1.14	8.75	v,w
Pivalolactone	6.92	2.89[6]	4.03	1.01	4.30	x
Butene-1 I	4.45	3.11	1.34	0.67		u
II	4.97	4.19	0.78	0.39	—	
4-methyl Pentene-1	2.46	2.07	0.39	0.20	—	y
Hexamethylene-adipamide α₂	19.08	16.37	2.71	0.18	25.75	z,aa
	19.21	13.56	5.56	0.38	15.05	
Ether ether ketone	18.53	6.43	12.10	1.72	—	bb

Table 6.5. *(cont.)*

Polymer		$\Delta S_{\mathrm{u}}^{(1)}$	ΔS_v	$(\Delta S_v)_v^{(3)}$	$\Delta S_{\mathrm{c}}/\mathrm{bond}^{(4)}$	$\Delta S_{\mathrm{c}}^{(5)}$ (calc)	References
Ester amides	6-6	42.2	15.70	26.50	1.30	28.70	cc
	12-2	47.4	17.80	29.60	1.40	33.30	
	12-6	57.1	20.10	37.00	1.40	39.70	
	12-12	71.3	25.60	45.70	1.40	50.70	

[a] Units cal mol^{-1}

(1) From Tables 6.1, 6.2 and 6.3. (2) Average between virgin sample and melt crystallized. (3) From Eq. (6.2). (4) Phenyl ring taken as single bond. (5) Calculated from rotational isomeric state theory. (6) Estimate.

a. Quinn, F. A., Jr. and L. Mandelkern, *J. Am. Chem. Soc.*, **80**, 3178 (1958).
b. Starkweather, H. W., Jr. and R. H. Boyd, *J. Phys. Chem.*, **64**, 410 (1960).
c. Tonelli, A. E., *J. Chem. Phys.*, **53**, 4339 (1970).
d. Sundararajan, P. R., *J. Appl. Polym. Sci.*, **22**, 1391 (1978).
e. Tsujita, Y., T. Nose and T. Hata, *Polym. J.*, **3**, 587 (1972).
f. Roberts, D. E. and L. Mandelkern, *J. Am. Chem. Soc.*, **77**, 781 (1955).
g. Tonelli, A. E., *Anal. Calorimetry*, **3**, 89 (1974).
h. Mandelkern, L., F. A. Quinn, Jr. and D. E. Roberts, *J. Am. Chem. Soc.*, **78**, 926 (1956).
i. Naoki, M. and T. Tomamatsu, *Macromolecules*, **13**, 322 (1980).
j. Fatou, J. G., *Eur. Polym. J.*, **7**, 1057 (1971).
k. Fortune, G. C. and G. N. Malcolm, *J. Phys. Chem.*, **71**, 876 (1967).
l. Tonelli, A. E., *Macromolecules*, **5**, 563 (1972).
m. Malcolm, G. N. and G. L. D. Ritchie, *J. Phys. Chem.*, **66**, 852 (1962).
n. Tsujita, Y., T. Nose and T. Hata, *Polym. J.*, **6**, 51 (1974).
o. Dedeurwaerder, R. and J. F. M. Oth, *J. Chim. Phys.*, **56**, 940 (1959).
p. Starkweather, H. W., Jr., P. Zoller, G. A. Jones and A. J. Vega, *J. Polym. Sci.: Polym. Phys. Ed.*, **20**, 751, (1982).
q. Hobbs, S. Y. and F. W. Billmeyer, Jr., *J. Polym. Sci., Pt. A-2*, **8**, 1387 (1970).
r. Tonelli, A. E., *J. Chem. Phys.*, **54**, 4637 (1971).
s. Roberts, R. C., *Polymer*, **10**, 113 (1969).
t. Allen, G., *J. Appl. Chem.*, **14**, 1 (1964).
u. Starkweather, H. W., Jr. and G. A. Jones, *J. Polym. Sci.: Polym. Phys. Ed.*, **24**, 1509 (1986).
v. Jones, L. D. and F. E. Karasz, *J. Polym. Sci.: Polym. Lett.*, **4**, 803 (1966).
w. Crescenzi, V., G. Manzini, G. Calzolari and C. Borri, *Eur. Polym. J.*, **8**, 449 (1972).
x. Borri, C., S. Brückner, V. Crescenzi, G. Della Fortuna, A. Mariano and P. Scarazzato, *Eur. Polym. J.*, **7**, 1515 (1971).
y. Zoller, P., H. W. Starkweather, Jr. and G. A. Jones, *J. Polym. Sci.: Polym. Phys. Ed.*, **26**, 257 (1988).
z. Starkweather, H. W., Jr., P. Zoller and G. A. Jones, *J. Polym. Sci.: Polym. Phys. Ed.*, **22**, 1615 (1984).
aa. Tonelli, A. E., *J. Polym. Sci.: Polym. Phys. Ed.*, **15**, 2015 (1977).
bb. Zoller, P., T. A. Kehl, H. W. Starkweather, Jr. and G. A. Jones, *J. Polym. Sci.: Polym. Phys. Ed.*, **27**, 993 (1989).
cc. Manzoni, G., V. Crescenzi, A. Ciana, L. Ciceri, G. Della Fortuna and L. Zotteri, *Eur. Polym. J.*, **9**, 941 (1973).

disordered with respect to chain packing.(235) Natta and Corradini have concluded from conformational analysis, and consistent with the x-ray diffraction analysis, that although the chains in the crystal maintain the same approximate shape there are statistical fluctuations in the ordered conformation.(236–238) These factors will contribute to the total entropy of fusion and a reduced value of ΔS_c. The value of ΔS_c for poly(hexamethylene adipamide) is extremely low, irrespective of which data set is used. The data are for the α_2 form which has a less ordered structure relative to the other crystalline forms.(239) The sensitivity of ΔS_c to disorder within the crystal structure is also manifested by the results for poly(4-methyl pentene-1).(240) Calorimetric and wide-angle x-ray measurements confirm that there is a structural change between the glass temperature and melting temperature of this polymer. The low value of ΔS_c for poly(butene-1) can also be attributed to the polymorphism of this polymer.

The change in the conformational entropy of a chain on fusion, at constant volume, can be evaluated from the partition function of the disordered chain, if it is assumed that there are no contributions from the ordered structure. Thus, the conformational entropy on fusion is identified with the entropy of the isolated chain in the pure melt. This entropy can be written as

$$S_1 = R[\ln Z + (T/Z)(dZ/dT)] \tag{6.6}$$

where Z is the partition function that describes the conformational characteristics of the isolated chain.

The partition function can be evaluated by adopting the rotational isomeric state model.(219,241). This method has been eminently successful in a variety of applications.(219) It has been adapted to the present problem by Tonelli.(232,242) In this procedure, each bond in the backbone of the chain is allowed to adopt a small number of discrete rotational states. The probability of the occurrence of a given state will usually depend on the rotational state of the adjoining bonds. A statistical weight matrix for each bond can be constructed from the Boltzman factors of the energies, $u_{\alpha,\beta}$ that are involved. Thus, $u_{\alpha,\beta} = \exp(-E_{\alpha,\beta}/RT)$, where $E_{\alpha,\beta}$ is the energy difference appropriate to the pairwise rotational states α and β. As an example, consider a system with three rotational states designated as α, β and γ, respectively. The statistical weight matrix U_i for the ith bond in the chain can then be written as

$$
U_i =
\begin{array}{c|ccc}
 & \alpha & \beta & \gamma \\
\hline
\alpha & u_{\alpha,\alpha} & u_{\alpha,\beta} & u_{\alpha,\gamma} \\
\beta & u_{\beta,\alpha} & u_{\beta,\beta} & u_{\beta,\gamma} \\
\gamma & u_{\gamma,\alpha} & u_{\gamma,\beta} & u_{\gamma,\gamma}
\end{array}
\tag{6.7}
$$

Here the element $u_{\beta,\alpha}$ represents the statstical weight of the ith bond in the α state following the $i-1$ bond in the β state, and so on. The sum of the statistical weights over all possible chain conformations gives the chain partition function Z. This sum can be obtained by sequential matrix multiplication.(242) For a chain of n bonds, each with ν rotational states

$$Z = J^* \left(\prod_{i=2}^{n-1} U_i \right) J \tag{6.8}$$

Here U_i is the appropriate statistical weight matrix for a specific bond type, J^* and J are $1 \times \nu$ and $\nu \times 1$ row and column vectors respectively. They can be represented as

$$J^* = |1000\ldots0| \qquad \text{and} \qquad J = \begin{vmatrix} 1 \\ 1 \\ 1 \\ \cdot \\ \cdot \\ \cdot \\ 1 \end{vmatrix} \tag{6.9}$$

The temperature coefficient dZ/dT can be expressed as

$$\frac{dZ}{dT} = G^* \left(\prod_{i=2}^{n-1} \hat{U}_{T,i} \right) G \tag{6.10}$$

Here G^* and G are the $1 \times 2\nu$ and $2\nu \times 1$ row and column vectors given by

$$G^* = |J^* 0 \ldots 0|; \qquad G = \begin{vmatrix} 0 \\ 0 \\ 0 \\ \cdot \\ \cdot \\ \cdot \\ J \end{vmatrix} \tag{6.11}$$

The $2\nu \times 2\nu$ matrix $\hat{u}_{T,i}$ is given by

$$\hat{u}_{T,i} = \begin{vmatrix} U_i & U'_{T,i} \\ 0 & U_i \end{vmatrix} \tag{6.12}$$

where $U'_{T,i} = dU_i/dT$. Thus, using the matrix formulation the conformational entropy partition function of the disordered chain can be calculated as long as the values of the statistical weights are available. These values can be obtained from the parameters that have been used to calculate a variety of configurational properties of chains.(219) The values of ΔS_c calculated in this manner are listed in the next to last column in Table 6.5. They can be compared with the experimental values, given in the fourth column.

For most of the polymers listed there is remarkably good agreement between the values of ΔS_c, calculated by invoking the methods of rotational isomeric state theory, and the entropy change at constant volume deduced from experiment. This agreement validates the approximation of separating the observed entropy of fusion into the two specific contributions, and identifying the entropy of fusion at constant volume with the average chain conformation in the melt. In this way the chain structure can be related to the melting temperature. The same parameters used in this calculation have also been used to calculate the characteristic ratio of many polymers.(219) Very good agreement has been obtained with experiment. The melting temperature depends on ΔS_u and thus on the volume change as well as the conformational contribution.

For a small number of polymers, the calculated values of ΔS_c are greater than those expected from the experimental data. As was indicated, these polymers show some element of disorder in the crystalline state prior to melting. The entropy in the crystalline state of these polymers cannot be neglected in the calculation.

Based on the role of chain conformation it is possible to envisage the development of three-dimensional order from the disordered state. Hypothetically, crystallization can be thought of as a two-step process. The first of these involves the cooperative intramolecular ordering of the individual chains. Successive bonds adopt a set of rotational angles that represent a low energy state and are perpetuated along the chain. As examples, the perpetuation of the trans bonds in linear polyethylene leads to a planar zigzag chain; or in the case of isotactic poly(propylene) the gauche–trans sequence leads to a helical chain structure. The crystallization process is completed by the further decrease in free energy that occurs as the chain atoms and substituents from the different molecules are suitably juxtaposed relative to one another so that order is developed in the lateral direction. The low energy form of the chain can be tempered and modified by the intermolecular interactions.(243–245) Chain conformation in the melt, prior to crystallization, can also be modified.(246) There is the general expectation, therefore, that there will be both intramolecular and intermolecular contributions to the free energy of fusion. However, a significant contribution from intramolecular interactions is to be expected.

6.4 Polymorphism

Polymers, in analogy with low molecular weight substances, can crystallize in different structural modifications. Different crystal structures can develop during crystallization from the pure melt by variations in temperature, pressure, and deformation in tension or shear. Different crystalline structures can also evolve from polymer–diluent mixtures. In this case the crystalline structure will depend on the nature of the diluent and its concentration. Polymorphism is not limited to synthetic polymers. It is also observed in proteins (247) and synthetic polypeptides.(248) The widespread observation of polymorphism in polymers precludes a discussion of each specific situation.(249) Rather, we shall set forth the main principles that are involved and use selected examples to illustrate these concepts. Polymorphism in polymers can be divided into two broad categories. In one group, the chain assumes a distinctly different conformation in the unit cell. In the other, the chain conformation and repeat distance along the chain axis is unaltered. However, the ordered chains are packed differently in the unit cell.

A classical example of chain molecules that have the same ordered conformation, but pack differently, is found in the crystallization of the *n*-alkanes. Either monoclinic, orthorhombic or hexagonal unit cells can be observed with many alkanes, depending on the chain length and crystallization conditions. However, the all trans planar zigzag chain structure is maintained despite the transformation from one type unit cell to another. In a similar manner, isotactic poly(propylene) exhibits three different, well-defined crystallographic forms. The chain conformation in each, however, is the 3/1 helix. The difference in the crystallography is the manner in which the chains are packed in the unit cell. The crystallographic habits of the three polymorphs have been described in detail.(250) The most commonly observed crystalline form has a monoclinic, or α, unit cell. The β, or hexagonal form, is found either after crystallization under stress or by adding specific nucleating agents to quiescent melts. This form transforms to α upon heating. In the third polymorph, the γ, the chains are packed in an orthorhombic unit cell. The structure of the unit cell for the γ form is very unusual for a crystalline polymer.(250,251) The chain axes in the unit cell are not parallel to one another. Furthermore, this crystal structure only develops when small crystallizable sequence lengths are available.

Some aliphatic poly(amides) exhibit similar packing behavior. For example, in poly(hexamethylene adipamide) (216) and poly(hexamethylene sebacamide) (252) the asymmetric packing present in the basal plane of the triclinic cell shifts to hexagonal as the temperature is increased. The planar zigzag chain conformation is, however, maintained. A latent enthalpy change is observed when poly(hexamethylene adipamide) undergoes this transition.(253) This type of crystal–crystal transition has been observed in many even–even type poly(amides).(254) Poly(undecanamide) displays a similar polymorphism in that a transition from a triclinic to hexagonal structure occurs at an elevated temperature with the ordered chain conformation being maintained.(255) The even poly(amides) also have different crystal structures. These are usually monoclinic and hexagonal.(254,256) Although the details of the transition from one form to the other have not been completely clarified, the ordered chain conformation is maintained.

An interesting example of polymorphism without a change in chain structure, is given by poly(p-phenylene vinylene).(257) This polymer becomes conducting when oxidized, or doped, with appropriate reagents. In the neutral state the unit cell is monoclinic. It is transformed into the orthorhombic form in the conducting state. It has been suggested that this process represents a first-order crystal–crystal transition. Of particular interest is the fact that although major changes take place in the crystal structure and electrical properties, the repeat distance and the ordered conformation remain unaltered. The change in the crystal structure involves alterations in the lateral dimensions to accommodate the dopant.

Polymorphism that reflects different ordered chain conformations in the unit cell is also well documented. This type of polymorphism is exemplified by the trans dienes and the α and β ordered structures that are observed in the polypeptides and fibrous proteins. The basic reason for the formation of different ordered chain structures in a given polymer is the existence of more than one minimum in the conformational energy surface.(219) Ordered chain conformations in a three-dimensional unit cell almost always correspond, or are close, to one of the low energy minima. As examples, the polymorphs of syndiotactic poly(propylene) and of syndiotactic poly(styrene) correspond to the repeat of a set of dihedral angles that correspond to low energy minima in each case.(258) These choices result in either extended or helical forms. Another example is given by poly(butene-1). Here the different ordered conformations correspond to the same broad minimum in the conformation energy surface. For this polymer, three different ordered helical structures are known, each of which have similar dihedral angles and form different unit cells.(259,260)

The polymorphism in poly(1,4-trans-isoprene) (gutta percha) has been studied in detail. Based on a detailed analysis of chain stereochemistry Bunn (261) predicted the possibility of four different crystalline modifications of this polymer, each with a different chain structure. Two of these, crystallized solely by cooling the polymer to an appropriate temperature, have been identified and their crystal structures determined.(261–263) A third form, that crystallizes upon stretching, has also been identified.(264) However, its structure has been questioned.(264)

Similarly, isotactic poly(4-methyl pentene) has been crystallized with several different unit cells with different ordered conformations in some cases.(264)

Polymorphic transitions are also observed in polymers as a result of a crystal–crystal transition. This behavior is similar to that observed in low molecular weight systems. A classical, and important, example is found in linear polyethylene. A new

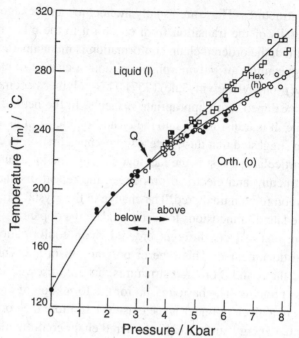

Fig. 6.13 Composite phase diagram for linear polyethylene. (From S. Rastogi *et al.* (271))

ordered phase initially appears at about 3 kbar and 215 °C.(265–268) Only the 100 and 110 reflections are found in the wide-angle x-ray pattern of this phase, indicating a hexagonal structure.(266,267) The suggestion has been made that this structure represents a continuity of the rotator phase that is found in low molecular weight *n*-alkanes.(269) Nuclear magnetic resonance studies of the hexagonal phase show a very high chain mobility and rapid axial reorientation similar to that observed in the "rotator" phase of the *n*-alkanes.(270)

A composite phase diagram for high molecular weight linear polyethylene, illustrating this polymorphism, is given in Fig. 6.13.(271–274) At low temperatures and pressures the conventional orthorhombic to liquid transition is observed. At higher temperatures and pressures the hexagonal form appears. The temperature–pressure curve that defines the transition from orthorhombic to hexagonal phase is indicated. The transition from the hexagonal form to the melt is also defined. There is a relatively narrow region in pressure–temperature space where the hexagonal structure is the stable form. A well-defined triple point appears in the diagram at about 215–220 °C and 3.3 kbar. It represents the pressure and temperature where the orthorhombic, hexagonal and liquid phases co-exist. Using statistical mechanical methods, Priest has been able to reproduce the pressure–temperature relation for the transition from the hexagonal to orthorhombic structure.(275) The entropy change between the hexagonal and orthorhombic form is calculated to be 1.70 e.u. mol^{-1}

at 5 kbar. The corresponding enthalpy change is 880 cal mol^{-1}. These quantities are comparable to the experimentally observed values.(274)

According to Raman spectroscopy the chains in the hexagonal form are conformationally disordered with an appreciable concentration of gauche bonds.(276,277) These results are consistent with the interpretation of the wide-angle diffraction pattern of this phase which requires an element of conformational disorder.(278,279) The gauche content of the pure melt of low molecular weight alkanes is known to increase with increasing applied hydrostatic pressure.

Different crystal structures can also be developed in polymers by the application of an external stress. The classical example is the reversible α to β transformation in fibrous proteins and synthetic polypeptides.(280) By the application of a tensile force the α-helical ordered structure is transformed into one of the extended β forms. This reversible process is accompanied by dimensional changes that reflect the difference in repeat length between the two crystalline forms. Similar transformations are also observed with synthetic polymers. For example, the poly(lactones) with the structural formula $(CH_2CR_1R_2COO)_n$, exhibit a polymorphic transformation when undergoing a tensile deformation that is analogous to the $\alpha \rightarrow \beta$ transition of the fibrous proteins.(281) In the α form, synthetic polymer is characterized by a helical structure, while in the β form a planar zigzag conformation is assumed. The poly(β-hydroxy alkanoates) show a similar transition under tensile deformation.(282) Many of the aromatic polyesters display polymorphism upon stretching.(283–288) For example, the uniaxial extension of poly(butylene terephthalate) is accompanied by a reversible crystal–crystal phase transition.(283–285) In the undeformed state the chain repeat distant is about 10% shorter than the repeat in the stretched state. These dimensional changes are manifested in macroscopic dimensional change.

Syndiotactic poly(styrene) displays a complex polymorphic behavior that reflects the specific role played by solvents. Four crystalline forms have been reported.(289,290) The α and β forms can be obtained from the melt (or glass), depending on the crystallization conditions.(291) Both structures comprise planar zigzag chains that have the same identity period of 5.1 Å. The α form has a trigonal unit cell while the β form is orthorhombic. The β form can also be produced by crystallization from solution.(292,293) The γ and δ structures develop after interaction with solvent. In contrast to the all trans bond orientation of the α and β structures, the chains in the γ and δ crystals adopt a ttggttgg sequence of bond orientation. Thus a helical ordered structure evolves. This structure is similar to the crystalline chain conformation of syndiotactic poly(propylene).(294) The difference between the γ and the δ polymorphs is that in the former the sample is completely dried, while the solvent is included in the δ form. It therefore represents a clathrate type structure. The formation of these structures is, thus, solvent specific.(292,293,295,296) The

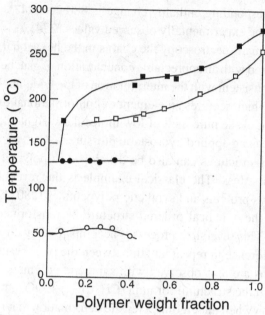

Fig. 6.14 Pseudo phase diagram of syndiotactic poly(styrene) in trans decalin. ○ crystal-lization of helix conformation; □ crystallization of zigzag conformation; ● melting of helix conformation; ■ melting of zigzag conformation. (From Deberdt and Berghmans (292))

crystal–crystal phase transition between the helical and planar zigzag structures is irreversible and takes place in the vicinity of 190–200 °C.

An example of the role of a solvent in polymorphism, a pseudo phase diagram of syndiotactic poly(styrene) in trans decalin is given in Fig. 6.14.(292) Here the crystallization temperatures of two of the polymorphs, obtained at a cooling rate of 5 °C min^{-1}, are represented by the open symbols. The filled symbols represent the melting of the β polymorph at high temperature and the helical δ form at lower temperatures. The more stable β structure can form over the complete composition range, while the formation of δ only occurs over a limited concentration range in this solvent. The invariance in both of the melting temperatures in the less concentrated polymer region could be indicative of liquid–liquid phase separation. Alternatively, the invariance could result from the coexistence of the two phases and solvent. Figure 6.14 is a pseudo phase diagram because, among other reasons, although the δ phase is transformed to the β phase on heating, it is not regenerated on cooling. The γ phase can only be developed by removing the solvent. It cannot be obtained by quenching the pure polymer to low temperature.

The examples that were described illustrate the different conditions under which polymorphism can occur. It is important to understand the underlying thermody-namic basis for polymorphism. Changes in the temperature, pressure, stress, type and solvent concentration can favor the development of one form and also affect the conversion of one crystalline structure to another. The transformation can occur either by direct conversion of one to the other or by the melting of one polymorph and the subsequent recrystallization of the other from the melt. These two processes may not be easily distinguishable by direct experimental observation. The determination of the free energy of fusion for each of the forms, as a function of the intensive variable involved, is necessary to decide their relative stabilities, and the thermodynamic basis for the transformation. However, it does not necessarily follow that the interconversion from one form to the other will follow the equilibrium path prescribed. The crystalline modification that is actually observed is a result of crystallization conditions and will be governed to a large extent by kinetic factors.

Figure 6.15 is a schematic representation of two possible modes for the transformation of one polymorphic structure to another. The temperature is taken as

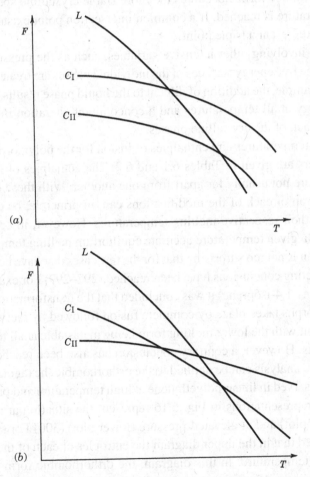

Fig. 6.15 Schematic diagram of the free energy (at constant pressure) as a function of temperature for two crystalline modifications and the liquid state of the same polymer.

the sole intensive variable in this example. Therefore, the free energy of each of the phases can be represented by curves in the planar diagram. The liquid phase is designated by L and the two crystalline phases by C_1 and C_{II}, respectively. In Fig. 6.15a, form II is higher melting than form I (as evidenced by the intersections of their free energy curves with that of the liquid phase) and has the lower free energy at all temperatures below its melting temperature. Hence, it is the thermodynamically more stable crystal structure at all temperatures. Form I must, therefore, be a metastable variety that will melt below form II. The system depicted does not undergo a crystal–crystal transformation. Figure 6.15b describes a different situation. Although form I is now the higher melting polymorph, the inverse situation of Fig. 6.15a does not exist. Rather, at low temperatures, form II is most stable (now the lower melting form). As the temperature is increased, its free energy curve first intersects that for form I so that a crystal–crystal transformation occurs. At intermediate temperatures form I becomes the more stable crystalline species until its melting temperature is reached. If a common intersection point exists for all three phases, it will represent a triple point.

For systems involving other intensive variables, such as the pressure, force, or composition, the free energy surfaces of the individual phases are treated in a similar manner. For example, the addition of diluent to the liquid phase results in a decrease in its free energy at all temperatures and a concomitant alteration in the stability conditions of each of the crystalline phases.

The melting temperatures and enthalpies of fusion for the polymorphic forms of several polymers are given in Tables 6.1 and 6.2. The enthalpies of fusion of the two structures are not usually far apart from one another. With these data the free energies of fusion of each of the modifications can, in principle, be calculated in the vicinity of their respective melting temperatures. However, in order to make comparisons at a given temperature accurate equilibrium melting temperatures are needed. Hence, it is not too surprising that for the few cases that have been analyzed in detail, conflicting conclusions have been reached.(297–299) For example, in the case of poly(trans-1,4-isoprene) it was concluded that the transformation of the low melting polymorphs takes place by complete fusion followed by recrystallization. This is consistent with the lower melting form being metastable at all temperatures at which it exists. However, a contrary conclusion has also been reached.(299)

A similar type analysis can be applied to the orthorhombic–hexagonal polymorphism that is observed in linear polyethylene at high temperature and pressure. The two schematic representations in Fig. 6.16 represent the situation at atmospheric pressure (upper plot) and at elevated pressure (lower plot).(300) For simplicity, it has been assumed that in the upper diagram the entropies of each of the phases are independent of temperature. In this diagram, the orthorhombic form is the most stable one up to its melting temperature, T_m^{or}. In order for the hexagonal form to

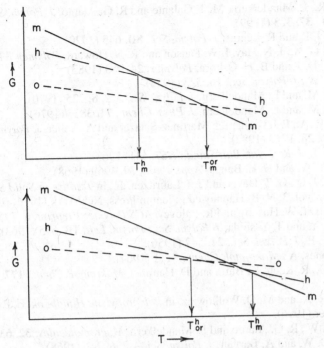

Fig. 6.16 Schematic diagram of the effect of pressure on the free energy functions and melting temperatures of two crystalline modifications of linear polyethylene. Upper graph, atmospheric pressure. Lower graph, elevated pressure. (From Asai(300))

appear, its free energy curve must intersect that of the orthorhombic structure at a temperature T_{or}^h less than that of T_m^h. This condition can be satisfied by the application of hydrostatic pressure as is schematically represented by the lower diagram in Fig. 6.16. Here $T_{or}^h < T_m^{or} < T_m^h$. Utilizing the available, but sparse, thermodynamic data, Asai was able to reproduce T_{or}^h and T_m^h at 5 kbar.(300)

References

1. Flory, P. J. and A. Vrij, *J. Am. Chem. Soc.*, **85**, 3548 (1963).
2. Mandelkern, L. and R. G. Alamo, in *American Institute of Physics Handbook of Polymer Properties*, J. E. Mark ed., American Institute of Physics Press (1996) p. 119.
3. Starkweather, H. W., Jr., P. Zoller, G. A. Jones and A. J. Vega, *J. Polym. Sci: Polym. Phys. Ed.*, **20**, 751(1982).
4. Jain, P. C., B. Wunderlich and D. R. Chaubey, *J. Polym. Sci: Polym. Phys. Ed.*, **15**, 2271 (1977).
5. Hobbs, S. W. and F. W. Billmeyer, *J. Polym. Sci., Pt. A2*, **8**, 1387 (1970).
6. Okui, N., H. M. Li and J. H. Magill, *Polymer*, **19**, 411 (1978).
7. Magill, J. M. private communication.
8. Fatou, J. G., *Eur. Polym. J.*, **7**, 1057 (1971).

9. Isasi, J. R., L. Mandelkern, M. T. Galante and R. G. Alamo, *J. Polym. Sci: Polym. Phys. Ed.*, **37**, 323 (1999).
10. Mercier, J. P. and R. Legras, *J. Polym. Sci.*, **8B**, 645 (1970).
11. Admans, G. N., J. N. Hay, I. W. Panson and R. N. Howard, *Polymer*, **17**, 51 (1976).
12. Blundell, D. J. and B. N. Osborn, *Polymer*, **24**, 953 (1983).
13. Bunn, C. W., *J. Polym. Sci.*, **16**, 323 (1955).
14. Rijke, A. M. and L. Mandelkern, *J. Polym. Sci. A-2*, **8**, 225 (1970).
15. Gopalan, M. and L. Mandelkern, *J. Phys. Chem.*, **71**, 3833 (1976).
16. Chivers, R. A., P. J. Barham, I. Martinez-Salazar and A. Keller, *J. Polym. Sci: Polym. Phys. Ed.*, **28**, 1717 (1982).
17. Weeks, J. J., *J. Res. Nat. Bur. Stand.*, **A67**, 441 (1963).
18. Huseby, T. W. and H. E. Bair, *J. Appl. Phys.*, **39**, 4969 (1968).
19. Hoffman, J. D., G. T. Davis and J. I. Lauritzen, Jr., in *Treatise in Solid State Chemistry*, vol. 3, N. B. Hannay ed., Plenum Press, New York (1976) p. 497.
20. Bair, H. E., T. W. Huseby and R. Salovey, *ACS Polymer Preprint*, **a**, 795 (1968).
21. Fujiwara, Y. and T. Yoshida, *J. Polym. Sci.: Polym. Lett*, **1B**, 675 (1963).
22. Reding, F. P., *J. Polym. Sci.*, **21**, 547 (1956).
23. Turner-Jones, A., *Makromol. Chem.*, **71**, 1 (1964).
24. Trafara, G., R. Koch, K. Blum and D. Hammel, *Makromol. Chem.*, **177**, 1089 (1976).
25. Phillips, R. A. and M. D. Wolkowicz, in *Polypropylene Handbook*, E. P. Moore Jr. ed., Hanser (1996).
26. Huang, T. W., R. G. Alamo and L. Mandelkern, *Macromolecules*, **32**, 6374 (1999).
27. Aubrey, D. W. and A. Barnatt, *J. Polym. Sci. A-2*, **6**, 241 (1968).
28. Kaufman, H. S., A. Sacher, T. Alfrey and I. Fankuchen, *J. Am. Chem. Soc.*, **70**, 3147 (1948).
29. Greenberg, S. and T. Alfrey, *J. Am. Chem. Soc.*, **76**, 6280 (1954).
30. Jordan, E. F., Jr., D. W. Feldeiser and A. N. Wrigley, *J. Polym. Sci: Polym. Chem. Ed.*, **9**, 1835 (1971).
31. Shibayev, V. P., B. S. Petrukhin, N. A. Platé and V. A. Kargin, *Polym. Sci. USSR*, **12**, 160 (1970).
32. Rogers, S. S. and L. Mandelkern, *J. Phys. Chem.*, **61**, 985 (1957).
33. Bacskai, R., J. E. Goodrich and J. B. Wilkes, *J. Polym. Sci. A-1*, **10**, 1529 (1972).
34. Raine, H. C., *Trans. Plastics Inst.*, **28**, 153 (1966).
35. Dunham, K. R., J. Vanderberghe, J. W. H. Faber and L. E. Contois, *J. Polym. Sci. A*, **1**, 751 (1963).
36. Campbell, T. W. and A. C. Haven, Jr., *J. Appl. Polym. Sci. A*, **1**, 73 (1959).
37. Faucher, J. A. and F. P. Reding, in *Crystalline Olefin Polymers*, vol. 1, Interscience (1965) p. 690.
38. Kaufman, M. H., H. F. Mark and R. B. Mesrobian, *J. Polym. Sci.*, **13**, 3 (1954).
39. Vansheidt, A. A., E. P. Melnikova, M. G. Krakovyak, L. V. Kukhareva and G. A. Gladkouskii, *J. Polym. Sci.*, **52**, 179 (1961).
40. Isoda, S., K. Kawagucht and K. I. Katayama, *J. Polym. Sci: Polym. Phys. Ed.*, **22**, 669 (1984).
41. Miller, K. J., H. B. Hollinger, J. Grebowicz and B. Wunderlich, *Macromolecules*, **23**, 3855 (1990).
41a. Greiner, A., S. Mang, O. Schäfer and P. Simon, *Acta Polym.*, **48**, 1 (1977).
41b. Steiger, D., C. Weder and P. Smith, *Macromolecules*, **32**, 5391 (1999).
42. Fox, T. G., B. S. Garrett, W. E. Goode, S. Gratch, J. Kincaid, A. Spell and J. D. Stroupe, *J. Am. Chem. Soc.*, **80**, 1768 (1958).

43. Miller, R. G. J., B. Mills, P. A. Small, A. Turner-Jones and G. M. Wood, *Chem. Ind.*, 1323 (1958).
44. de Boer, A., R. A. Van Ekenstein and G. Challa, *Polymer*, **16**, 931 (1975).
45. Mark, J. E., R. A. Wessling and R. E. Hughes, *J. Phys. Chem.*, **70**, 1895 (1966).
46. DeRosa, C., F. Auriemma, V. Vinti and M. Galimberti, *Macromolecules*, **31**, 6206 (1998).
47. Natta, G., L. Porri, P. Corradini and D. Morero, *Atti Accad. Naz. Lincei, Mem., Cl. Sci.*, **20**, 560 (1956).
48. Natta, G. and P. Corradini, *Atti Accad. Naz. Lincei, Mem., Cl. Sci.*, **19**, 229 (1955).
49. DeRosa, C., V. Venditto, G. Guerra and P. Corradini, *Macromolecules*, **25**, 6938 (1992).
50. DeRosa, C., V. Venditto, G. Guerra, B. Pirozzi and P. Corradini, *Macromolecules*, **24**, 5645 (1991).
51. Fujii, K., *J. Polym. Sci. Macromol. Rev.*, **5D**, 431 (1971).
52. Ohgi, H. and T. Sato, *Macromolecules*, **26**, 559 (1993).
52a. Tubbs, R. K., *J. Polym. Sci: Pt.A*, **3**, 4181 (1965).
53. Dall' Astra, G. and P. Scaglione, *Rubber Chem. Tech.*, **42**, 1235 (1960).
54. Caplizzi, A. and G. Gianotti, *Makromol. Chem.*, **157**, 123 (1972).
55. Wilkes, C. E., M. J. Pekló and R. J. Minchak, *J. Polym. Sci.*, **43C**, 97 (1973).
56. Marvel, C. S. and C. H. Young, *J. Am. Chem. Soc.*, **73**, 1066 (1951).
57. Lyman, D. J., *J. Polym. Sci.*, **55**, 507 (1961).
58. Natta, G. and P. Corradini, *Nuovo Cimento*, **15**, Suppl. 1, 9 (1960).
59. Natta, G., G. Dall' Astra, I. W. Bassi and D. Carella, *Makromol. Chem.*, **91**, 87 (1966).
60. Natta, G., I. W. Bassi and G. Fagherazzi, *Eur. Polym. J.*, **5**, 239 (1969).
61. Kell, R. M., B. Bennett and P. B. Stickney, *Rubber Chem. Technol.*, **31**, 499 (1958).
62. Allcock, H. R. and R. A. Arcus, *Macromolecules*, **12**, 1130 (1979).
63. Noether, H. D., *Textile Res. J.*, **28**, 533 (1958).
64. Lopez, L. C. and G. W. Wilkes, *J. Macromol. Sci. Rev. Macromol. Chem. Phys.*, **C29**, 83 (1989).
65. Bunn, C. W., in *Fibers from Synthetic Polymers*, R. E. Hill ed., Elsevier (1953) p. 310.
66. Hill, R. and E. E. Walker, *J. Polym. Sci.*, **3**, 609 (1948).
67. Fuller, C. S. and C. J. Frosch, *J. Am. Chem. Soc.*, **61**, 2575 (1939).
68. Bhaumik, D. and J. E. Mark, *Makromol. Chem.*, **187**, 1329 (1986).
69. Veberreiter, K., V. H. Karl and A. Altmeyer, *Eur. Polym. J.*, **14**, 1045 (1978).
70. Morgan, P. W., *Macromolecules*, **3**, 536 (1970).
71. Goodman, I., *Angew. Chem.*, **74**, 606 (1962).
72. Smith, J. G., C. J. Kibler and B. J. Sublett, *J. Polym. Sci., Pt. A-1*, **4**, 1851 (1966).
73. Kip, H. K. and H. L. Williams, *J. Appl. Polym. Sci.*, **20**, 1209 (1976).
74. Goodman, I., *Encyclopedia of Polymer Science and Technology*, **11**, 62 (1969).
75. Bello, P., A. Bello and E. Riande, *Macromolecules*, **32**, 8197 (1999).
76. Doak, K. W. and H. N. Campbell, *J. Polym. Sci.*, **18**, 215 (1955).
77. Guzman, J. and J. G. Fatou, *Eur. Polym. J.*, **14**, 943 (1978).
78. Coffman, D. D., G. J. Berchet, W. R. Peterson and E. W. Spanagel, *J. Polym. Sci.*, **2**, 306 (1947).
79. Ke, B. and A. W. Sisko, *J. Polym. Sci.*, **50**, 87 (1961).
80. Kinoshita, Y., *Makromol. Chem.*, **33**, 1 (1959).
81. Saotome, K. and H. Komoto, *J. Polym. Sci., Pt. A-1*, **4**, 1463 (1966).
82. Franco, L. and J. Puiggali, *J. Polym. Sci.: Pt. B: Polym. Phys.*, **33**, 2065 (1995).

82a. Ehrenstein, M., S. Dellsberger, C. Kocher, N. Stutzman, C. Weder and P. Smith, *Polymer*, **41**, 3531 (2000).
83. Puiggali, J., J. E. Aceituni, L. Franco, J. Lloueras, A. Prieto, A. Xenopoules, J. M. Fernandez-Santin and J. A. Subirana, *Prog. Coll. Polym. Sci.*, **87**, 35 (1992).
84. Aceituni, J. E., V. Tereshko, B. Lotz and J. A. Subirana, *Macromolecules*, **29**, 1886 (1996).
85. Franco, L., E. Navanno, J. A. Subirana and J. Puiggali, *Macromolecules*, **27**, 4284 (1994).
86. Cofmann, D. D., N. L. Cox, E. L. Martin, W. E. Mochel and F. J. Van Natta, *J. Polym. Sci.*, **3**, 851 (1948).
87. Schroeder, L. R. and S. L. Cooper, *J. Appl. Phys.*, **47**, 4310 (1976).
88. Garcia, D. and H. W. Starkweather, Jr., *J. Polym. Sci.: Polym. Phys. Ed.*, **23**, 537 (1985).
89. Skrovanek, D. J., S. E. Howe, P. C. Painter and M. M. Coleman, *Macromolecules*, **18**, 1676 (1985).
90. Skrovanek, D. J., P. C. Painter and M. M. Coleman, *Macromolecules*, **19**, 699 (1986).
91. Aelion, R., *Ann. Chim. (Paris)*, **3**, 5 (1948).
92. Shalby, S. W., R. J. Fredericks and E. M. Pearce, *J. Polym. Sci., Pt. A-2*, **10**, 1699 (1972).
93. Shalby, S. W., R. J. Fredericks and E. M. Pearce, *J. Polym. Sci.: Polym. Phys. Ed.*, **11**, 939 (1973).
94. Maglio, G., E. Musco, R. Palumbo and F. Riva, *J. Polym. Sci.: Polym. Lett.*, **12B**, 129 (1974).
95. Levesque, G. and J. C. Gressner, *J. Polym. Sci.: Polym. Lett.*, **17B**, 281 (1979).
96. Edgar, O. B. and R. Hill, *J. Polym. Sci.*, **8**, 1 (1951).
97. Morgan, P. W. and S. L. Kwolek, *Macromolecules*, **8**, 104 (1975).
98. Gaymans, R. J. and S. Harkema, *J. Polym. Sci.: Polym. Phys. Ed.*, **15**, 587 (1977).
99. Bell, A., J. G. Smith and C. J. Kibler, *J. Polym. Sci., Pt A*, **3**, 19 (1965).
100. Ridgway, J. S., *J. Polym. Sci.: Polym. Lett.*, **13B**, 87 (1975).
101. Ridgway, J. S., *J. Polym. Sci.: Polym. Chem. Ed.*, **12**, 2005 (1974).
102. MacKnight, W. J., M. Yang and T. Kajiyama, in *Analytical Calorimetry*, R. S. Porter and J. F. Johnson eds., Plenum Press (1968) p. 99.
103. Kajiyama, T. and W. J. MacKnight, *Polym. J.*, **1**, 548 (1970).
104. Wittbecker, E. L. and M. Katz, *J. Polym. Sci.*, **50**, 367 (1959).
105. Manzini, G., V. Crescenzi, A. Ciana, L. Ciceri, G. Della Fortuna and L. Zotteri, *Eur. Polym. J.*, **9**, 941 (1973).
106. Hill, J. W. and W. H. Carothers, *J. Am. Chem. Soc.*, **54**, 1569 (1932); *ibid.*, **54**, 5023 (1933).
107. Yoda, N., *Makromol. Chem.*, **55**, 174 (1962); *ibid.*, **56**, 10 (1962).
108. Yoda, N., *J. Polym. Sci., Pt A*, **1**, 1323 (1963).
109. Hergenrother, P. M., N. T. Wakelyn and S. J. Havens, *J. Polym. Sci.: Pt. A: Polym. Chem.*, **25**, 1093 (1987).
110. Hergenrother, P. M. and S. J. Havens, *Macromolecules*, **27**, 4659 (1994).
111. Cheng, S. Z. D., D. P. Heberer, H.-S. Lien and F. W. Harris, *J. Polym. Sci.: Pt. B: Polym. Phys.*, **28**, 655 (1990).
112. Muellerleile, J. T., B. G. Risch, D. E. Rodriques, G. L. Wilkes and D. M. Jones, *Polymer*, **34**, 789 (1993).
113. Korshak, V. V., T. M. Babchinitser, L. G. Kazaryan, V. A. Vasilev, Y. V. Genin, A. Y. Azriel, Y. S. Vygodsky, N. A. Churochkina, S. V. Vinogradova and D. Y. Tsuankin, *J. Polym. Sci.: Polym. Phys. Ed.*, **18**, 247 (1980).

114. Allen, G., C. Booth and C. Price, *Polymer*, **8**, 414 (1967).

115. Kambara, S. and A. Takahadi, *Makromol. Chem.*, **58**, 226 (1962).

116. Vandenberg, E. J., *J. Polym. Sci., Pt. A-1*, **7**, 525 (1969).

117. Bello, A., E. Perez and J. G. Fatou, *Macromolecules*, **19**, 2497 (1986).

118. Parris, J. M., R. H. Marchessault, E. J. Vandenberg and J. C. Mills, *J. Polym. Sci.: Pt. B: Polym. Phys.*, **32**, 749 (1994).

119. Alamo, R., A. Bello and J. G. Fatou, *J. Polym. Sci.: Polym. Phys. Ed.*, **28**, 907 (1990).

120. Lal, J. and G. S. Trick, *J. Polym. Sci.*, **50**, 13 (1961).

121. Chirico, A. D. and L. Zotteri, *Eur. Polym. J.*, **11**, 487 (1975).

122. Marco, C., S. Lazcono and J. G. Fatou, *Makromol. Chem.*, **191**, 1151 (1990).

123. Bello, A., S. Lazcono, C. Marco and J. G. Fatou, *J. Polym. Sci.: Polym. Chem. Ed.*, **22**, 1197 (1984).

124. Blundell, D. J. and B. N. Osborn, *Polymer*, **24**, 953 (1983).

125. Kelsey, D. R., L. M. Robeson, R. A. Clendinny and C. S. Blackwell, *Macromolecules*, **20**, 1204 (1987).

126. Rueda, D. R., M. G. Zolotukhin, M. E. Cagiao, F. J. Balta Callja, D. Villers and M. Dosiere, *Macromolecules*, **29**, 7016 (1996).

126a. van Dort, H. M., C. A. M. Halfs, E. P. Magré, A. J. Schopf and K. Yntema, *Eur. Polym. J.*, **4**, 275 (1968).

127. Zimmerman, H. J. and K. Könnecke, *Polymer*, **32**, 3162 (1991).

128. Harris, J. E., L. M. Robeson, *J. Polym. Sci.: Pt. B: Polym. Phys.*, **25**, 311 (1987).

129. Carlier, V., J. Devaux, R. Legras and P. T. McGrail, *Macromolecules*, **25**, 6646 (1992).

130. Wilski, H., *Koll. Z. Z. Polym.*, **238**, 426 (1970).

131. Berendswaard, W., V. M. Litvinov, F. Soeven, R. I. Scherrenberg, C. Gondard and C. Colemonts, *Macromolecules*, **32**, 167 (1999).

132. Mandelkern, L. and R. G. Alamo, in *American Institute of Physics Handbook of Polymer Properties*, J. E. Mark ed., American Institute of Physics Press (1996). Table 11.3, p. 132.

133. Okuda, A., *J. Polym. Sci. Polym. Chem. Ed.*, **2**, 1749 (1964).

134. Baker, W. O. and C. S. Fuller, *Ind. Eng. Chem.*, **38**, 272 (1946).

135. Kargin, V. A. and G. L. Slonimskii, *Usp. Khim.*, **24**, 785 (1955).

136. Kargin, V. A., *J. Polym. Sci.*, **30**, 247 (1958).

137. Kargin, V. A., *High Molecular Weight Compounds*, **2**, 466 (1960).

138. Malm, C. J., J. W. Mench, D. C. Kendall and G. D. Hiatt, *Ind. Eng. Chem.*, **43**, 684, 688 (1951).

139. Glasser, W. G., G. Samaranayake, M. Dumay and V. Dave, *J. Polym. Sci.: Pt B: Polym. Phys.*, **33**, 2045 (1995).

140. Lee, J. L., E. M. Pearce and T. W. Kwei, *Macromolecules*, **30**, 8233 (1997).

141. Godovsky, Y. K. and V. S. Papkov, *Adv. Polym. Sci.*, **88**, 129 (1989).

142. Weir, C. E., W. H. Leser and L. A. Wood, *J. Res. Nat. Bur. Stand.*, **44**, 367 (1950).

143. Lee, C. L., O. K. Johannson, O. L. Flaningam and P. Hahn, *Polym. Prepr.*, **10**, 1311 (1969).

144. Friedrich, J. and J. F. Rabolt, *Macromolecules*, **20**, 1975 (1987).

145. Aranguren, M. I., *Polymer*, **39**, 4897 (1998).

146. Varma-Nair, M., J. P. Wesson and B. Wunderlich, *J. Therm. Anal.*, **35**, 1913 (1989).

147. Lebedev, B. V., N. N. Mukhina and T. G. Kulagin, *Polym. Sci. USSR*, **20**, 1458 (1979).

148. Godovsky, Y. K., N. N. Makarova, V. S. Popkov and N. N. Kuzmin, *Makromol. Chem., Rapid Comm.*, **6**, 443 (1985).
149. Wunderlich, B. and J. Grebowicz, *Adv. Polym. Sci.*, **83**, 6016 (1984).
150. Wunderlich, B., M. Möller, J. Grebowicz and H. Baur, *Adv. Polym. Sci.*, **87**, 1 (1988).
151. Natta, G. and P. Corradini, *J. Polym. Sci.*, **39**, 29 (1959).
152. Beatty, C. L. and F. E. Karasz, *J. Polym. Sci.: Polym. Phys. Ed.*, **13**, 971 (1975).
153. Papkov, V. S., Y. K. Godovsky, V. S. Svistunov, V. M. Lituinov and A. A. Zhdanov, *J. Polym. Sci.: Polym. Chem. Ed.*, **22**, 3617 (1984).
154. Tsvankin, D. Y., V. S. Papkov, V. P. Zhukov, Y. K. Godovsky, V. S. Svistunov and A. A. Zhdanov, *J. Polym. Sci.: Polym. Chem. Ed.*, **23**, 1043 (1985).
155. Kögler, G., A. Hasenhindl and M. Möller, *Macromolecules*, **22**, 4190 (1989).
156. Möller, M., S. Siffrin, G. Kögler and D. Aelfin, *Macromol. Chem. Makromol. Symp.*, **34**, 171 (1990).
157. Out, G. J. J., A. A. Turetskii, M. Möller and D. Aelfin, *Macromolecules*, **27**, 3310 (1994).
158. Rim, P. B., A. A. Husan, A. A. Rasoul, S. M. Hurley, E. B. Orler and K. M. Schlosky, *Macromolecules*, **20**, 208 (1987).
159. Falender, J. R., G. S. Y. Yeh, D. S. Chiu and J. E. Mark, *J. Polym. Sci.: Polym. Phys. Ed.*, **18**, 389 (1980).
160. Harkness, B. R., M. Tachikawa and I. Mita, *Macromolecules*, **28**, 1323 (1995); *ibid.*, **28**, 8136 (1995).
161. Meier, D. J. and M. K. Lee, *Polymer*, **34**, 4882 (1993).
162. Miller, R. D. and J. Michl, *Chem. Rev.*, **89**, 1359 (1989).
163. Ungar, G., *Polymer*, **34**, 2050 (1993).
164. Newburger, N., I. Baham and W. L. Mattice, *Macromolecules*, **25**, 2447 (1992).
165. Godovsky, Y. K. and V. S. Papkov, *Macromol. Chem. Macromol. Symp.*, **4**, 71 (1980).
166. Godovsky, Y. K., N. N. Makarova, I. M. Petrova and A. A. Zhdanov, *Macromol. Chem. Rapid. Comm.*, **5**, 427 (1984).
167. Papkov, V. S., V. S. Svistunov, Y. K. Godovsky, A. A. Zhdanov, *J. Polym. Sci.: Polym. Phys. Ed.*, **25**, 1858 (1987).
168. Molenberg, A. and M. Möller, *Macromolecules*, **30**, 8332 (1997).
169. Okui, N., H. M. Li and J. H. Magill, *Polymer*, **19**, 411 (1978).
170. Cotts, P. M., R. D. Miller, P. T. Trefonas, III, R. West and G. N. Ficker, *Macromolecules*, **20**, 1046 (1987).
171. Cotts, P. M., S. Ferline, G. Dagli and D. S. Pearson, *Macromolecules*, **24**, 6730 (1991).
172. Wesson, J. P. and T. C. Williams, *J. Polym. Sci.: Polym. Chem. Ed.*, **17**, 2833 (1979).
173. Weber, P., D. Guillon, A. Skoulios and R. D. Miller, *J. Phys. France*, **50**, 793 (1989).
174. Weber, P., D. Guillon, A. Skoulios and R. D. Miller, *Liq. Cryst.*, **8**, 825 (1990).
175. Varma-Nair, M., J. Cheng, Y. Jin and B. Wunderlich, *Macromolecules*, **24**, 5442 (1991).
176. Lovinger, A. J., D. D. Davis, F. C. Schilling, F. J. Padden, Jr. and F. A. Bovey, *Macromolecules*, **24**, 132 (1991).
177. Kleman, B., R. West and J. A. Koutsky, *Macromolecules*, **26**, 1042 (1993); *ibid.*, **29**, 198 (1996).
178. Bukalov, S. S., L. A. Leiter, R. West and T. Asuke, *Macromolecules*, **29**, 907 (1996).
179. Allcock, H. R., *Chem. Rev.*, **72**, 315 (1972).
180. Allcock, H. R., *Angew. Chem. Int. Ed. Engl.*, **16**, 147 (1977).
181. Singler, R. E., N. S. Schneider and G. L. Hagnauer, *Polym. Eng. Sci.*, **15**, 321 (1975).

182. Russell, T. P., D. P. Anderson, R. S. Stein, C. R. Desper, J. J. Beres and N. S. Schneider, *Macromolecules*, **17**, 1795 (1984).

183. Masuko, T., R. L. Simeone, J. H. Magill and D. J. Plazek, *Macromolecules*, **17**, 2857 (1984).

184. Schneider, N. S., C. R. Desper and R. E. Singles, *J. Appl. Polym. Sci.*, **20**, 3087 (1976).

185. Schneider, N. S., C. R. Desper and J. J. Beres, in *Liquid Crystalline Order in Polymers*, A. Blumstein ed., Academic Press (1978) p. 299.

186. Sun, D. C. and J. H. Magill, *Polymer*, **28**, 1243 (1987).

187. Allen, G., C. J. Lewis and S. M. Todd, *Polymer*, **11**, 44 (1970).

188. Allcock, H. R., S. R. Puches and A. G. Scopelianos, *Macromolecules*, **27**, 1071 (1994).

189. Kojima, M. and J. M. Magill, *Polymer*, **30**, 579 (1989).

190. Schelten, J., D. G. H. Ballard, G. D. Wignall, G. Longman and W. Schmatz, *Polymer*, **17**, 751 (1976).

191. Tasumi, M. and S. Krimm, *J. Polym. Sci., Pt. A-2*, **6**, 995 (1968).

192. Bank, M. I. and S. Krimm, *J. Polym. Sci. Pt. A-2*, **7**, 1785 (1969).

193. Bates, F. S., L. G. Fetters and G. D. Wignall, *Macromolecules*, **21**, 1086 (1988).

194. Dorset, D. L., H. L. Strauss and R. G. Snyder, *J. Phys. Chem.*, **95**, 938 (1991).

195. Stehling, F. C., E. Ergoz and L. Mandelkern, *Macromolecules*, **4**, 672 (1971).

196. English, A. D., P. S. Smith and D. F. Axelson, *Polymer*, **26**, 1523 (1985).

197. Bates, F. S., H. D. Keith and D. B. McWhan, *Macromolecules*, **20**, 3065 (1987).

198. Crist, B., W. W. Graessley and G. D. Wignall, *Polymer*, **23**, 1561 (1982).

199. Fischer, E. W., private communication.

200. Ballard, D. G. H., P. Cheshire, G. W. Longman and J. Schelten, *Polymer*, **19**, 379 (1978).

201. Cais, R. E. and J. M. Kometani, *Macromolecules*, **17**, 1887 (1984).

202. Nandi, A. K. and L. Mandelkern, *J. Polym. Sci.:Pt. B*, **29**, 1287 (1991).

203. Buckingham, A. D. and H. G. E. Hentschel, *J. Polym. Sci.: Polym. Phys. Ed.*, **18**, 853 (1980).

204. Okamura, S., K. Hayashi and Y. Kitamishi, *J. Polym. Sci.*, **58**, 925 (1962).

205. Mateua, R., G. Wegner and G. Lieser, *J. Polym. Sci.: Polym. Lett.*, **11B**, 369 (1973).

206. Wegner, G., *Farady Discuss. Chem. Soc.*, **68**, 494 (1979).

207. Flory, P. J. and R. R. Garrett, *J. Am. Chem. Soc.*, **80**, 4836 (1958).

208. Gustavson, K. H. *The Chemistry and Reactivity of Collagen*, Academic Press (1956) Chapter 9.

209. Garrett, R. R., see P. J. Flory in *Protein Structure and Function, Brookhaven Symp. Biol.* No. 13 (1960) p. 229.

210. Nemethy, G. and H. A. Scheraga, *J. Phys. Chem.*, **66**, 1773 (1962).

211. Holmgren, S. K., K. M. Taylor, L. E. Bretscher and R. T. Raines, *Nature*, **392**, 666 (1998).

212. Rigby, B. J., *Symposium on Fibrous Proteins, Australia*, Plenum Press (1968) p. 217.

213. Rigby, B. J., *Nature*, **219**, 166 (1968).

214. Mandelkern, L. *Chem. Rev.*, **56**, 903 (1956).

215. Dole, M., *Adv. Polym. Sci.*, **2**, 221 (1960).

216. Brill, R., *J. Prakt. Chem.*, **161**, 49 (1942).

217. Wilhoit, R. C. and M. Dole, *J. Phys. Chem.*, **57**, 14 (1953).

218. Flory, P. J., H. D. Bedon and E. H. Keefer, *J. Polym. Sci.*, **28**, 151 (1958).

219. Flory, P. J., *Statistical Mechanics of Chain Molecules*, Wiley-Interscience (1969).

220. Erman, B., P. J. Flory and J. P. Hummel, *Macromolecules*, **13**, 484 (1980).
221. Takahashi, Y. and J. E. Mark, *J. Amer. Chem. Soc.*, **98**, 756 (1976).
222. Bhaumick, D. and J. E. Mark, *Macromolecules*, **14**, 162 (1981).
223. Flory, P. J. and A. D. Williams, *J. Polym. Sci. A-2*, **5**, 399 (1967); *ibid.*, *A-2*, **5**, 417 (1967).
224. Williams, A. D. and P. J. Flory, *J. Polym. Sci. A-2*, **6**, 1945 (1968).
225. Oriani, R. A., *J. Chem. Phys.*, **19**, 93 (1951).
226. Kubaschewski, O., *Trans. Faraday Soc.*, **45**, 930 (1949).
227. Allen, G., *J. Appl. Chem.*, **14**, 1 (1964).
228. Karasz, F. E., P. R. Couchman and D. Klempner, *Macromolecules*, **10**, 88 (1977).
229. Naoki, M. and T. Tomomatsu, *Macromolecules*, **13**, 322 (1980).
230. Abe, A., *Macromol. Symp.*, **118**, 23 (1997).
230a. Abe, A., T. Takeda, T. Hiejima and H. Feruya, *Polym. J.*, **31**, 728 (1999).
230b. Abe, A., T. Takeda and T. Hiejima, *Macromol. Symp.*, **152**, 255 (2000).
231. Starkweather, H. W., Jr. and R. H. Boyd, Jr., *J. Phys. Chem.*, **64**, 410 (1960).
232. Tonelli, A. E., *Anal. Calorimetry*, **3**, 89 (1974).
233. Starkweather, H. W., Jr., P. Zoller, G. A. Jones and A. J. Vega, *J. Polym. Sci.: Polym. Phys. Ed.*, **20**, 751 (1982).
234. Lau, S. F., H. Suzuki and B. Wunderlich, *J. Polym. Sci.: Polym. Phys. Ed.*, **22**, 379 (1984).
235. Nyburg, S. C., *Acta Cryst.*, **7**, 385 (1954).
236. Natta, G. and P. Corradini, *J. Polym. Sci.*, **20**, 251 (1956).
237. Natta, G. and P. Corradini, *Angew. Chem.*, **68**, 615 (1956).
238. Corradini, P. *J. Polym. Sci.*, **50C**, 327 (1975).
239. Starkweather, H. W., P. Zoller and G. A. Jones, *J. Polym. Sci.: Polym. Phys. Ed.*, **22**, 1615 (1984).
240. Karasz, F. E., H. E. Bair and J. M. O'Reilly, *Polymer*, **8**, 547 (1967).
241. Volkenstein, M. V., *Configurational Statistics of Polymer Chains*, Interscience (1963).
242. Tonelli, A. E., *J. Chem. Phys.*, **53**, 4339 (1970).
243. Sorensen, R. A., W. B. Liau and R. H. Boyd, *Macromolecules*, **21**, 194 (1988).
244. Sorensen, R. A., W. B. Liau, L. Kesner and R. H. Boyd, *Macromolecules*, **21**, 200 (1988).
245. Leon, S., J. J. Navas and C. Aleman, *Polymer*, **40**, 7351 (1999).
246. Smith, G. D., R. L. Jaffe and D. Y. Yoon, *Macromolecules*, **26**, 298 (1993).
247. Astbury, W. T. and H. J. Woods, *Phil. Trans. R. Soc. London Ser. A*, **232**, 333 (1934).
248. Bamford, C. H., W. E. Hambey and F. Happly, *Proc. R. Soc. London Ser. A*, **205**, 30 (1951).
249. Corradini, P. and G. Guerra, *Adv. Polym. Sci.*, **100**, 183 (1992).
250. Bruckner, S., S. V. Meille, V. Petraccone and B. Pirozzi, *Prog. Polym. Sci.*, **16**, 361 (1991).
251. Alamo, R. G., M. H. Kim, M. J. Galante, J. R. Isasi and L. Mandelkern, *Macromolecules*, **32**, 4050 (1999).
252. Slichter, W. P., *J. Polym. Sci.*, **35**, 82 (1959).
253. Starkweather, H. W., Jr. and G. A. Jones, *J. Polym. Sci.: Polym. Phys. Ed.*, **19**, 467 (1981).
254. Ramesh, C., *Macromolecules*, **32**, 372 (1999); *ibid.*, 5074 (1999).
255. Mathias, L. J., D. G. Powell, J. P. Autran and R. S. Porter, *Macromolecules*, **23**, 963 (1990).

256. Vogelsong, D. C., *J. Polym. Sci. Pt A*, **1**, 1053 (1963).
257. Masse, M. A., J. B. Schlenoff, F. E. Karasz and E. L. Thomas, *J. Polym. Sci.: Pt B: Polym. Phys.*, **27**, 2045 (1989).
258. Corradini, P., *J. Polym. Sci.*, **51C**, 1 (1975).
259. DeRosa, C. and D. Scaldarella, *Macromolecules*, **30**, 4153 (1997).
260. DeRosa, C., A. Borriello, V. Venditto and P. Corradini, *Macromolecules*, **27**, 3864 (1994).
261. Bunn, C. W., *Proc. R. Soc. London Ser. A*, **180**, 40 (1942).
262. Natta, G., P. Corradini and L. Porri, *Atti Accad. Naz. Lincei*, **20**, 728 (1956).
263. Takahasi, Y., T. Sato and H. Tadokoro, *J. Polym. Sci.: Polym. Phys. Ed.*, **11**, 233 (1973).
264. Fischer, D., *Proc. Phys. Soc. London*, **66**, 7 (1953).
265. Bassett, D. C. and B. Turner, *Nature*, **240**, 146 (1972).
266. Bassett, D. C., S. Block and G. Piermarini, *J. Appl. Phys.*, **45**, 4146 (1974).
267. Yusiniwa, M., C. Nakafuku and T. Takemua, *Polym. J.*, **4**, 526 (1974).
268. Bassett, D. C., *Polymer*, **17**, 460 (1976).
269. Ungar, G., *Macromolecules*, **19**, 1317 (1986).
270. deLaugon, M., H. Luigjes and K. O. Prius, *Polymer*, **41**, 1193 (2000).
271. Rastogi, S., M. Hikosaka, H. Kawabata and A. Keller, *Macromolecules*, **24**, 6384 (1991).
272. Hikoska, M. and S. Tamaki, *Phys. Sci. Jpn*, **50**, 638 (1981).
273. Bassett, D. C. and B. A. Khalifa, *Polymer*, **14**, 390 (1973); *ibid.*, **17**, 275 (1976).
274. Yasuniwa, M., R. Enoshoto and T. Takemura, *Jpn. J. Appl. Phys.*, **15**, 1421 (1976).
275. Priest, R. G., *Macromolecules*, **18**, 1504 (1985).
276. Tanaka, H. and T. Takemura, *Polymer J.*, **12**, 355 (1980).
277. Wunder, S. L., *Macromolecules*, **14**, 1024 (1981).
278. Yamamoto, T., H. Miyaji and K. Asai, *Jpn. J. Appl. Phys.*, **16**, 1890 (1977).
279. Yamamoto, T., *J. Macromol. Sci.*, **B16**, 487 (1979).
280. Astbury, W. T., *Proc. R. Soc. London Ser. A*, **134**, 303 (1947).
281. Prud'homme, R. E. and R. H. Marchessault, *Macromolecules*, **7**, 541 (1974).
282. Orts, W. J., R. H. Marchessault, T. L. Blumm and G. K. Hamer, *Macromolecules*, **23**, 5368 (1990).
283. Boye, C. A., Jr. and J. R. Overton, *Bull. Amer. Phys. Soc.*, **19**, 352 (1974).
284. Jakeways, R., I. M. Ward, M. A. Wilding, I. H. Hall, I. J. Desborough and M. G. Pass, *J. Polym. Sci.: Polym. Phys. Ed.*, **13**, 799 (1975).
285. Brereton, M. G., G. R. Davies, R. Jakeways, T. Smith and I. M. Ward, *Polymer*, **19**, 17 (1978).
286. Yamadera, R. and C. Sonoda, *J. Polym. Sci.: Polym. Lett.*, **3B**, 411 (1965).
287. Hall, I. H. and B. A. Ibrahim, *J. Polym. Sci.: Polym. Lett.*, **18B**, 183 (1980).
288. Stanbaugh, B., J. B. Lando and J. L. Koenig, *J. Polym. Sci.: Polym. Phys. Ed.*, **17**, 1063 (1979).
289. Guerra, G., V. M. Vitagliano, C. DeRosa, V. Petraccone and P. Corradini, *Macromolecules*, **23**, 1539 (1990).
290. Lin, R. H. and E. M. Woo, *Polymer*, **41**, 121 (2000).
291. Ho, R. M., C. P. Lin, H. Y. Tsai and E. M. Woo, *Macromolecules*, **33**, 6517 (2000).
292. Deberdt, F. and H. Berghmans, *Polymer*, **34**, 2193 (1993).
293. Deberdt, F. and H. Berghmans, *Polymer*, **35**, 1694 (1994).
294. Natta, G., I. Pasqmon, P. Corradini, M. Peraldo, M. Pegoraro and A. Zambelli, *Atti Accad. Naz. Lincei, Cl. Sci. Fis. Mat. Nat. Rend.*, **28**, 539 (1960).

295. Immirzi, A., F. DeCandia, P. Iannelli, V. Vittoria and A. Zambelli, *Macromol. Chem. Rapid Commun.*, **9**, 761 (1988).
296. Chatani, Y., Y. Shimane, Y. Inone, T. Inagaki, T. Ishioka, T. Ijitsu and T. Yukinari, *Polymer*, **33**, 488 (1992).
297. Mandelkern, L., F. A. Quinn, Jr. and D. E. Roberts, *J. Am. Chem. Soc.*, **78**, 926 (1956).
298. Lovering, E. G. and D. C. Wooden, *J. Polym. Sci., Pt. A-2*, **7**, 1639 (1969).
299. Flanagan, R. D. and A. M. Rijke, *J. Polym. Sci., Pt. A-2*, **10**, 1207 (1972).
300. Asai, K., *Polymer*, **23**, 391 (1982).

7

Fusion of cross-linked polymers

7.1 Introduction

Chain units that are involved in forming intermolecular cross-links require special attention as far as crystallization is concerned. When a sufficient number of intermolecular cross-links are imposed on a collection of linear polymer chains, a three-dimensional network structure reaching macroscopic dimensions is developed. Such structures are termed infinite networks. According to theory (1,2) the initial formation of a network occurs when the fraction of cross-linked units ρ exceeds a critical value ρ_c that is expressed as

$$\rho_c = \frac{1}{\bar{y}_w - 1} \cong \frac{1}{\bar{y}_w} \tag{7.1}$$

where $\bar{y}_w$ is the weight-average degree of polymerization of the initial polymers. At this critical value called the gel point, not all the polymer chains are attached to the insoluble network. Depending on details of the initial molecular weight distribution, the further introduction of cross-links into the system results in the incorporation of the remaining chains into the network.(3) In the usual cases of interest, complete network formation requires that only a small percentage of the chain units be involved in intermolecular cross-linkages. If the only effect of the cross-linkages on crystallization was through their concentration, then the previous discussion of copolymers could be generalized to include such structural variables. However, further theoretical insight and experimental observations do not justify this conclusion.

In contrast to other types of structural irregularities, chain units involved in inter-molecular cross-linkages act in a unique manner since they actually join together portions of different chains. There is, therefore, the distinct possibility that the cross-linked units could be restricted from participating in the crystallization for steric reasons. In addition, the fact that a network structure is formed can lead to alterations in the crystallization pattern as compared to that of a collection of individual polymer chains.

In the theoretical treatment for the formation of networks, it is customary to assume that the points of cross-linkage are randomly distributed over the complete volume of the sample. It is not necessary, however, to make any restrictive assumptions with regard to the disposition of the polymer chains at the time of network formation. Networks are commonly formed from randomly coiled chains. However, this represents a special case among several possibilities. Networks can also be formed from either deformed systems or systems where the chains are in ordered or partially ordered array when the cross-links are introduced. Theory has shown(4) that properties of a network are strongly influenced by the nature of the chain arrangement when the cross-links are introduced. Therefore, in discussing the properties of networks in general, and their crystallization behavior in particular, careful distinction must be made as to their mode of formation.

Intermolecular cross-linkages can be introduced into a collection of polymer chains by either chemical reaction, as, for example, the vulcanization of natural rubber,(5) or in favorable cases by the action of high-energy ionizing radiation.(6,7,8) In many cases the efficacy of the cross-linking process depends on the state of the polymer when the linkages are introduced.(7) Many naturally occurring macromolecular systems develop a sufficient number of intermolecular cross-links during the course of their synthesis so that in the molten state they display the characteristics of an infinite network.

A network in the liquid or amorphous state can be given a quantitative description(4,9,10) by defining a chain as that portion of the molecule which traverses from one cross-linked unit to a succeeding one. It is convenient to characterize each chain by a vector $\mathbf{r}$ which connects the average position of its terminal units, namely, the cross-linked units. The number of chains ν must be equal to the number of intermolecularly cross-linked units. If N_0 is the total number of chain units in the network, then ρ is equal to ν/N_0. The network can then be characterized by the number of chains and their vectorial distribution. When the network is deformed, a common assumption made is that the chain vector distribution is altered directly as the macroscopic dimensions. An affine transformation of the average position of the coordinates of the cross-links occurs. It is also usually assumed that the individual chains obey Gaussian statistics.

A reference state for the network is conveniently taken as one which represents the isotropic network with mean-square vector components, $\overline{x_0^2} = \overline{y_0^2} = \overline{z_0^2} = \overline{r_0^2}/3$. The reference state is chosen so that the mean-square chain vector length $\overline{r_0^2}$ is identical with the corresponding unperturbed length of the free chain. For any given state of the network, where $\overline{x^2}$, $\overline{y^2}$, $\overline{z^2}$ are the average squares of the Cartesian components of the chains, the entropy of the network contains a term

$$\frac{kv}{2}\left[-\frac{3}{r_0^2}(\overline{x^2} + \overline{y^2} + \overline{z^2})\right] \tag{7.2}$$

for the internal configurations.(4,9) The contribution to the total entropy of the random distribution of cross-linkages over the volume V of the sample is(4,11)

$$\frac{kv}{2}\ln V + \text{const} \tag{7.3}$$

for tetrafunctional cross-links (four chains emanating from a junction). Therefore, the entropy difference between a given specified state and the reference state can be expressed as

$$\Delta S = \frac{3kv}{2}\left[-\frac{(\overline{x^2} + \overline{y^2} + \overline{z^2})}{r_0^2} + 1 + \ln\langle\alpha\rangle\right] \tag{7.4}$$

where $\langle\alpha\rangle = (\overline{x^2}\,\overline{y^2}\,\overline{z^2}/\overline{x_0^2}\,\overline{y_0^2}\,\overline{z_0^2})^{1/6}$. The parameter $\langle\alpha\rangle$ measures the geometric mean of the linear dilation in the actual state relative to that in the reference state. It is not necessary that the volume of the reference state and the actual state under consideration be the same.

For a network formed from polymer chains in the isotropic randomly coiled state, $\langle\alpha\rangle$ equals unity at the same temperature and network volume as prevailed during cross-linking. On the other hand, for a network formed by the cross-linking of chains that are not randomly arranged, the value of $\langle\alpha\rangle$ depends on the details of the chain organization. For the network at its initial volume and temperature, $\langle\alpha\rangle$ may be either less than or greater than unity. With this brief description of the formation and characterization of networks, attention can now be given to the melting of crystallizable networks.

7.2 Theory of the melting of isotropic networks

The melting temperature of an unstressed isotropic network, T_m^i, can be expressed quite generally as the ratio of the enthalpy of fusion to the entropy of fusion. The entropy of fusion can be treated conveniently as the additive contribution of three terms. These are ΔS^0, the entropy of fusion in the absence of the constraints imposed on the chain conformation by the cross-linkages; ΔS_x^0, the alteration of the chain configurational entropy in the reference state, $\langle\alpha\rangle = 1$, that results from the presence of the cross-linkages; and ΔS_{el}^i the entropy change that occurs in going from the reference state to the real isotropic state. In the latter, $\langle\alpha\rangle$ assumes a value characteristic of the network structure. The melting temperature of the isotropic network can therefore be expressed as (4)

$$\frac{1}{T_{m}^{i}} = \frac{\Delta S^{0} + \Delta S_{x}^{0} + \Delta S_{el}^{i}}{\Delta H} \qquad (7.5)$$

and depends on the network constitution and the mode of its formation.

If the network structure is such that the crystallization of the cross-linked units is not restricted, ΔH and ΔS^{0} can be taken to be independent of the fraction of units cross-linked. Under these conditions, ΔS^{0} is identified with the entropy of fusion of the pure non-cross-linked polymer, and the ratio of ΔS^{0} to ΔH is identified with the equilibrium melting temperature T_{m}^{0} of the pure polymer. If, however, steric requirements are such that cross-linked units are excluded from the crystalline regions, an alteration will occur in these quantities. The presence of cross-linked units in the molten phase and not in the crystalline phase results in an increase in ΔS^{0} (when compared with the non-cross-linked polymer) of an amount $R\rho$ per mole of chain units. The melting temperature must accordingly be depressed for this reason, as long as ΔH is unaffected by the presence of cross-links.

For networks formed from randomly coiled chains, ΔS_{x}^{0} must be essentially zero since the units cross-linked are selected at random. This type of cross-linking process does not influence the configurational entropy characteristic of random non-cross-linked chains. For this case ΔS_{el}^{i} must also be zero. However, if polymer chains, initially arranged in parallel array, are cross-linked to form a network, the above factors must be greatly modified. If $\langle \alpha \rangle$ is known, ΔS_{el}^{i} can be calculated from theory.(4) Its contribution to Eq. (7.5) is shown to be small. However, because of the nature of the chain disposition at the time of network formation, a certain element of the high degree of order that is initially present will be imposed on the network. This element of order will be maintained throughout any subsequent transformations that the network may undergo as long as the initially imposed cross-links are not severed.

When cross-linking occurs in this initial state of axial order, it is required that the unit of a molecule that is being cross-linked be joined to a neighboring pre-determined unit. Thus, even though the cross-links are randomly distributed in space, units to be paired can no longer be selected at random. Since this pairing of units is maintained even in the liquid state, a decrease in the configurational entropy of the liquid occurs as a consequence of the introduction of cross-links in the prescribed manner. This conclusion should be contrasted with the random cross-linking of random chains. This decrease in the configurational entropy of the liquid manifests itself in a decrease in the total entropy of fusion, which is embodied in the term ΔS_{x}^{0}.

To calculate ΔS_{x}^{0} for networks formed from perfectly axially ordered chains, it is necessary to compute the probability that the units involved in cross-linkages will occur in suitable juxtaposition. The results of such a calculation can be expressed as (4)

$$\Delta S_x^0 = kv \left(\frac{\ln C}{2} - \frac{9}{4} + \frac{3}{4} \ln \frac{v}{N_0} \right) \tag{7.6}$$

Here C is a dimensionless quantity of the order of unity and N_0 is the number of statistical elements in the network.[1] A similar expression has been derived by Schellman for the effect of intramolecular cross-linkages in stabilizing ordered polypeptide chains.(12)

From the foregoing, quantitative expressions can be developed for the isotropic melting temperature of various types of crystallizable networks. For networks formed from random chains, if the cross-linked units participate in an unrestricted manner in the crystallization, i.e. if the cross-linked and non-cross-linked units are indistinguishable, then no change in the melting temperature in comparison with the non-cross-linked polymer should be observed. On the other hand, if the cross-linked units do not enter the crystal lattice, then

$$\frac{1}{T_m^i} - \frac{1}{T_m^0} = \frac{R}{\Delta H_u}(1 - \rho) \tag{7.7}$$

For small values of ρ, Eq. (7.7) becomes

$$\frac{1}{T_m^i} - \frac{1}{T_m^0} = \frac{R\rho}{\Delta H_u} \tag{7.8}$$

and a decrease in the melting temperature should occur. Equation (7.7) is recognized as the limiting form of Eq. (5.42), that describes the melting temperature of a random copolymer containing the fraction ρ of noncrystallizable chain units.

When a network is formed from perfectly axially ordered chains and the cross-linked units participate as equals in the crystallization process the isotropic melting temperature can be expressed as

$$\frac{1}{T_m^0} - \frac{1}{T_m^i} = \frac{R\rho}{\Delta H_u} \left(\frac{9}{4} - \frac{3}{4} \ln \rho k_0 \right) \tag{7.9}$$

to a good approximation. Here k_0 is the number of chemical repeating units that can be identified with a statistical element. According to Eq. (7.9), the isotropic melting temperature of such a network should increase relative to the melting temperature of the initially non-cross-linked system. If the cross-linked units are restricted from participating in the crystallization, however, this effect would be partially offset by the necessity of introducing into Eq. (7.9) a term equivalent to the right-hand side

[1] A statistical element of the network bears the same relation to a chain unit as the statistical element of an "equivalent statistical chain" does to the repeating unit of a real chain. See Ref. (9), pp. 410 ff.

of Eq. (7.7). Experimentally determined melting temperatures for networks formed under different conditions can be examined in terms of the above analysis.

It is important when treating semi-crystalline polymers that the state of the system at the time the cross-links are introduced be specified. This point cannot be overemphasized since the network properties are affected in a very significant way.

7.3 Melting temperature of networks formed from random chains

The simplest case of network formation is the random introduction of cross-linkages into a system of randomly arranged chains. This type network can be illustrated by several examples. These include, among others, the usual vulcanization of natural rubber by chemical means, the cross-linking of natural rubber at room temperature by high-energy ionizing radiation, and the irradiation cross-linking of polyethylene at temperatures above its melting temperature. Such networks can be crystallized from the melt merely by cooling and the isotropic melting temperatures subsequently determined. A summary of results for some typical networks formed by these methods is given in Fig. 7.1. Here the networks were formed by either

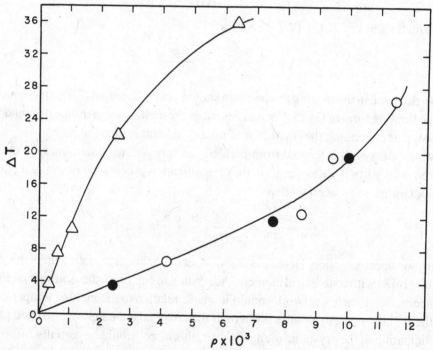

Fig. 7.1 Plot of melting point depression (ΔT) as a function of fraction of units cross-linked, ρ, for polymer networks formed from random chains. △ molten polyethylene cross-linked by ionizing radiation; ○ natural rubber cross-linked with sulfur; ● natural rubber cross-linked with di-t-butyl peroxide. (Data from Refs. (13) and (14))

irradiating linear polyethylene above its melting temperature (13) or by cross-linking natural rubber by chemical means.(14) The latter type networks were formed by reaction with either sulfur or di-t-butyl peroxide. The melting temperatures were determined by utilizing slow heating rates subsequent to crystallization in all cases, in an effort to approach equilibrium values. The depression of the melting temperature, relative to that of non-cross-linked polymer, is plotted in the figure as a function of the fraction of units cross-linked.

For the two different natural rubber networks illustrated, the melting point depression depends only on the fraction of units cross-linked and not on the chemical process by which the cross-links were introduced. The most significant observation here, however, is the fact that a substantial depression of the melting temperature occurs when only a very small number of chain units are cross-linked. For the natural rubber networks the melting temperature is depressed 20°C when 1% of the units are involved in cross-linkages. For the polyethylene networks a depression of 30°C is observed when only about 0.5% of the chain units are cross-linked. Comparable results have been reported for other polyethylene networks (15–17) as well as those formed by both cis and trans poly(butadiene) (18–20) and poly(tetrahydrofuran) (21,22) among others. Melting point depressions larger than expected are typical of networks formed from random chains.

If the melting point depression results solely from the fact that cross-linked units are excluded from the crystalline phase, then Eq. (7.7) should be applicable to the data of Fig. 7.1. According to this equation, when values of ΔH_u for natural rubber and polyethylene are utilized, a melting point depression of, at most, only about 2 to 3°C is expected. This theoretical expectation is clearly not in harmony with the experimental observations. Thus, the applicability of Eq. (7.7), and those that follow from it, can be seriously questioned. In deriving this equation, the inherent assumption is made that equilibrium conditions prevail. It is, therefore, implied that the development of crystallinity in the chain direction is impeded only by the random distribution of the noncrystallizing cross-linked units. The lateral development of crystallinity is unrestricted. If these conditions are not fulfilled, so that a less perfect crystalline state is generated, Eq. (7.7) cannot be applied, and a more severe melting point depression results. Several lines of evidence indicate that the crystallization of the type of networks under discussion results in a state that does not adhere to the rigid specifications set forth. The results with copolymers, that were discussed previously, anticipate such a conclusion.

Wide-angle x-ray diffraction studies of crystalline networks of both polyethylene and natural rubber show that with increasing cross-linking density there is a progressive broadening of the reflections from various crystalline planes.(13,14,23) This broadening can be attributed to either a decrease in crystallite size, the development of further imperfections in the crystals, or to strain. Irrespective of which of these effects causes the broadening of the reflections, they each can make a contribution to the melting point depression. Therefore, one of the major reasons for the large melting point depression that is observed in polymer networks is that the perfection of the crystallinity that can be developed is severely restricted, even after careful annealing procedures are adopted. Permanent type cross-links act to prevent the lateral accretion of polymer chains, a necessary step in the formation of larger crystallites.

Units that neighbor those cross-linked may also be prevented from crystallizing. Hence the longitudinal development of crystallinity is restricted to an extent greater than would be expected solely by the concentration of cross-linked units themselves. To account for this effect, empirical modifications have been made to Eq. (7.7).(15,20) The fraction of cross-linked units ρ is replaced by the quantity $K\rho$, where K represents the number of chain units per cross-link unit that are excluded from the crystallization process. Thus Eq. (7.7) can be rewritten as

$$\frac{1}{T_m^i} - \frac{1}{T_m^0} = \frac{R}{\Delta H_u}(1 - K\rho) \tag{7.10}$$

The fractions of the chain units that are cross-linked can be expressed in terms of M_c, the molecular weight between points of cross-links as

$$\rho = M_0/2M_c \tag{7.11}$$

where M_0 is the molecular weight of the repeating unit. This expression can then be substituted into Eqs. (7.7) and (7.10).

Figure 7.2 is a plot of the extrapolated equilibrium melting temperatures against the molecular weight between crosslinks for poly(tetrahydrofuran) networks.(21) Unimodal and bimodal networks are represented by the filled and open circles respectively. There is, thus, no obvious effect of network architecture. The solid line represents Eq. (7.10) (M_c substituted for ρ) with the best fit value of K. This value of K turns out to be 33. Thus, according to this analysis there are 33 noncrystallizable monomer units per cross-link. For a tetrafunctional junction this corresponds to 8 adjoining units. For trans poly(1,4-butadiene), K was found to be 10 by a similar analysis.(20) However, here observed melting temperatures were used. The K value for linear polyethylene was found to be 60.(15) This analysis indicates that a significant number of repeating units are restrained from participating in the crystallization due to their proximity to the intermolecular cross-links. The K values thus estimated may represent an overestimate because of other factors. For example, the perfection of the crystallites could also cause a depression of the melting temperature from that expected. The analysis can be generalized to include other types of imperfections.(15)

It follows from Eq. (7.1) that only a small fraction of intermolecular cross-linked units are required for network formation at the gel point. For example, for a

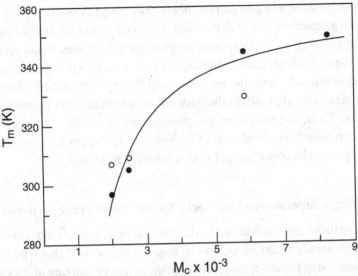

Fig. 7.2 The extrapolated equilibrium melting temperature of poly(tetrahydrofuran) networks as a function of molecular weight between cross-links. Filled and open circles represent unimodal and bimodal networks respectively. Solid curve calculated from Eq. 7.10. (From Roland and Buckley (21))

polyethylene having $M_w = 14\,000$, $\rho_c = 10^{-4}$. The further introduction of cross-links results in the partitioning of the system into sol and gel. With the introduction of intermolecular cross-links branch points must develop in both the pre and post gelation system. Such branching will also contribute to a reduction in the melting temperature. Studies of chemically cross-linked polyethylene (in an undefined initial state) show that there are no major differences in melting temperatures between the gel and sol portions and the nonextracted whole polymer.(23) There is essentially no difference in melting temperatures between the three at small cross-linking levels, after isothermal crystallization. For increased levels of cross-linking, M_0 between 3600 and 1900, the difference is no more than 1 to 2 °C.

Since the melting temperatures of networks are depressed well beyond that expected, based on the concentration of cross-linkages, it can be anticipated that the level of crystallinity will be influenced in a similar manner. This expectation is fulfilled due to the crystallization restraints placed on units adjacent to network junction points. In polyethylene networks there is a twofold decrease in the enthalpy of fusion for the networks studied.(15) This can be directly related to the decrease in the level of crystallinity. Similar results were found with poly(tetrahydrofuran) networks.(21)

Kuhn and Majer (18,24–26) have shown that in a polymeric network swollen with diluent the freezing point of the monomeric liquid component is significantly depressed when compared to that of the pure liquid. In both natural rubber networks swollen with benzene and poly(acrylic acid)–poly(vinyl alcohol) networks swollen in water, the magnitude of the depression is related to the fraction of units cross-linked. The freezing point depression progressively increases as the cross-linking density of the network increases. As much as a 21 °C depression has been observed. This depression results from the limited size of the crystals formed. The restriction on crystal size is attributable to the network structure and the presence of cross-linked units. Thus, not only does the presence of cross-links in relatively small concentration retard crystallization of the network itself, but the crystallization of the diluent present in a swollen gel is also severely restricted.

7.4 Melting temperature of networks formed from axially ordered chains

In contrast to those just studied, networks can also be prepared from chains that are initially in an axially oriented, or fibrous state. A network of this type can be obtained by subjecting fibrous natural rubber to the action of ionizing radiation.(14,27) After network formation the sample can be retracted, or relaxed, and then crystallized merely by cooling. On subsequent heating the isotropic melting temperature is obtained. A comparison can be made between the melting temperatures of networks thus formed and those when the chains are initially in the randomly coiled state utilizing the same cross-linking process.(14) The dependence of the melting temperature on the fraction of units cross-linked for these two extreme types of natural rubber networks is given in Fig. 7.3. The lower curve represents the results for the networks formed from random chains. The melting point depression is similar to that previously described for natural rubber networks formed by chemical cross-linking. The upper curve, however, shows only a very small melting temperature depression over an appreciable cross-linking range. The latter data represents networks formed from axially oriented chains. Consequently, at any cross-linking level, T_m^i is greater for these type networks in comparison with those formed from random chains. Moreover, the difference in melting temperatures between the two networks becomes greater as the cross-linking level is increased. These results are definitive examples of the greater stability that is ultimately imparted to the system by imposing the cross-links on ordered chains. The reason for the enhanced stability is the decreased configurational entropy in the liquid state that accompanies the mode of network formation, although the cross-links are still randomly distributed.

According to Eq. (7.6), if the cross-linked units participated in the crystallization, a progressive increase in the isotropic melting temperature should result. Since this expectation is not observed it can be concluded that in this case also the cross-linkages still impede the crystallization process. However, if it is assumed that the nonconfigurational effects of the cross-links are the same for the two different

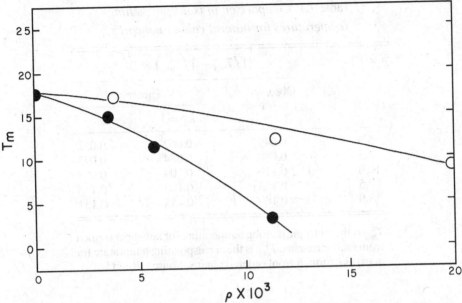

Fig. 7.3 Plot of isotropic melting temperature of natural rubber networks formed by irradiation. ●, random chains at time of network formation; ○, chains axially oriented at time of network formation.(14)

types of networks, i.e. the number of units restrained from crystallizing in the isotropic state, a quantitative comparison of the differences in melting temperature can still be made. By invoking Eq. (7.6), the differences in expected melting temperatures can be calculated for different values of ρ. The results of such a calculation are given in Table 7.1 for networks formed from random and oriented chains, respectively. A value of $\Delta H_u = 1050$ cal/mol was used in the calculation, and the parameter k_0 was assigned values of 1 and 3. When comparison is made at the same values of ρ, the differences in the observed melting temperature are in good accord with theoretical expectations over the cross-linking range in which crystallizable networks can be prepared. Excellent agreement is obtained when 1% or less of the units are cross-linked. The slight deviations that occur for the higher values for ρ can be attributed to the large depression of the melting point observed for networks formed from random chains in this range. Quantitative support is, therefore, given to the concept that a significant decrease in the configurational entropy in the liquid state occurs when networks are formed from axially oriented chains. A partially ordered liquid can be said to have been developed. It can be surmised that other interesting changes in liquid state properties could be accomplished, that would then be reflected in the crystalline state that was formed subsequently. The importance of the chain disposition at the time of network formation is emphasized by these results.

Table 7.1. *Comparison of isotropic melting*
temperatures for natural rubber networks

$\rho \times 10^3$	$(1/T_{m,r}^i - 1/T_{m,o}^i) \times 10^{3\,a}$		
	Observed	Theoretical	
		$k_0 = 1$	$k_0 = 3$
5.0	0.049	0.062	0.052
7.5	0.080	0.084	0.073
10.0	0.110	0.109	0.093
12.5	0.163	0.133	0.113
15.0	0.180	0.155	0.131

[a] $T_{m,r}^i$ is the isotropic melting temperature for networks formed from random chains. $T_{m,o}^i$ is the corresponding temprature for networks formed from ordered chains. Source: Ref. (14).

7.5 Melting temperature of networks formed from randomly arranged crystallites

Since networks with unique properties result from cross-linking highly axially oriented polymer chains, the question naturally arises as to what limits there are, if any, to the degree of intermolecular order required to observe these effects. Partially crystalline undeformed polymers possess a large amount of intermolecular order, since significant portions of the polymer molecules are constrained to lie parallel to one another in three-dimensional array. This order is only on a microscopic scale since the crystalline regions are randomly arranged relative to one another. The question as to whether the presence of such order influences the properties of the resultant isotropic network is a matter to be decided by experiment.

Investigations have shown that the isotropic melting temperatures of networks formed from crystalline, but nonoriented, linear polyethylene are different from those of networks of the same polymer but formed when the chains are initially in the molten state.(13,28) As was illustrated in Fig. 7.1 for the latter type network, a large and continuous decrease in T_m^i is observed with cross-linking. On the other hand, the melting temperatures of a set of crystalline linear polyethylenes, that were cross-linked by high energy radiation, are only depressed by about 6.5 °C relative to the non-cross-linked polymer. The melting temperatures remain independent of the fraction of units cross-linked up to relatively high cross-linking levels.(13) These are isotropic melting temperatures, determined following melting and recrystallizing, after the initial introduction of cross-links. The observed melting temperatures of both types of networks are summarized in columns (a) of Tables 7.2 and 7.3. Each of

Table 7.2. *Properties of polyethylene networks formed at 17°C by the action of high-energy ionizing radiation*[a]

(a)		(b)	
v_1 (130°C)	T_m^i	T_m^*	v_1 (at T_m^*)
0.97	131	102	0.88
0.95	131	104	0.83
0.83	131	105.3	0.77
0.74	131	106.2	0.69
0.61	131	110	0.58
0.39	131	114	0.36
0.28	132		

[a] From Ref. (14).

Table 7.3. *Properties of polyethylene networks formed at 175°C by the action of high-energy ionizing radiation*[a]

(a)		(b)	
v_1 (130°C)	T_m^i	T_m^*	v_1 (at T_m^*)
0.93	134	101	0.84
0.91	132.5	100.2	0.85
0.87	130	100	0.80
0.76	128	100.6	0.71
0.70	115.5	94	0.68
0.49	107.5		
	101		

[a] From Ref. (13).

the networks is characterized by the volume fraction of xylene imbibed at swelling equilibrium at 130°C, (v_1 130°C) in the tables. Decreasing values of v_1 indicate a progressively increasing value of ρ. For the least swollen network formed at 17°C (in the highly crystalline state), it is estimated that approximately 2.5% of the units are involved in cross-linkages.(13) Despite the large variation in ρ, T_m^i does not change after the slight initial decrease from the value of the pure polymer.(Table 7.2) In other works a slight but steady increase in the melting temperature with cross-linking is observed.(28,29) Irrespective of the overall differences between these studies the major conclusion is clear. Networks formed by cross-linking in the crystalline, but nonoriented, state are quite different than networks formed from completely random chains. In the latter case a large decrease in the isotropic melting temperature is observed with increasing cross-linking density.

The greater stability of the crystalline state of networks formed from unoriented but crystalline chains compared with networks formed from amorphous polymers, can be explained in the same way as for networks formed from axially oriented natural rubber. Although prior to network formation the crystallites are randomly arranged relative to one another, portions of chains are still constrained to lie in parallel array. The cross-linking of the predominantly crystalline polymer cannot, therefore, involve the random selection of pairs of units. The units that can be paired are limited by the local chain orientation imposed by the crystalline structure. An increase in the isotropic melting temperature of such networks would therefore be expected. It can be concluded that orientation on a macroscopic scale is not required for partial order in the liquid state to develop. Concomitantly a decrease in the entropy of fusion will result, which reflects the increase in molecular order in the melt. This is an important concept that must be kept in mind when studying the properties of networks formed in this manner. This conclusion has important implications in studying the properties of networks formed from unoriented crystalline polymers.

The manifestation of molecular order in the liquid state, after cross-linking in the crystalline state, is substantiated by direct microscopic observation of the polymer melt.(13,30) Figure 7.4 illustrates the intense birefringence observed in the spherulitic pattern of linear polyethylenes that persists at temperatures above the melting temperature. The pattern on the left is that of the sample after cross-linking by irradiation, but before heating. The pattern on the right is the same sample after heating above the melting temperature, to 150 °C. The persistence of the birefringence in the melt is quite striking. It can be directly attributed to the unique, partially ordered, liquid structure that has been developed. In contrast, when polyethylene is cross-linked in the molten state no birefringence or structure is observed above the melting temperature.

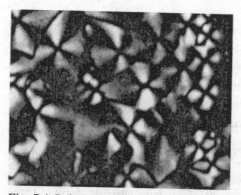

Fig. 7.4 Polarized optical micrographs of linear polyethylene cross-linked at room temperature. (From Hammer, Brandt and Peticolas (30))

The major influence of high energy ionizing radiation on crystalline polyethylene is the introduction of intermolecular cross-links into the system. Although the cross-links are proportioned between the crystalline and noncrystalline regions a sufficient number of chains within the interior of the crystallite are cross-linked.(7) Consequently the configuration entropy is reduced. With a sufficiently high radiation dose, a hexagonal structure that only has lateral order is developed prior to complete melting.(31) This is a manifestation of the reduced configurational entropy and the persistence of molecular order after complete melting.

The intensity of the electron beam used to examine thin crystalline polymer films by electron microscopy is usually of sufficient intensity to induce cross-linking. It is not surprising, therefore, after initial examination in the crystalline state, that thin films of poly(amides) and polyethylene display ordered structures when subsequently examined in the molten state by this technique. These observations are to be expected. They cannot be construed as evidence that, in general, the liquid state in polymers is an ordered one.(33) The partially ordered liquid represents an interesting, unique situation that results from the nature of the chain arrangement at the time of network formation.

7.6 Melting of network–diluent mixtures

Polymer networks can also crystallize when in contact with a monomeric liquid or diluent. The simplest case to analyze is when the crystalline network is in contact with a large excess of a one-component liquid phase. This case corresponds to a thermodynamic open system. Upon melting, the network in the amorphous state imbibes large quantities of the surrounding fluid. The amount of swelling that occurs depends on the network structure, the temperature, and the polymer–diluent thermodynamic interaction parameter. Conversely, on crystallization, diluent is expelled from the network. At the melting temperature, the crystalline polymer phase is in equilibrium with a mixed phase composed of amorphous polymer and imbibed liquid. In turn, the latter phase is in equilibrium with the pure solvent. Three distinct phases must co-exist in equilibrium at the melting temperature. Therefore, the equilibrium requirements are

$$\mu_u^c - \mu_u^0 = \mu_u^m - \mu_u^0 \tag{7.12}$$

$$\mu_1^l - \mu_1^0 = 0 \tag{7.13}$$

for the components common to each of the phases. In these equations μ_u^c and μ_u^m represent the chemical potentials of the polymer unit in the crystalline and mixed phases, while μ_u^0 represents the chemical potential of the pure molten polymer unit. Equation (7.13) specifies the equality of the chemical potential of the solvent component in the mixed and supernatant phases.[2]

The free energy change ΔG for the formation of the mixed phase from its pure components, the pure solvent and the pure isotropic amorphous network, consists of two parts. One is the free energy of mixing ΔG_M and the other is the elastic free energy ΔG_{el} that results from the expansion of the network structure because of swelling (4,34)

From the Flory–Huggins theory (34,35) of polymer solutions

$$\Delta G_M = kT(n_1 \ln v_1 + \chi_1 n_1 v_2) \tag{7.14}$$

where n_1 is the number of solvent molecules. Utilizing an idealized theory for rubber elasticity, where it is assumed that the deformation process accompanying the swelling involves no internal energy change attributable to interactions between chains, ΔG_{el} can be written as (4)

$$\Delta G_{el} = \frac{kTv}{2}\left[(3\langle\alpha\rangle^2\alpha_s^2 - \ln\alpha_s^3 - 3 - 3\ln\langle\alpha\rangle)\right] \tag{7.15}$$

where α_s represents the linear swelling factor for the network–diluent mixture. From the sum of Eqs. (7.14) and (7.15),

$$\frac{\mu_u^m - \mu_u^0}{RT} = -\frac{V_u}{V_0}(v_1 - \chi_1 v_1^2) + \frac{\rho}{2}\left[1 - v_2\langle\alpha\rangle(v_2^{1/3} - v_2^{-2/3})\right] \tag{7.16}$$

since $\alpha_s^3 = v_2^{-1}$. As for non-networks, $\mu_u^c - \mu_u^0$ can be expressed as

$$\mu_u^c - \mu_u^0 = -\Delta H_u\left(1 - \frac{T}{T_m}\right) \tag{7.17}$$

where T_m is now identified with the equilibrium melting temperature of a given network. At equilibrium, $T = T_m^*$ so that

$$\frac{1}{T_m^*} - \frac{1}{T_m^0} = \frac{R}{\Delta H_u}\frac{V_u}{V_1}(v_1 - \chi_1 v_1^2) + \frac{R\rho}{2\Delta H_u}\left\{2\langle\alpha\rangle\left[v_2^{-2/3} - v_2^{-1/3} - (1 - v_2)\right]\right\} \tag{7.18}$$

The first two terms on the right-hand side of Eq. (7.18) are identical to those obtained for the non-cross-linked polymer–diluent mixture at the same composition. The remaining terms represent the contribution of the elastic free energy of the mixed phase. For an open system, the composition of the mixed phase v_2 is determined from Eq. (7.13), which specifies the swelling equilibrium.(34) Therefore v_2 is an equilibrium quantity and should be so designated. It can be identified with the reciprocal of the equilibrium swelling ratio at $T = T_m^*$. For a closed system, where

---

[2] The treatment can be generalized to include a multi-component supernatant phase and a partitioning of components between it and the mixed phase. The case where diluent enters the crystalline phase can also be treated.

the composition of the mixed phase is fixed and the supernatant phase is absent, Eqs. (7.9) and (7.17) suffice to specify the melting point relations.

A comparison of Eq. (7.18) with Eq. (3.2) indicates that the melting point depression should be greater for a network than for just a collection of polymer chains of the same constitution at the same concentration. This is due to the contribution from the elastic free energy to Eq. (7.18). However, since the values of ρ usually encountered are of the order of 0.01 to 0.02 or less, this effect is quite small. It manifests itself only when v_2 of the mixed phase becomes less than 0.5.

The melting temperatures of polyethylene networks immersed in a large excess of xylene have been measured.[13] The results can be examined in terms of the equilibrium theory. For networks formed from either random chains or from nonoriented crystalline chains, a depression of the isotropic melting temperature relative to the undiluted system is observed. However, as an examination of columns (b) in Tables 7.2 and 7.3 reveals, the melting temperatures T_m^* of the two different networks depend quite differently on the cross-linking density. When immersed in xylene, the networks formed in the crystalline state display a continuous and significant increase of melting temperature with increased cross-linking. However, in contrast, the melting temperatures of the networks formed from random chains display a slight decrease in T_m^* with increased cross-linking. The results for the networks formed from the crystalline chains immersed in an excess of diluent are in sharp contrast with the melting points of the undiluted networks of natural rubber or polyethylene. Irrespective of the chain disposition prior to network formation in the latter cases, a decrease in T_m with cross-linking is invariably observed.

The melting temperatures of the network–diluent mixtures depend on the nature of the initial network and on the volume fraction of liquid that is imbibed subsequent to fusion. T_m for the undiluted networks is constant with cross-linking for the networks formed at 17 °C. However, the equilibrium swelling at T_m^* (v_1 at T_m^*) in Tables 7.2 and 7.3 continuously decreases. The melting point depression must be progressively diminished, with the net result that T_m^* increases. However, for the networks formed from random chains, the rate of decrease of T_m is not compensated by the concentration changes in the mixed phase, so that a decrease in T_m^* results.

An attempt to quantitatively examine these observations in terms of theory is given in Fig. 7.5. The solid line in this plot is computed from Eq. (7.18) with $\chi_1 = 0$ and neglect of the elastic contribution. For the low values of v_1, where neither the thermodynamic interaction term nor the elastic term make an appreciable contribution to the melting point depression, the data follows the simplest theoretical expectation. As the polymer concentration in the mixed phase decreases, a contribution to the melting point depression of the omitted terms is expected. Small deviations from the simplified theory are observed. A small positive value of χ_1, believed to be appropriate for this system, brings the observed and calculated values very close

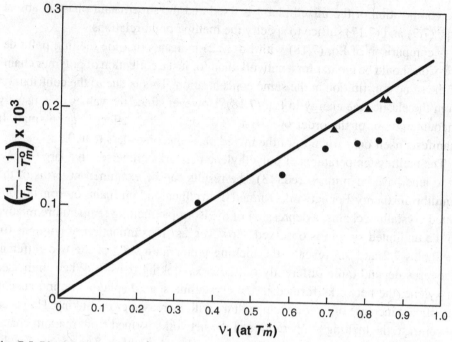

Fig. 7.5 Plot of $1/T_m - 1/T_m^0$ against volume fraction of xylene imbibed upon melting, v_1. ● networks from crystalline chains; ▲ networks from molten chains; —— theoretical plot according to Eq. (7.18) with $\chi_1 = 0$ and $\rho = 0$. (Data from ref. (13))

to one another. It is noteworthy that the existing theory can encompass the apparently diverse manner in which the melting points of the two different types of network–diluent systems vary with increasing amounts of cross-linking. The importance of accounting for the concentration of the mixed phase in an open system is emphasized.

7.7 Fibrous proteins

Certain of the fibrous proteins, such as collagen (37) and α-keratin from various layers of epidermis,(38) display an increase in melting temperature as the number of intermolecular cross-links are increased. For collagen, increases in melting temperature of up to 35 °C have been observed after the introduction of cross-links by means of specific tanning processes.(37) In various layers of the epidermis of cow's lip the melting temperature progressively decreases from the outer to inner layer, while the cystine content (which can be related to the number of intermolecular cross-links) also decreases.(38) These examples represent typical findings in the fibrous proteins.

The fibrous proteins are naturally occurring axially ordered systems, with the cross-links imposed on the ordered structure. Therefore, according to theory,(4) if the cross-linkages do not impede the crystallization, a continuous increase of melting temperature with cross-linking is expected. However, the melting of the fibrous proteins is almost invariably determined when they are immersed in a suitable liquid medium. Consequently, at equilibrium the polymer concentration in the mixed phase must also increase with increasing cross-linking density. An elevation of the melting temperature is also expected from this cause, in analogy to the results for the polyethylene network–diluent mixtures. Thus both an alteration in the entropy of fusion and a compositional change of the mixed phase result from the introduction of cross-links into a fibrous protein system. Both these effects act to raise the melting temperature and favor the stability of the crystalline phase. As would then be anticipated, when experiments are carried out in the presence of a large excess of liquid, a strong correlation exists between the melting of collagen and the swelling capacity in the mixed phase.(39)

References

1. Flory, P. J., *J. Am. Chem. Soc.* **63**, 3097 (1941).
2. Stockmayer, W. J., *J. Chem. Phys.*, **12**, 125 (1944).
3. Flory, P. J., *J. Am. Chem. Soc.*, **69**, 30 (1947).
4. Flory, P. J., *J. Am. Chem. Soc.*, **78**, 5222 (1956).
5. Craig, D., *Rubber Chem. Technol.*, **30**, 1291 (1957).
6. Charlesby, A., *Atomic Radiation and Polymers*, Pergamon Press (1960).
7. Kitamaru, R. and L. Mandelkern, *J. Am. Chem. Soc.*, **86**, 3529 (1964).
8. Dole, M., *Radiation Chemistry of Macromolecules*, Academic Press (1972).
9. Flory, P. J., *Principles of Polymer Chemistry*, Cornell University Press (1953) p. 464.
10. Treloar, L. R. G., *The Physics of Rubber Elasticity*, Oxford University Press (1947).
11. Flory, P. J., *J. Chem. Phys.*, **18**, 108 (1950).
12. Schellman, J. G., *Compt. Rend. Trav. Lab. Carlsberg, Ser. Chim.*, **29**, 223 (1955).
13. Mandelkern, L., D. E. Roberts, J. C. Halpin and F. P. Price, *J. Am. Chem. Soc.*, **82**, 46 (1960).
14. Roberts, D. E. and L. Mandelkern, *J. Am. Chem. Soc.*, **82**, 1091 (1960).
15. de Boer, A. P. and S. J. Pennings, *Faraday Discuss Chem. Soc.*, **68**, 345 (1979).
16. de Boer, A. P. and S. J. Pennings, *Polymer*, **23**, 1944 (1982).
17. Jäger, E., J. Muller and B.-J. Jungnickel, *Prog. Coll. Polym. Sci.*, **71**, 145 (1985).
18. Kuhn, W. and J. Majer, *Angew. Chem.*, **68**, 345 (1956).
19. Trick, G. S., *J. Polym. Sci.*, **41**, 213 (1959).
20. Akana, Y. and R. S. Stein, *J. Polym. Sci.: Polym. Phys. Ed.*, **13**, 2195 (1975).
21. Roland, C. M. and G. S. Buckley, *Rubber Chem. Technol.*, **64**, 74 (1991).
22. Shibayama, M., H. Takahashi, H. Yamaguchi, S. Sakurai and S. Nomura, *Polymer*, **35**, 2945 (1994).
23. Lambert, W. S. and P. J. Phillips, *Polymer*, **31**, 2077 (1990).
24. Kuhn, W., E. Poterli and H. Majer, *Z. Electrochem.*, **62**, 296 (1958).
25. Kuhn, W. and H. Majer, *Z. Physik. Chem.*, **3**, 330 (1955).
26. Kuhn, W. and H. Majer, *Ric. Sci. Suppl. A.*, **3** (1955).
27. Roberts, D. E. and L. Mandelkern, *J. Am. Chem. Soc.*, **80**, 1289 (1958); D. E. Roberts, L. Mandelkern and P. J. Flory, *J. Am. Chem. Soc.*, **79**, 1515 (1957).

28. Gielenz, G. and B.-J. Jungnickel, *Coll. Polym. Sci.*, **260**, 742 (1980).
29. Bhateja, S. K., E. H. Andrews and R. S. Young, *J. Polym. Sci.: Polym. Phys. Ed.*, **21**, 523 (1983).
30. Hammer, C. F., W. W. Brandt and W. L. Peticolas, *J. Polym. Sci.*, **24**, 291 (1957).
31. Ungar, G. and A. Keller, *Polymer*, **21**, 1273 (1980).
32. Orth, H. and E. W. Fischer, *Makromol. Chem.*, **88**, 188 (1965).
33. Kargin, V. A., *J. Polym. Sci.*, **30**, 247 (1958); V. A. Kargin and G. L. Slonimskii, *Usp. Khim.*, **24**, 785 (1955).
34. Flory, P. J., *Principles of Polymer Chemistry*, Cornell University Press (1953) p. 578.
35. Huggins, M. L., *J. Phys. Chem.*, **46**, 151 (1942); *Ann. N.Y. Acad. Sci.*, **41**, 1 (1942).
36. Flory, P. J., *J. Chem. Phys.*, **10**, 51 (1942).
37. Gustavson, K. H., *The Chemistry and Reactivity of Collagens*, Academic Press, Inc. (1956) p. 227.
38. Rudall, K. M., *Symposium on Fibrous Proteins, J. Soc. Dyers Colourists*, 15 (1946); *Adv. Protein Chem.*, **7**, 253 (1952).
39. Theis, F. R., *Trans. Faraday Soc.*, **42B**, 244 (1946).

8

Oriented crystallization and contractility

8.1 Introduction

A characteristic property of amorphous polymers is the ability to sustain large strains. For cross-linked three-dimensional networks the strain is usually recoverable and the deformation process reversible. The tendency toward crystallization is greatly enhanced by deformation since chains between points of cross-linkages are distorted from their most probable conformations. A decrease in conformational entropy consequently ensues. Hence, if the deformation is maintained, less entropy is sacrificed in the transformation to the crystalline state. The decrease in the total entropy of fusion allows crystallization, and melting, to occur at a higher temperature than would normally be observed for the same polymer in the absence of any deformation. This enhanced tendency toward crystallization is exemplified by natural rubber and polyisobutylene. These two polymers crystallize very slowly in the absence of an external stress. However, they crystallize extremely rapidly upon stretching.

It is a widely observed experimental fact that crystallites produced by stretching usually occur with their chain direction preferentially oriented parallel to the axis of elongation. The extent of the orientation will depend on the type and amount of the deformation. This is particularly true for crystallization at large deformations.[1] These observations contrast with the crystalline texture that results when the transformation is induced in the absence of an external stress merely by cooling. In the latter case the crystallites are, on the average, randomly arranged relative to one another. When a portion of a deformed chain is incorporated into a crystallite, the average stress that it exerts at its end points is reduced. This conclusion can be reached either by the application of Le Châtelier's principle or from a more detailed molecular analysis.(2,3) According to the molecular theory of rubber elasticity,

---

[1] Certain exceptions to this generalization can be noted. These usually result from nonisothermal crystallization at small deformations. In these instances the chain axes are more preferentially oriented normal to the stretching directions.(1)

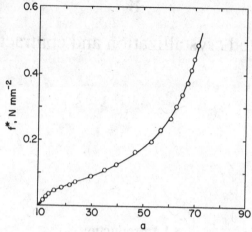

Fig. 8.1 Stress–elongation curve for natural rubber in the vicinity of room temperature. (From Mark (5))

the force exerted by the fixed chain ends is inversely proportional to the number of statistical elements contained in the chain and the magnitude of its end-to-end distance.(2,4)[2] Since only the remaining amorphous units contribute to the resulting retractive force, the former quantity is decreased somewhat as a result of oriented crystallization. Moreover, the distance traversed by the remaining amorphous units is severely reduced because of the disproportionately greater distance taken up by the crystalline units. Consequently, the retractive force exerted is diminished by the crystallization process. The conclusion is reached that orientation imposed by stretching promotes crystallization and that crystallization in an oriented polymer diminishes the stress.

A stress–strain isotherm for the uniaxial deformation of natural rubber, at ambient temperature, that was cross-linked in the liquid state is shown in Fig. 8.1.(5) Here f^* is the nominal stress defined as the tensile force, f, in the stretching direction divided by the initial cross-section, and α is the extension ratio. Using the most rudimentary form of molecular rubber elasticity theory f^* can be expressed as (6–9)

$$f^* = \left(\frac{vkT}{V}\right)(\alpha - \alpha^{-1}) \tag{8.1}$$

where k is the Boltzman constant and V is the volume. The initial portion of the stress–strain isotherm in Fig. 8.1 is that expected from Eq. (8.1). For large values of α, f^* approaches linearity. However, at larger deformation ratios, $\alpha \geq 5$, the large increase in f^* that is observed cannot be attributed to the deformation of

⸱[2] For a further detailed discussion of rubber elasticity theory, see Refs. (6,7,8,9).

disordered chain units. Rather, crystallization has been induced by the stretching, as demonstrated by direct measurements. The oriented crystallites have a much higher modulus than the disordered chain and introduce an element of rigidity to the system. In addition, the crystallites act as physical cross-links which will also act to increase the modulus of the system. On further stretching, the chain segments in the amorphous regions will be oriented much more than normal. A proportionately larger decrease in the entropy ensues, resulting in an increase in the retractive force. Since further crystallization will occur with subsequent elongation these effects will be enhanced and the increase in force will be accelerated.

These factors, due to oriented crystallization, explain the large upsweep of f^* with α in Fig. 8.1. The results shown for natural rubber are typical of different elastomers of reasonably regular structure.(10–13) However, for a structurally irregular chain, as for example a poly(cis-1,4-butadiene) that only contains about 37% of the cis 1,4 units, the large upsweep in the stress–strain curve is not observed.(11) These results support the contention that many of the unusually high modulus values that are reported are not due to interchain interactions at the high chain extension, but to the reinforcing effect of crystallization induced by stretching. These results should not be taken as a shortcoming of rubber elasticity theory. The theory is based on the deformation of disordered chains. It might appear that these findings are in contradiction to the discussion that was just given above. However, the two processes that have been discussed are quite different. In the case just discussed crystallites are found during isothermal stretching. In the previous case an equilibrium process was considered.

For any stress likely to be borne by amorphous chains, the length of the randomly coiled molecule projected on the fiber axis is considerably less than its length in the crystalline state. This statement is in accord with the known crystal structures of polymers. Hence for axially oriented systems, melting results in contraction and crystallization in elongation. Macroscopic dimensional changes, as well as changes in the exerted stress, can be coupled with and related to the crystal–liquid phase transition. This behavior, which reflects one of the unique properties of polymer chains, results from their conformational versatility. It is not limited to the simpler types of chain molecules but should apply equally well to the fibrous proteins and other macromolecules of biological interest. Many polymers in the latter category are characterized by the prevalence of an ordered crystalline arrangement in the native state. Cognizance must be taken of the existence of this state when attention is given to such properties as thermoelastic behavior and to the mechanism by which major changes in length are incurred.

In order to properly analyze the melting of an oriented system, it must be ascertained whether the process is reversible, i.e. whether oriented crystallites are formed on recrystallization. This concern exists since it is possible that the original oriented crystalline state will not be regenerated. The possible nonequilibrium aspects of the melting of an oriented polymer and the complications that result have been discussed in connection with the melting of "stark" rubber.(14) When natural rubber is stored in temperate climates, it frequently becomes hard and inelastic because of the development of significant amounts of crystallinity. Upon initial heating, the melting point is significantly higher than that assigned to the equilibrium melting temperature of natural rubber (in the absence of any external force). This apparent contradiction is resolved when it is observed that in "stark" rubber the crystalline regions are preferentially oriented despite the absence of any external force. The maintenance of this orientation during fusion results in an elevated melting temperature. After the initial melting and subsequent recrystallization, melting points that are normal for natural rubber are observed since oriented crystallization does not redevelop.

Many polymers can be rendered fibrous, i.e. made to possess a high axial orientation of the crystallites, by suitable mechanical means. This condition can, in many cases, be maintained below the melting temperature without application of an external force. On melting, in addition to the usual changes in properties, an axial contraction is observed. This transformation temperature has, therefore, been designated as the shrinkage temperature. However, only under certain unique conditions (see following) can this temperature be identified with the equilibrium melting temperature. In general, in the absence of an equilibrium tensile force, the original crystalline state is not regenerated merely by reversing the melting process. Even if a tensile force is applied to the system, a distinction must be made between the shrinkage temperature T_s and the equilibrium melting temperature T_m. The latter temperature requires the co-existence of amorphous and crystalline phases along the fiber length, whereas in a well-oriented highly crystalline fiber a significant amount of superheating may be required to initiate melting and observe shrinkage. Hence, error may arise by failure to discriminate between T_m and T_s. The shrinkage temperature by itself is not an appropriate quantity for thermodynamic analyses.

Although irreversible melting is commonly associated with oriented crystalline polymers, the possibility of conducting the transformation under reversible conditions that approach equilibrium cannot be disregarded. In fact, the treatment of this problem as one of phase equilibria lead to important relations between crystallization, deformation, and dimensional changes.(3,4)

8.2 One-component system subject to a tensile force

Consider a cross-linked fibrous system, composed of highly axially oriented crystalline regions co-existing with amorphous zones; the latter being devoid of any vestiges of crystalline order. The fiber is subject to a uniform tensile force, f, acting along its axis. The fibers are assumed to be homogeneous and uniform with respect to chemical composition, structure, and cross-section, apart from such differences as may exist in cross-section because of the interspersion of crystalline and amorphous regions along the length.[3]

According to the first law of thermodynamics, the change in internal energy E of any system can be written with complete generality as

$$dE = dQ - dW \tag{8.2}$$

where dQ is the heat absorbed by the system and dW is the work performed by the system on its surroundings. If x_i represents the extensive variables characterizing the system and y_i the conjugate intensive variables,

$$dW = -\sum_i y_i \, dx_i \tag{8.3}$$

and

$$dE = dQ - \sum_i y_i \, dx_i \tag{8.4}$$

For a one-component system the intensive–extensive pairs p, V and f, L are those of interest. Here, p and V are the pressure and volume, respectively, and L is the length of the fiber. For a process that is conducted reversibly, $dQ = T \, dS$, where S is the entropy. Thus

$$dE = T \, dS - p \, dV + f \, dL \tag{8.5}$$

Defining the Gibbs free energy by

$$G = E + pV - TS = H - TS \tag{8.6}$$

where H is the enthalpy, from Eqs. (8.5) and (8.6)

$$dG = -S \, dT + V \, dp + f \, dL \tag{8.7}$$

For present purposes, it is convenient to choose p, T and f as the independent variables. It is advantageous to utilize the equivalent relation

$$d(G - fL) = -S \, dT + V \, dp - L \, df \tag{8.8}$$

For the system to be in equilibrium at constant p, T and f, the function $G - fL$ must be a minimum with respect to all permissible displacements. In particular,

---

[3] A tensile force, or stress, is not the only kind that can be applied to a polymeric system.(6,8,9) Other types of deformation could be treated equally well with, however, more complexity in the analysis. The case being considered here serves quite well in illustrating the principles involved.

it must be a minimum with respect to changes in the fraction of the fiber that is crystalline. Thus

$$\left[\frac{\partial(G - fL)}{\partial\lambda}\right]_{p,T,f} = 0 \tag{8.9}$$

if equilibrium is to be maintained between the two phases. The total free energy of the fibrous system can be expressed as

$$G = \lambda G^a + (1 - \lambda)G^c \tag{8.10}$$

where G^a and G^c are the free energies of the fiber when totally amorphous and totally crystalline, respectively, under the conditions specified by p, T and f. The other extensive properties can be expressed in a similar manner. Accordingly, the requirement for equilibrium becomes

$$G^a - fL^a = G^c - fL^c \tag{8.11}$$

or

$$d(G^a - fL^a) = d(G^c - fL^c) \tag{8.12}$$

From Eq. (8.8) it follows that

$$\left(\frac{\partial f}{\partial T}\right)_{p,eq} = \frac{-\Delta S}{\Delta L} \tag{8.13}$$

at constant pressure. Here ΔS and ΔL are the changes in entropy and length that occur upon fusion of the entire fiber at constant T, p and f. For the reversible process being treated, the heat absorbed is expressed as

$$Q = T\,\Delta S = \Delta E + \Delta W = \Delta E + p\,\Delta V - f\,\Delta L \tag{8.14}$$

so that

$$\Delta S = \frac{\Delta H - f\,\Delta L}{T} \tag{8.15}$$

Combination of Eq. (8.15) with Eq. (8.13) yields

$$\left(\frac{\partial f}{\partial T}\right)_{p,eq} = \frac{f}{T} - \frac{\Delta H}{T\,\Delta L} \tag{8.16}$$

or, in more compact form,

$$\left[\frac{\partial(f/T)}{\partial(1/T)}\right]_{p,eq} = \frac{\Delta H}{\Delta L} \tag{8.17}$$

These equations have been derived by Gee (15) and by Flory.(4) Equations (8.16) and (8.17) are variants of the Clapeyron equation applied to a unidimensional system of axially oriented crystalline and amorphous phases. The temperature T may be regarded as the melting temperature T_m under a force f and a pressure p. The analogy between this problem in phase equilibrium and the vapor–liquid or solid–liquid equilibrium of monomeric substances becomes apparent when it is realized that in Eqs. (8.16) and (8.17) – f and L correspond to the pressure and volume in the more conventional formulation of the Clapeyron equation. At the temperature of vapor–liquid equilibrium for a one-component system, the pressure is independent of the volume of the system, i.e. independent of the relative abundance of each phase. Similarly, it is implicit in the above formulation that for a one-component fibrous system, with uniform properties throughout, the equilibrium force f must be independent of the length over the two-phase region at constant T and p. It will ordinarily be expected that $\Delta L < 0$, whereas $\Delta H > 0$. Therefore from Eq. (8.17), f/T will increase with T. In other words, the melting temperature increases with an increase in the applied tensile force at constant pressure.

The integration of Eq. (8.17) between specified limits leads to a relation between the equilibrium tensile force, f_{eq}, and the melting temperature. This is analogous to integrating the Clapeyron equation for vapor–liquid equilibrium. In this case, if the equation of state relating the pressure and volume of the liquid is known, the dependence of the pressure on temperature is obtained. For the present problem the equation of state relating the applied force to the length of the network is required. This information can be obtained from the theory of rubber elasticity.(6–9)

When a one-component amorphous network, composed of chains whose distribution of end-to-end distances is Gaussian, is subject to a simple tensile force, the relation between the force and length is expressed by(4)

$$f = BTL_a\left(1 - \frac{L_i^3}{L_a^3}\right) \tag{8.18}$$

with

$$B = k\nu\left(\frac{\langle\alpha\rangle}{L_i}\right)^2 \tag{8.19}$$

L_i is the length of the isotropic amorphous network, i.e. the length under zero force, L_a is the length in the amorphous state under the equilibrium tensile force f and $\langle\alpha\rangle$ is a parameter which measures the geometric mean of the linear dilation of the actual network relative to that in the isotropic state.[4] The relation between f and L is completely general and applies equally to networks formed from polymer molecules in random configuration and to those formed from highly oriented chains.

---

[4] Equation (8.18) is derived for a Gaussian network from the relation $f = (\partial\Delta G_{el}/\partial L)_{P,T\langle\alpha\rangle}$ and the assumption $\Delta F_{el} = -T\,\Delta S_{el}$; the expression for ΔS_{el} is well known.

The macroscopic isotropic length of the sample can be related to the number of chains v and their mean-square end-to-end distance by

$$L_i = \frac{v}{\sigma'}\left(\frac{\overline{r_0^2}}{3}\right)^{1/2}\langle\alpha\rangle \tag{8.20}$$

where σ' is the number of chain vectors traversing a plane transverse to the axis of the sample. For networks formed from highly ordered chains, which are of particular interest in the present context, L_i increases as $v^{1/2}$. For such a system, σ' can be identified with the number of chains in a cross-section and hence is independent of v. However, $\overline{r_0^2}$, the mean-square end-to-end distance, is proportional to the number of units in the chain and varies inversely as v. Consequently L_i varies as $v^{1/2}\langle\alpha\rangle$. Hence, from Eq. (8.18), B is independent of v for networks formed in this manner. It is convenient to introduce the quantity L_m which represents the length of the amorphous fiber at its maximum extension. Then (4)

$$B = \frac{3kvn'}{L_m^2} \tag{8.21}$$

where n' is the number of statistical elements in the chain. If L^a is sufficiently greater than L_i, the retractive force can be expressed as

$$f \cong 3kTvn'\frac{L}{L_m^2} \tag{8.22}$$

Upon substitution of Eq. (8.18) into (8.16) one obtains

$$(L^a - L^c)d\left[L^a - \frac{L_i^3}{(L^a)^2}\right] = \frac{\Delta H}{B}d\left(\frac{1}{T}\right) \tag{8.23}$$

when it is recalled that $\Delta L = L^a - L^c$. Integration of this equation between the limits of L_i and L_a and T_m and T_m^i yields

$$2(L^a - L^c)\left[L^a - \frac{L_i^3}{(L^a)^2}\right] - \left[(L^a)^2 + \frac{2L_i^3}{L^a} - 3L_i^2\right] = \frac{2\Delta H}{B}\left(\frac{1}{T_m} - \frac{1}{T_m^i}\right) \tag{8.24}$$

where T_m^i is the equilibrium melting temperature at zero force and T_m is the melting temperature at a force f such that the amorphous length is L^a. The implicit relationship between T_m and the applied tensile force can also be obtained by utilizing Eqs. (8.18) and (8.19) to eliminate L^a in the above. For networks formed from highly ordered chains, where B is given by Eq. (8.21),

$$2(L^a - L^c)\left[L^a - \frac{L_i^3}{(L^a)^2}\right] - \left[(L^a)^2 + \frac{2L_i^3}{L^a} - 3L_i^2\right] = \frac{2L_m^2 \Delta h'}{3R}\left(\frac{1}{T_m} - \frac{1}{T_m^i}\right)$$

(8.25)

where $\Delta h'$ is the heat of fusion per mole of equivalent statistical elements. For large deformations, where $(L_i/L^a)^3 \ll 1$, so that Eq. (8.22) can be employed, the above simplifies to

$$\frac{(L^a)^2 - 2L^a L^c}{L_m^2} \cong \frac{2\Delta h'}{3R}\left(\frac{1}{T_m} - \frac{1}{T_m^i}\right)$$

(8.26)

Alternatively, the integration can be carried out between the limits L^c and L^a, with the result that

$$(L^c - L^a)^2\left[1 + \frac{2L_i^3}{L^c(L^a)^2}\right] = \frac{2\Delta H}{B}\left(\frac{1}{T_m} - \frac{1}{T_m^c}\right)$$

(8.27)

where T_m^c is the melting point when $L^a = L^c$. Using the previous expression for B

$$\left(\frac{L^c - L^a}{L_m}\right)^2\left[1 + \frac{2L_i^3}{L^c(L^a)^2}\right] = \frac{2\Delta h'}{3R}\left(\frac{1}{T_m} - \frac{1}{T_m^c}\right)$$

(8.28)

For large deformations, this expression further simplifies to

$$\frac{(L^c - L^a)^2}{L_m^2} \cong \frac{2\Delta h'}{3R}\left(\frac{1}{T_m} - \frac{1}{T_m^c}\right)$$

(8.29)

When $T_m < T_m^c$, Eqs. (8.27) and (8.28) yield two solutions for L^a, one less than and the other greater than L^c. No real solutions exist when $T_m > T_m^c$. Thus, T_m^c plays the role of a critical temperature above which the crystalline phase cannot exist.

If the deformation is sufficiently large so that (8.27) and (8.29) can be used, then L^a can be eliminated from each by means of Eq. (8.22). This manipulation leads to the results

$$\left(\frac{f}{T}\right)_{eq} \cong \frac{3kvn'}{L_m}\left[\frac{L^c}{L_m} \pm \sqrt{\frac{2\Delta h'}{3R}\left(\frac{1}{T_m} - \frac{1}{T_m^c}\right)}\right]$$

(8.30)

$$\left(\frac{f}{T}\right)_{eq} \cong \frac{3kvn'}{L_m}\left[\frac{L^c}{L_m} \pm \sqrt{\left(\frac{L^c}{L_m}\right)^2 + \frac{2\Delta h'}{3R}\left(\frac{1}{T_m} - \frac{1}{T_m^i}\right)}\right]$$

(8.31)

The approximate results, embodied in Eqs. (8.30) and (8.31) allow for a concise graphical representation of the phenomenon, as is illustrated in Fig. 8.2.(4) If the deformation process is initiated at a temperature at which the network is in the amorphous state and if the equation of state is given by Eq. (8.22), f/T will

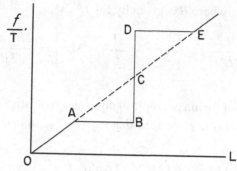

Fig. 8.2 Plot of f/T against length L for polymer networks undergoing a crystal–liquid transformation according to Eqs. (8.30) and (8.31). (From Flory (4))

increase linearly with L until crystallization sets in at point A. This point represents the melting temperature of the network under the specified force and elongation. As crystallinity develops, the length of the specimen increases along the line AB. For a one-component system the force must remain invariant until the transformation is complete at point B. The lesser of the two roots of Eq. (8.30) and (8.31) is applicable to this equilibrium. The stress is then assumed to rise almost vertically in the inelastic highly rigid crystalline state that was developed at point B. If it is possible to attain a state in which $L^a > L^c$, the amorphous phase will be reconstituted along the line DE. The equilibrium force corresponds to the larger of the two roots in this case. With increasing temperature, the points A and E are displaced toward C and a temperature is reached where the equilibrium lines AB and DE vanish. This temperature corresponds to T_m^c, the critical temperature above which crystallization cannot occur. The regeneration of the amorphous phase along the line DE seems scarcely to be a physically realizable situation. It is highly unlikely that a polymer chain could sustain the large deformation required for L^a to exceed L^c. Attention should therefore be focused primarily on the path OABD for real systems. Utilization of the less restrictive equation of state, Eq. (8.18), would not affect the salient features of Fig. 8.2. The linear stress–strain curve for the amorphous network that passes through the origin would be replaced by a curve starting at $L = L_i$ corresponding to $f/T = 0$ and which would be asymptotic to a line through the origin. The force–temperature–length relations expressed above in analytical and graphical form are general in concept. They do not depend on any details of the crystallographic structure of the ordered phase. Modification of these relations can be anticipated, however, with additional refinements in the statistical mechanical development of rubber elasticity theory.

The experimental investigations of Oth and Flory (16) substantiate the major conclusions of the theory outlined above. Their studies of the force–length–temperature relations for fibrous natural rubber, that was cross-linked in the oriented state, give

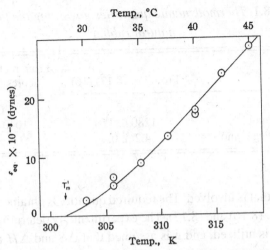

Fig. 8.3 Plot of force required for phase equilibrium against the temperature, for cross-linked fibrous natural rubber. $\rho = 1.56 \times 10^{-2}$ and $T_m^i = 302$ K. (From Oth and Flory (16))

strong support to the concept that this problem is a classical one in phase equilibria. Their basic experiments involved the determination of the equilibrium force f_{eq} required to maintain the two co-existing phases in equilibrium at temperatures above the isotropic melting temperature. These intricate experiments were accomplished by initiating melting at a temperature above the shrinkage temperature. The completion of the transformation was prevented by increasing the force or by lowering the temperature or by performing both operations simultaneously. Equilibrium is approached from several directions in these experiments and f_{eq} thus established. In accord with theory, it was found that, as long as the two phases coexist, f_{eq} was independent of the specimen length and increased with increasing temperature. The variation of the equilibrium force with the melting temperature is shown in Fig. 8.3 for a natural rubber network.(16) Even for the relatively small temperature interval, within which equilibrium was established, substantial forces were required. Based on the cross-section of the fiber, large stresses, of the order of 3 to 4 kg cm^{-2}, had to be imposed to maintain the equilibrium. These stresses would become much larger if equilibrium were established at still higher temperatures. When the curve of Fig. 8.3 is extrapolated to zero force, the isotropic melting temperature T_m^i is obtained. In this case, T_m^i is 6 °C lower than the shrinkage temperature that is observed in the absence of an external force. This result demonstrates the nonequilibrium character of the latter temperature.

In analogy with monomeric substances, where an analysis of the change in the transformation temperature with pressure yields either the latent heat of vaporization or of fusion, the variation of the transformation temperature with force yields

Table 8.1. *Thermodynamic quantities governing the fusion of natural rubber*

	From Eq. (8.17) (16)	From melting point depression (17)
T_m^i (K)	302	301
ΔH_u (cal mol^{-1})	1280 ± 150	1040 ± 60
ΔS_u (cal deg^{-1} mol^{-1})	4.2 ± 0.4	3.5 ± 0.2

the heat of fusion that is involved. The required quantity is obtained by the graphical integration of Eqs. (8.16) and (8.17). The experimentally determined change in ΔL with temperature is utilized, and it is assumed that ΔS and ΔH are constant over the small temperature range of interest. The latent changes ΔS and ΔH computed by this method can be ascribed entirely to the melting of that fraction of the polymer which is crystalline. Hence, the changes on fusion for the hypothetically completely crystalline fiber are obtained by dividing the calculated quantity by the degree of crystallinity. A comparison of the thermodynamic quantities governing the fusion as determined by this method and those obtained from an analysis of the melting point depression of natural rubber by monomeric diluents is given in Table 8.1. The results obtained from the two methods are in good agreement and further substantiate the analysis.

A compilation of the results for fibrous natural rubber can be represented graphically as in Fig. 8.4. The equilibrium force is plotted as a function of the length of the specimen at the indicated temperatures above the isotropic melting temperature. For the particular network represented by Fig. 8.4 the latter temperature is 302 K. The horizontal solid lines represent the stresses necessary to maintain the two phases in equilibrium. The length of the sample upon the completion of melting, at a given force and temperature, is indicated by the solid circles. The dashed lines represent the dependence of the force on length at each temperature in the amorphous state, as calculated from rubber elasticity theory. The force–length relation in the crystalline state at 303.2 K is indicated by the vertically rising straight lines. A similar behavior would be expected at other temperatures as long as the observations are restricted to the crystalline state. The set of isotherms in Fig. 8.4, which encompass the axially oriented crystalline and the liquid states, correspond to the isotherms in a p–v diagram describing vapor–liquid condensation in monomeric substances. It is to be noted in Fig. 8.4 that in the two-phase region the force is independent of the length.

The relations between the force, length and temperature when conditions are varied so that the fiber traverses the two-phase region, are also illustrated in this figure. Consider, for example, a network that is maintained under conditions

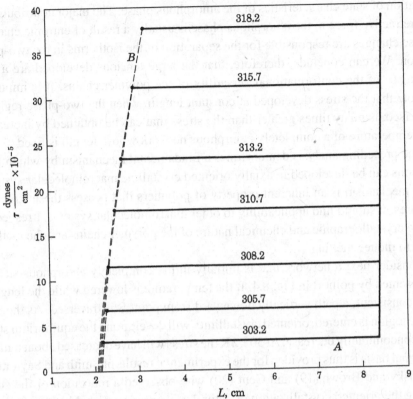

Fig. 8.4 Composite plot of tension–length relation at various temperatures for fibrous natural rubber. (From Oth and Flory (16))

specified by point A in Fig. 8.4. This condition corresponds to a sample 8 cm long in phase equilibrium at 303.2 K, under a tension slightly less than 4×10^5 dyn cm^{-2}. If a process is prescribed where the temperature is increased while the length is maintained constant, a path described by a vertical line upward from point A will be followed. In order to maintain a constant length, it is clear that an external force must be applied to balance the retractive force developed by the crystalline network. This additional stress is needed to prevent melting as the temperature is increased. Otherwise, the original length would not be preserved. If the temperature is raised to 318.2 K during this process, a tension of 4 kg cm^{-2} is developed. This tension is of the same order of magnitude as that developed by the muscle fiber system in tetanic contraction. For the fibrous natural rubber, a still greater tension could be developed by merely increasing the temperature. The stress will continuously increase with temperature as long as the two-phase region is maintained, i.e. until the critical temperature T_m^c is reached. The development of such large tensions with increasing temperature is obviously due to the separation of the horizontal portions of the isotherms in Fig. 8.4. This separation in turn is a reflection of the equation of state characteristics of the amorphous phase. The major contribution to the retractive force in the amorphous phase is usually a result of entropic changes. These changes are responsible for the separation of the isotherms in the two-phase region. We can conclude, therefore, that the large tensions developed are a consequence of the conformational versatility of the polymer chains. It is important to note that the stress developed at constant length, when the two-phase region is traversed, is many times greater than the stress that can be obtained by increasing the temperature of a completely amorphous network whose length is fixed.

The preceding analysis demonstrates a fundamental mechanism by which large tensions can be developed in axially oriented crystalline macromolecular systems. This mechanism is an inherent property of polymers that possess these structural features. It should find applicability to other macromolecular systems, irrespective of the crystallographic and chemical nature of the polymer chains and the methods used to induce melting.

Consider next a network that is initially in the completely amorphous state as represented by point B in Fig. 8.4. If the temperature is lowered while the length is held consistent, a path vertically downward from point B is traversed. As the two-phase region is entered, oriented crystallinity will develop and the equilibrium stress will concomitantly decrease. At 303.2 K the stress will have decreased about tenfold. A formal basis is thus provided for the experimental results of Smith and Saylor,[18] Tobolsky and Brown,[19] and Gent [20] who observed a relaxation of the stress during the oriented crystallization of natural rubber networks held at fixed length.

Processes can also occur where the stress rather than the length is held constant as the temperature is varied. Consider the system to again be in the crystalline state at point A of Fig. 8.4. If the stress is now maintained constant while the temperature is raised, a horizontal path will be followed which will terminate at the appropriate dashed curve representing the completely amorphous state. Accompanying the transformation, in this example, will be a fourfold diminution in length. This process is reversible as long as the equilibrium stress is maintained. Thus, by returning to the original temperature a spontaneous elongation will accompany the transformation from the amorphous to the crystalline state. A spontaneous increase in length during the crystallization of deformed natural rubber networks held has been observed.[18]

It was pointed out in Chapter 6 that the polymorphic transitions from one crystalline form to another can be induced by the application of an external stress on an axially oriented crystalline system. Anisotropic dimensional changes usually accompany the transformation. Typical examples are the classical α–β transition of the keratins,[21] and the crystal–crystal transition in poly(1,4-trans-butadienes).[22] The dimensional changes in these cases reflect the different axial or fiber repeat distances of the two polymorphs. The dimensional change would be expected to be appreciably less than what occurs during a crystal–liquid transformation. Moreover, since the elastic equations of state of the two crystalline forms will be similar to one another a large separation of the isotherms in the two-phase region is not expected. Thus, the development of a large retractive force is not anticipated. However, the force will still be independent of length in the two-phase crystallite region. Contractility and tension development will still be observed to some degree. The force–length relations in the two crystalline states will depend on details of the structures of each but can be expected to be very steep, i.e. to represent high moduli.

The melting–crystallization cycle of an oriented network that is conducted under equilibrium conditions results in a reversible contractile system when the force is held fixed. Alternatively, large changes in the tension are observed when the length is held constant. These two complementary observations are inherent properties of all types of macromolecular systems. The above analysis has been limited to a pure one-component homopolymer of uniform cross-section. However, it can be extended, in a straightforward manner to include inhomogeneous fibers, copolymers, and polymer–diluent mixtures.(4)

Variations in either chemical structure or cross-section along the fiber length result in a broadening of the transition between the crystalline and amorphous states. The primary effect of varying chemical structure, as in a copolymer, is manifested in a change of the melting point at a given force. Alterations in the cross-section will affect the stress. Since the equilibrium depends directly on the stress, different values of the critical stress occur in various cross-sections. It is, therefore, possible for the transition to occur over a range in tensile forces, at constant temperature and pressure, in nonhomogeneous axially oriented polymers. A more detailed analysis indicates that relations similar to Eqs. (8.24) to (8.31) hold for inhomogeneous fibers, provided that they are interpreted to apply to the particular element of the fiber in phase equilibrium.(4) Consequently, the lines AB and BD of Fig. 8.2 are replaced by sigmoidal curves.

Because the two states available to a polymer network can co-exist in a macroscopic sample, unique thermoelastic coefficients are observed. The coefficients of interest are those of force–temperature and of length–temperature. These are related to each other by the identity

$$\left(\frac{\partial f}{\partial T}\right)_{p,L} = -\left(\frac{\partial f}{\partial L}\right)_{p,T}\left(\frac{\partial L}{\partial T}\right)_{p,f} \tag{8.32}$$

Since $(\partial f/\partial L)_{p,T}$ is always positive, the sign of $(\partial f/\partial L)_{p,L}$ is opposite to that of $(\partial L/\partial T)_{p,f}$. Both coefficients are zero at the same length or force. The dependence of the length on the temperature at constant force is schematically illustrated in Fig. 8.5a for an idealized homogeneous fiber. At large L a small, positive thermal expansion coefficient typical of a crystalline solid is indicated. The melting

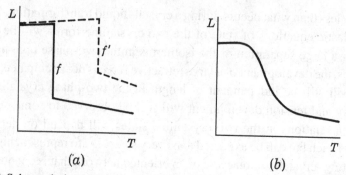

Fig. 8.5 (a) Schematic length–temperature relations for an idealized homogeneous fiber. $f' > f$. (b) Same for an inhomogeneous fiber. (From Flory (3))

point appears as a discontinuity in the diagram, and subsequently the molten fiber exhibits a moderate negative thermal expansion expected for a rubber-like substance. The melting point, of course, increases with increased force, as indicated by the dashed lines. The melting range is broadened for fibers that are inhomogeneous with respect to chemical structure or cross-sectional area. The sharp discontinuity in length is now smoothed to a continuous curve, as indicated in Fig. 8.5b. The length–temperature coefficient is still slightly negative for shrunken fibers and positive at large extensions in the highly crystalline states. At intermediate degrees of crystallinity, however, the coefficient is strongly negative. It reaches a maximum negative value with increasing crystallinity and then assumes the normal positive values. The force–temperature coefficients can be described in a similar manner by utilizing Eq. (8.32). A wide variation in the behavior of the thermoelastic coefficients with extension can be expected as a consequence of the phase change that occurs and the diffuse melting of inhomogeneous fibers. Thermoelastic behavior of the type described has, in fact, been observed for many of the fibrous proteins.(23–25)

The increase in melting temperature that occurs with the extent of deformation, is readily discerned by the locus of the solid points in Fig. 8.4. The development of a theoretical relation between the equilibrium melting temperature and the extension ratio, that agrees with experiment over the complete deformation range, has been very elusive. However, there are theoretical analyses that quantitatively account for portions of the deformation.(2,26–29) Since our interest here is in the equilibrium condition we must limit ourselves to the formation of extended chain crystallites. The state of equilibrium is reached by deforming the network at sufficiently high temperature so that the system is in the completely amorphous state. The temperature is then lowered, while the network is held at fixed length. Crystallization then ensues, equilibrium is approached and the melting temperature is then determined at constant length. This requirement is to be distinguished from the nonequilibrium case where crystallization occurs while the network is being stretched. Equilibrium requires that extended chain crystallites be formed.[5] The extended chain crystallites need to be sufficiently long to satisfy the requirements of equilibrium. Since there is a kinetic aspect to the crystallization process described,(30) this requirement is not automatically satisfied. A further assumption that is commonly made is that the network chains obey Gaussian statistics in both the isotropic and deformed states.

It might be expected that the analysis of this problem merely involves equating the free energy of the deformed amorphous network with the free energy of fusion in order to obtain the relation between the equilibrium melting temperature and the extension ratio.(29) However, the problem being considered presents a very unique situation. The structure in the crystalline state, particularly the crystallite orientation, will affect the free energy of fusion. The basic analysis of the problem has been developed by Flory.(2) Statistical mechanical procedures, similar to those used in the development of rubber elasticity theory, are employed. The crystalliza-tion occurs in two distinct steps. The network is first elongated to its final relative length α, and then allowed to crystallize. This procedure is best accomplished if the network is elongated at a sufficiently elevated temperature so that crystallization does not occur and then cooled to a temperature at which crystallization can ensue. In the analysis some important premises are made. A primary assumption is that the deformation is affine, i.e. the coordinates of the relative average position of the junction points change in proportion to the changes in the macroscopic dimensions of the sample. It is also assumed that a chain only passes through a crystallite once. The crystallites are taken to be oriented parallel to the stretching direction. A chain is assumed to traverse a crystallite in the direction of the displacement of its length with respect to the orientation axis. These assumptions have important ramifications when analyzing results of real systems. It is assumed the chain conformation can be approximated by a hypothetical one that is composed of a large number of segments joined together by bonds which permit freedom of notation.[6]

With this model two main changes in the entropy need to be considered. A segment entering the crystallite sacrifices its orientational and rotational disorder of the original state. Consequently, the entropy will decrease. This change in entropy is akin to that which takes place during crystallization of undeformed systems. A further entropy change, unique to this particular type of crystallization results from the change in the distance traversed by the remaining disordered portion of the chain. If we take the z-axis as the elongation axis then the z component of the chain displacement length is decreased. At the same time the number of amorphous

---

[5] Situations where crystallization under stress leads to some type of folded-chain lamellar structure are not considered at this point since this represents a nonequilibrium situation.

[6] Although in many applications this hypothetical equivalent chain can be replaced by the real chain by using rotational isomeric theory (31) this simplified concept is maintained at present for illustrative purposes.

segments that are available to traverse the required distance is diminished. The total entropy change accompanying the crystallization could be obtained by computing the separate entropy changes just described. Instead, following Flory, (2) we outline the calculation of the absolute configurational entropy of the stretched, crystalline polymers taking the hypothetical totally crystalline polymers as the reference state.

The relative number of conformations of the disordered chain is assumed to be a Gaussian function of the chain displacement length r. Accordingly,

$$W(xyz) = \left(\beta/\pi^{1/2}\right)^3 \exp[-\beta^2(x^2 + y^2 + z^2)] \qquad (8.33)$$

Here x, y, and z represent the coordinates of one end of the chain with respect to the other. The chain displacement length $r = (x^2 + y^2 + z^2)^{1/2}$ and $1/\beta$ is the most probable value of r. It is assumed in this model that the cross-linkages are introduced at random into the undeformed, isotropic polymer. The chains are thus free to assume random conformations. Hence Eq. (8.33) also represents the distribution of chain coordinates before stretching.(6,7) After stretching by a factor α, along the z-axis the distribution of chain coordinates becomes

$$v(xyz) = \sigma\left(\beta/\pi^{1/2}\right)^3 \exp[-\beta^2(\alpha x^2 + \alpha y^2 + z^2/\alpha^2)] \qquad (8.34)$$

assuming that the volume remains constant. Here σ is the total number of chains under consideration. For the hypothetical chain being considered, which has freely orienting segments, β can be expressed as

$$\beta = (3/2m)^{1/2}/l \qquad (8.35)$$

where l is the length of each segment and m is the number of segments per chain. When ζ of the m segments occur in a crystalline region, the relative number of configurations available to the remaining $m - \zeta$ segments becomes

$$W'(xyz') = \left(\beta'/\pi^{1/2}\right)^3 \exp[-(\beta')^2(x^2 + y^2 + z'^2)] \qquad (8.36)$$

where

$$\beta' = [\text{const}/(m - \zeta)l^2]^{1/2} = \beta[m/(m - \zeta)]^{1/2} \qquad (8.37)$$

and z' is the algebraic sum of the z displacement lengths of the two amorphous sections of the chain. The x and y displacements are unaffected by the formation of crystallites with axes parallel to the stretching direction. However, the z displacement will be altered by the amount ζl. The assumption that all the chains traverse the crystallite in the same direction as the z displacement is involved here.

The calculation of the configurational entropy with respect to the totally crystalline polymer is carried out in two hypothetical steps. The first is the melting of $m - \zeta$ segments from each of the σ chains each having m segments. In this step the ends of the chains are free to occupy most probable locations. The distribution of displacement lengths, x, y and z' of the amorphous portion is then given by

$$v'(xyz') = \sigma W'(xyz') \tag{8.38}$$

The second hypothetical step is the assignment of chain ends to the locations of the cross-linkages within the deformed polymer as is required by Eq. (8.34). The entropy change for the first step is given by

$$S_{\mathrm{a}} = \sigma(m - \zeta)\Delta S_{\mathrm{f}} \tag{8.39}$$

Here ΔS_{f} is the entropy of fusion per segment. The entropy change in the second step arises from the transformation of the chain length distribution in the amorphous portion given by Eq. (8.38) to that given by Eq. (8.2). From the Boltzmann relationship $S = k \sum \ln W$ this entropy change can be expressed as

$$S_{\mathrm{b}} = k \sum_{xyz} v(xyz) \ln m \, W'(xyz') - k \sum_{xyz'} v'(xyz) \ln m \, W'(xyz') \tag{8.40}$$

After substitution, replacing the sums by integrals, and performing the necessary integrations Eq. (8.40) becomes

$$S_{\mathrm{b}} = -\sigma k \left[(\zeta \beta l)^2 m/(m - \zeta) - 2\alpha (\zeta \beta l/\pi^{1/2})m/(m - \zeta) \right.$$
$$\left. + (\alpha^2/2 + 1/\alpha)m/(m - \zeta) - \frac{3}{2} \right] \tag{8.41}$$

The total conformational entropy involved is then given by the sum of Eqs. (8.39) and (8.41).

In order to calculate the free energy change it is assumed that the second step in the procedure occurs without any change in internal energy. For the first step the heat change accompanying the fusion of $m - \zeta$ segments per chain is $\sigma \Delta H_{\mathrm{f}}(m - \zeta)$, where ΔH_{f} is the heat of fusion per chain segment. Accordingly, the free energy change can be expressed as

$$G = \sigma RT \left[m\theta(1 - \lambda) + (m\beta l)^2 (1 - \lambda)^2/\lambda - (2\alpha m\beta l/\pi^{1/2}) \right.$$
$$\left. \times (1 - \lambda)/\lambda + (\alpha^2/2 + 1/\alpha)/\lambda - \frac{3}{2} - m\theta \right] \tag{8.42}$$

The perfectly ordered completed crystalline chain has been taken as the standard state. The fraction noncrystalline is given by

$$\lambda = (m - \zeta)/m \tag{8.43}$$

The temperature function

$$\theta = (\Delta H_{\mathrm{f}}/R)(1/T_{\mathrm{m}}^0 - 1/T) \tag{8.44}$$

can also be introduced. Here, $\Delta H_f/\Delta S_f = T_m^0$, the equilibrium melting temperature of the undeformed polymer network. The equilibrium condition with respect to the longitudinal length of the crystallite is given by $(\partial G/\partial \zeta)_\alpha = 0$, or $(\delta G/\delta \lambda)_\alpha = 0$. It is found, with appropriate substitution, that

$$\lambda_e = \left\{ \left[\frac{3}{2} - \varphi(\alpha) \right] \Big/ \left[\frac{3}{2} - \theta \right] \right\}^{1/2} \tag{8.45}$$

where

$$\varphi(\alpha) = (\sigma/\pi)^{1/2} 6/m^{1/2} - (\alpha^2/2 + 1/\alpha)/m \tag{8.46}$$

The equilibrium level of crystallinity is given by $1 - \lambda_e$. The dependence of the equilibrium melting temperature on the elongation ratio α is determined by setting $\lambda_e = 1$. It is then found that

$$1/T_m - 1/T_m^0 = \frac{R}{\Delta H_u} \varphi_F(\alpha) \tag{8.47}$$

The designation, $\varphi_F(\alpha) \equiv \varphi(\alpha)$, indicates that the Flory function is used here.

It is important to note that there is an inconsistency in Eq. (8.47) at low elongations.(2) The function $\varphi_F(\alpha)$ retains a small positive value when α becomes unity. Therefore, the theory predicts that at $\alpha = 1$, T_m will be less than T_m^0, rather than being identical to it. The reason for this anomaly lies in Eq. (8.41). From this equation it is found that at small degrees of crystallinity and low elongation S_b is positive. Obviously, S_b should always be zero or negative since the final state cannot have a higher entropy than the most probable one. The failure of Eq. (8.41) arises from the assumption made of complete axial orientation of the crystallites along the stretching direction. Thus, in the crystalline state the chain traverses a crystallite in the same direction as the z displacement component. It can be presumed that more accurate theories would replace $\varphi_F(\alpha)$ with a function that would equal zero at $\alpha = 1$ and thus remove the anomaly in Eq. (8.47). In order to overcome the recognized deficiencies in the Flory theory another approach was taken.

In the approach taken by Krigbaum and Roe it was assumed that the only contribution to the entropy of fusion in the deformed state is the conventional isotropic one.(29) They found that

$$\frac{1}{T_m} - \frac{1}{T_m^0} = -\frac{R}{2N_u\Delta H_u} \varphi_K(\alpha) \tag{8.48}$$

where

$$\varphi_K(\alpha) = \alpha^2 + (2/\alpha) - 3 \tag{8.49}$$

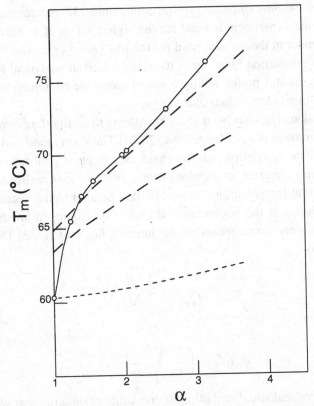

Fig. 8.6 Comparison of experimental extrapolated equilibrium melting temperatures of poly(chloroprene) at various elongation ratios with those predicted. ○ experimental results; (- - - -) according to Eq. (8.48); (– – –) according to Eq. (8.47) for two different values of number of repeating units per statistical segment. (From Krigbaum *et al.* (32))

Theoretical plots of T_m as a function of α are given in Fig. 8.6 for the two theories discussed up to this point.(32) The lower dashed curve, representing the Krigbaum–Roe theory, indicates that for this analysis $T_m = T_m^0$ at $\alpha = 1$. This agreement is a consequence of the basic assumption that was made with regard to the entropy of fusion. However, only a small increase in T_m is predicted at high deformation, relative to $\alpha = 1$. This is again a result of the entropy of fusion that was assumed. In this theory the melting temperature does not change much with deformation, a statement that is contrary to observation. The two upper dashed curves are theoretical plots based on the Flory theory for two different values of the parameter that relates the repeating unit to the statistical segment. The Flory theory predicts much higher melting temperatures for α greater than unity. Illustrated by the open circles in this figure are extrapolated equilibrium melting temperatures as a function of elongation for poly(trans chloroprene).(32) The melting temperatures obey the Flory relation for α values greater than about 1.5. Similar results are obtained with networks that have different type repeating units.(32) Agreement between the Flory theory and experiment is good for the higher values of α. However, significant improvement in theory is needed for the low values of α. The dependence of the crystallite orientation on α needs to be expressed in analytical form. Clearly, having the crystallites preferentially oriented along the stretching direction is an oversimplification at low values of α.(2)

A theoretical analysis has been given by Allegra to rectify the discrepancy in the melting temperatures at low elongations.(26,27) The main modification made was the removal of the restriction that the chain axes in the crystallites were oriented in the stretching direction at all elongations. In this more realistic approach, it was assumed that the crystallites were oriented parallel to the vector connecting the junction points at the beginning and end of the chain. In another variation, different constraints were imposed on the junction fluctuation.(33) The results can be expressed as

$$\frac{1}{T_M} - \frac{1}{T_m^0} = -\frac{3}{2}\frac{R}{\Delta H_u}\varphi_A(\alpha) \tag{8.50}$$

where

$$\varphi_A(\alpha) = \left(\frac{32}{3\pi}\right)^{1/2}\frac{\alpha}{m^{1/2}} - \frac{\alpha^2}{m} \tag{8.51}$$

Despite the more realistic distribution of crystallite orientation that was assumed, $\varphi_A(\alpha)$ still does not equal zero at $\alpha = 1$. Although the theory does not reduce to $T_m = T_m^0$ at $\alpha = 1$, the difference from experiment is fairly small. When this theory is compared with the observed melting temperature of poly(cis-1,4-butadiene) the calculated value is only about 3°C lower.(34) At higher elongations the agreement with experiment does not appear to be as good as is obtained with the Flory theory.

A more sophisticated and realistic rubber elasticity theory (applicable to the amorphous polymers) was also applied to the problem of strain induced crystallization.(28,35) The theory is based on the constrained junction model as developed by Flory and Erman.(36–38) A major premise of the theory is that local intermolecular entanglements and steric constraints on the junction fluctuations contribute to the modulus and network deformability. Two parameters are introduced. The parameter κ is a measure of the severity of the entanglement constraints and is proportional to the degree of chain interpenetration. The parameter ζ accounts for the possible nonaffine nature of the transformation with increasing strain of the constrained domains. The deformation process, prior to crystallization, is not taken as an affine process. The crystallite orientation was the same as originally used by Flory. In effect, this treatment generalizes the initial Flory treatment by taking into account a more realistic deformation process. The relation between the melting temperature and the extension ratio for this model can be expressed as (28,35)

$$\frac{1}{T_m} - \frac{1}{T_m^0} = -\frac{R}{\Delta H_u} \varphi_M(\Lambda_x, \Lambda_y, \Lambda_z) \tag{8.52}$$

where

$$\varphi_M(\Lambda_x, \Lambda_y, \Lambda_z) = \left(\frac{6}{\pi m}\right)^{1/2} \Lambda_z - \frac{1}{-2m\left(\Lambda_x^2 + \Lambda_y^2 + \Lambda_z^2\right)} \tag{8.53}$$

The quantities Λ_x, Λ_y and Λ_z represent the molecular deformation tensors in the three principal directions. For the undeformed network the function $\varphi_M(\Lambda_x, \Lambda_y, \Lambda_z)$ is identical to the Flory $\varphi_F(\alpha)$. Hence, according to this theory, at $\alpha = 1$, T_m does not equal T_m^0. This result is to be expected based on the crystallite orientation that was assumed. Plots comparing the theoretical melting temperature–elongation ratios deduced from both the constrained junction and the Flory model are given in Fig. 8.7.(9) The same parameters pertinent to poly(cis-1,4-butadiene) were used in the calculations. A value of $\kappa = 10$ was taken for the constrained junction model; while $\kappa = \infty$ corresponds to the Flory model. The two models give very similar results at the smaller values of α. There are however, significant differences in the melting temperatures that are predicted for the higher elongations.

The foregoing analysis of stress-induced crystallization has deliberately been limited to equilibrium concepts. By definition, therefore, only extended chain crystallites are being considered. Even with this restriction, theory and experiment are not in as complete harmony with one another as would be desired. Several

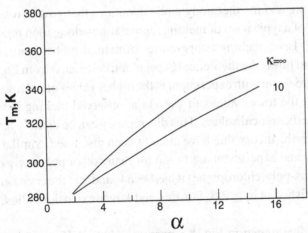

Fig. 8.7 Plot of the theoretical dependence of the melting temperature on elongation ratio for $\kappa = 10$ for constrained junction model and $\kappa = \infty$ for Flory model. (From Erman and Mark (9))

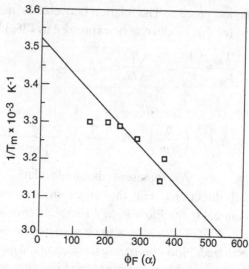

Fig. 8.8 Plot of reciprocal of extrapolated equilibrium melting temperature against $\varphi_F(\alpha)$ according to Eq. (8.47) for cis-poly(isoprene).(39)

shortcomings in theory are apparent. Despite the deficiencies, particularly at low elongations, it is still of interest to examine experimental data in terms of the theoretical base available. It is found in general, for the different networks studied, that the extrapolated equilibrium melting temperatures increase with the extension ratio. However, it is not surprising that when examined in detail there are distinct differences between networks of different type repeating units as well as those of the same type prepared in a different manner. Irrespective of the detailed theoretical analysis, the melting temperatures, at a fixed value of α, should depend on the network structure and the thermodynamic quantities that govern fusion.

An analysis of a typical set of melting temperature–elongation ratio data is given in Fig. 8.8.(39) Here, melting temperatures of natural rubber were obtained up to $\alpha = 5$. The solid line is the theoretical expectation calculated from Eq. (8.47). There is very good agreement with experiment at the higher values of α. However, as would be expected, at the lower values of $\varphi_F(\alpha)$ the observed melting temperatures are greater than the theoretical values. This discrepancy can be directly attributed to the shortcomings in the theory that have already been discussed. Similar comparisons between theory and experiment are found for many other polymeric networks. For networks of trans-poly(chloroprene) it has been found that the crystallite orientation condition is satisfied at $\alpha \geq 2.5$. For this and larger extension ratios, Eq. (8.47) is obeyed.

The straight line drawn in Fig. 8.8 corresponds to $\Delta H_u = 620\,\mathrm{cal\,mol^{-1}}$. This value is significantly lower than that determined by other methods. The ΔH_u value deduced by this method is sensitive to small errors in the experimentally determined melting temperatures. For example, in analyzing another set of data for natural rubber, Candia *et al.* only found an 18% deviation in ΔH_u. The value of ΔH_u deduced from deformation experiments for poly(ethylene oxide) networks is in good agreement with those obtained by the diluent method.(40)

Only crystallization induced by a tensile type deformation has been discussed here. Other types of deformation such as biaxial extension, shear and torsion should also be considered. Such deformations have been studied and analyzed for amorphous networks. However, there is a paucity of experimental data, as well as analysis, of the equilibrium aspects of crystallization induced by these deformations. In one available report the observed melting temperature of natural rubber networks increased substantially when subject to biaxial deformation.(41) An increase in melting temperature of about 50 °C was found for a biaxial stretching ratio of three. This increase is much larger than that observed for natural rubber when crystallized in simple extension.

8.3 Multicomponent systems subject to a tensile force

Melting and crystallization of oriented polymers also occurs when in contact with either diluent or a solution containing monomeric solutes. Although the introduction of additional components and phases results in some alterations of the analyses, the fundamental physical processes involved are not changed. These conditions hold for the fibrous proteins since the necessary experiments can only be carried out either in the presence of diluent or by appropriate chemical reactions.

The appropriate relation for a multicomponent system that corresponds to Eq. (8.7) is

$$d(G - fL) = -S\,dT + V\,dp - L\,df + \sum_i \mu_i\,dn_i \qquad (8.54)$$

The requirement for equilibrium between the crystalline and amorphous phases is that the function $G - fL$ be a minimum when δn_i moles of component i are transferred from one phase to another at constant p, T, f. Hence

$$\delta(G - fL) = \delta n_i \mu_i^a - \delta n_i \mu_i^c = 0 \qquad (8.55)$$

or

$$\mu_i^a = \mu_i^c \qquad (8.56)$$

for each of the components present in both polymer phases. When the simplifying assumption is made that the crystalline phase is pure, Eq. (8.56) becomes

$$\mu_u^c = \mu_u^a \qquad (8.57)$$

Similarly, when a supernatant phase (designated by the superscript s) comprised of monomeric components is present, then for the amorphous portion of the fiber to be in equilibrium with this phase it is required that

$$\mu_i^a = \mu_i^s \tag{8.58}$$

for each of the monomeric components present in the amorphous or mixed phase.

In the two polymer phases, the chemical potential of the polymer unit is a function of T, p, f and composition. If the crystalline phase is assumed pure then

$$d\mu_u^c = -S_u^c \, dT + V_u^c \, dp - L_u^c \, df \tag{8.59}$$

For the amorphous phase containing r components

$$d\mu^a = -\bar{S}_u^a \, dT + \bar{V}_u^a \, dp - \bar{L}_u^a \, df + \sum_{i=1}^{r-1} \left(\frac{\partial \mu_u}{\partial x_i} \right)_{T,p,f} dx_i \tag{8.60}$$

where $\bar{S}_u^a$, $\bar{V}_u^a$ and $\bar{L}_u^a$ are the respective partial molar quantities of the polymer unit in this phase and x_i is the mole fraction of the ith component. For equilibrium between the two phases, at constant p

$$d\mu_u^a = d\mu_u^c \tag{8.61}$$

and

$$\left(\bar{L}_u^a - L_u^c \right) df = -\left(\bar{S}_u^a - S_u^c \right) dT + \sum_{i=1}^{r-1} \left(\frac{\partial \mu_i}{\partial x_i} \right)_{T,p,f} dx_i \tag{8.62}$$

If the composition of the amorphous phase is held fixed, then

$$\left(\frac{\partial f}{\partial T} \right)_{p,n} = -\frac{\bar{S}_u^a - S_u^c}{\bar{L}_u^a - L_u^c} \tag{8.63}$$

The subscript n denotes that the concentrations of all components are held constant. The invariance in composition required by Eq. (8.63) implies not only a fixed polymer concentration but also a constant concentration of the monomeric constituents present in this phase. The entropy change per polymer unit that occurs on melting, at constant p, T, and n, is then given by

$$\bar{S}_u^a - S_u^c = \frac{\left(\bar{H}_u^a - H_u^c \right) - f \left(\bar{L}_u^a - L_u^c \right)}{T} \tag{8.64}$$

so that

$$\left[\frac{\partial (f/T)}{\partial (1/T)} \right]_{p,n} = \frac{\bar{H}_u^a - H_u^c}{\bar{L}_u^a - L_u^c} \tag{8.65}$$

It is convenient to multiply the numerator and denominator of the right-hand side of Eq. (8.65) by n_u, the total number of structural units in the fiber. Equation (8.65) then becomes

$$\left[\frac{\partial (f/T)}{\partial (1/T)} \right]_{p,n} = \frac{\Delta \bar{H}}{\Delta \bar{L}} \tag{8.66}$$

where $\Delta\bar{H} = \bar{H}^a - H^c$ and $\Delta\bar{L} = \bar{L}^a - L^c$. H^c and L^c are the enthalpy and length of the totally crystalline fiber at p, T, and f. $\bar{H}^a$ and $\bar{L}^a$ are the partial derivatives in the amorphous phase of the total entropy and length with respect to the fraction λ of the polymer in this phase. Thus, $\Delta\bar{H}$ consists of the heat of fusion plus the differential heat of dilution. The quantity $\Delta\bar{L}$ is similarly defined.

Two cases must now be distinguished. In one, the total quantity of the nonpolymeric components is fixed. In the other, the amorphous portion of the fiber is in equilibrium with a supernatant phase containing a large excess of the monomeric species. In the former case the fiber and its contents operate as a closed system. If only a one-component diluent is present, the system is bivariant at constant pressure. As melting progresses, the length of the fiber decreases. The composition of the amorphous phase changes since the polymer concentration increases while that of the diluent is fixed. The differential coefficient of Eq. (8.66) is for constant composition, a condition that can be identified with the constancy λ. Hence

$$\left[\frac{\partial(f/T)}{\partial(1/T)}\right]_{p,n} = \left[\frac{\partial(f/T)}{\partial(1/t)}\right]_{p,\lambda} = \frac{\Delta\bar{H}}{\Delta\bar{L}} \tag{8.67}$$

In contrast to the pure one-component polymer system, $\Delta\bar{S}$, $\Delta\bar{L}$, and $\Delta\bar{H}$ are now dependent on the composition. Therefore, the force–temperature derivative depends on λ. The force in this instance is not uniquely determined by the temperature, and total melting does not occur at constant force.

When a supernatant phase is present, the fiber and its contents operate as an open system since there can be an exchange of matter between the supernatant and amorphous polymer phase. If the supernatant consists of a single component, the system is univariant at constant pressure. The equilibrium force is thus uniquely determined by the temperature. Total melting now occurs at constant force, independent of the length of the specimen, in analogy to a pure one-component system. Since an excess of diluent is present in the supernatant phase, equilibrium swelling in the amorphous phase can be established at the given f and T for all values of L. Thus, as melting proceeds the composition of the mixed phase remains constant so that

$$\left(\frac{\partial f}{\partial T}\right)_{p,n} \equiv \left(\frac{\partial f}{\partial T}\right)_{p,\lambda} \equiv \left(\frac{\partial f}{\partial T}\right)_{p,L} = -\frac{\Delta\bar{\bar{S}}}{\Delta\bar{\bar{L}}} \tag{8.68}$$

and

$$\left(\frac{\partial(f/T)}{\partial(1/T)}\right)_{p,L} = \frac{\Delta\bar{\bar{H}}}{\Delta\bar{\bar{L}}} \tag{8.69}$$

The double-barred quantities represent the sum of the latent change that occurs on fusion of the polymeric component and the integral change for mixing the required amounts of each component to arrive at the equilibrium composition of the amorphous phase.

When the supernatant phase is multicomponent, the system is no longer univariant. Although the conditions of Eq. (8.68) must still be satisfied, this does not ensure that the composition of the amorphous phase will remain fixed with changes in λ. At constant pressure the equilibrium force need no longer depend solely on the temperature. Consequently, total melting does not have to occur at constant force, in analogy to the behavior of a closed system.

Since the crystal–liquid equilibrium can also be regulated by chemical processes, the force–length–temperature relations of axially oriented crystalline systems will be influenced accordingly. The formal analysis of the problem is similar to that for a nonreacting system with

$$\left(\frac{\partial f}{\partial T}\right)_{p,n} = -\frac{\left[\partial(\mu_u^a - \mu_u^c)/\partial T\right]_{p,f,n}}{\left[\partial(\mu_u^a - \mu_u^c)/\partial f\right]_{p,T,n}} \tag{8.70}$$

Attention must now be given to the changes in the chemical potential of the polymer unit caused by the specific chemical reaction and to the phase(s) in which the reaction occurs. When these conditions are specified, the differential coefficient $(\partial f/\partial T)_{p,n}$ can be evaluated.

For purposes of illustration, and for simplification, it will be assumed that the chemical reaction is restricted to the amorphous polymer phase, so that the crystalline phase remains pure. Furthermore, we shall assume that the composition of the amorphous phase is invariant with λ even if the supernatant phase is multicomponent. Then

$$\left(\frac{\partial f}{\partial T}\right)_{p,n} = -\frac{\bar{S}_u^a - \bar{S}_u^c}{\bar{L}_u^a - \bar{L}_u^c} = -\frac{\Delta \bar{\bar{S}}}{\Delta \bar{\bar{L}}} \tag{8.71}$$

and

$$\left[\frac{\partial(f/T)}{\partial(1/T)}\right]_{p,n} = \frac{\Delta \bar{\bar{H}}}{\Delta \bar{\bar{L}}} \tag{8.72}$$

The triple-barred quantities represent the sum of three terms: the fusion of the pure polymer; the integral mixing of components to the composition specified by n; and the change in the quantity resulting from the change in the chemical potential of the structural unit caused by the chemical reaction. Thus, for example, under the assumption of the constancy of composition of the amorphous phase with L, $\Delta \bar{\bar{H}}$ can be written as

$$\Delta \bar{\bar{H}} = \Delta H + \Delta H_M + \Delta H_R \tag{8.73}$$

where ΔH is the heat of fusion, ΔH_M is the integral heat of mixing, and ΔH_R the enthalpic change per structural unit caused by the chemical reaction at the total composition specified by n.

For a simple complexing reaction, of the type discussed in Chapter 3, the change in chemical potential can be expressed as (42)

$$\mu_u - \mu_u^0 = -RT \ln (1 + Ka) \qquad (8.74)$$

Hence

$$\Delta H_R = RT^2 \frac{a}{1 + Ka} \left(\frac{\partial K}{\partial T} \right)_{p,f} = T \frac{a}{1 + Ka} \Delta H^0 \qquad (8.75)$$

Where ΔH^0 is the standard state enthalpic change for the complexing reaction. Whether the corresponding term ΔL_R differs from zero depends on whether the equilibrium constant for the reaction is a function of the applied stress. Other possible chemical reactions can be treated in a similar manner (43) as long as the changes that occur in the chemical potential of the polymer unit can be specified.

When Eqs. (8.68) and (8.69) are integrated, relations similar to those for the one-component system are obtained. The $\Delta \bar{\bar{H}}$ term now includes the additive contributions of the heats of dilution and of reaction. The integration must be carried out at constant composition of the amorphous phase. The equation of state used must take cognizance of the polymer concentration in this phase. The integration constants L_i and T_m^i or L^c and T_m^0 refer to this fixed composition. Thus, not only are the enthalpy and length terms affected by changes in composition but the isotropic length and melting temperature are as well.

A study involving a multicomponent fibrous system can now be examined in terms of the above analysis. The tension required to maintain equilibrium between the crystalline and amorphous phases of cross-linked collagen has been determined.(44) In these experiments the fiber is immersed either in a large excess of pure water or in an aqueous KCNS solution. The experiments were conducted over a wide temperature range. The equilibrium force at a given temperature was approximately independent of the total sample length and consequently the extent of the transformation. When the supernatant phase consists solely of pure water, the system is univariant and the aforementioned result is to be expected. However, the results obtained when the supernatant phase contains two components indicates that the single-liquid approximation is also valid in this particular case.

Some results for the change in the equilibrium stress with temperature for this fibrous system are illustrated in Fig. 8.9. The change in the required stress with temperature is quite substantial. Extremely large stresses can be developed by this process, as has been previously noted for fibrous natural rubber. These changes in stress with temperature are in contrast with those observed during the deformation

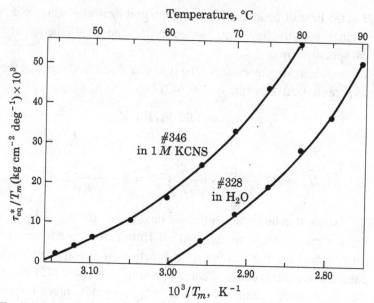

Fig. 8.9 Equilibrium stress τ_{eq}^* divided by T_m plotted against $1/T_m$ for collagen fibers immersed in pure water and in 1 M KCNS. (From Flory and Spurr (44))

of the completely amorphous collagen fiber.(44) When the length of the completely amorphous fiber is kept constant, only relatively small increases in the stress are observed with increasing temperature. Figure 8.9 also demonstrates the changes that occur at fixed temperature as the composition of the supernatant phase is varied. For example, in pure water at 70 °C a stress of 4.4 kg cm^{-2} is required to keep the two phases in equilibrium. However, if the supernatant phase is made 1 M in KCNS, the stress required is increased to about 11.5 kg cm^{-2}. Thus substantial changes in the tension are developed solely by changing the composition of the supernatant phase. Since the slopes of the two curves in Fig. 8.9 are approximately the same at all temperatures, the major reason for the increase in the equilibrium stress resides in the change in T_m^i from 60 °C in pure water to 43 °C in 1 M KCNS. These changes in T_m^i are a result of the specific chemical processes involved. It should also be noted that a collagen fiber immersed in a 2 M mercury–potassium iodide solution (a medium known to promote the melting of fibrous proteins) develops a tension of 100 kg cm^{-2} when the length is maintained constant.(45)

According to Eqs. (8.71) and (8.75), the quantity $\Delta\bar{\bar{H}}/\Delta\bar{\bar{L}}$ can be obtained from the slopes of the curves given in Fig. 8.9. If L^c is treated as a constant, $\Delta\bar{\bar{H}}$ can be calculated. Proper decomposition of $\Delta\bar{\bar{H}}$ into its constituent parts allows for an evaluation of ΔH_u, the heat of fusion of the polymer. This involves calculation of the integral heat of solution ΔH_{sol} and an estimation of ΔH_R, the contribution from the chemical reaction. With neglect of the latter term, the results

Table 8.2. *Thermodynamic parameters for the fusion of collagen*[a]

Supernatant phase	T_m^i (°C)	$\Delta \bar{H}_u$ (kcal mol^{-1})	ΔH_{sol} (kcal mol^{-1})	ΔH_u (kcal mol^{-1})	ΔS_u (cal deg^{-1} mol^{-1})
Water	60	1.2	−0.15	1.35	4.1
1 M KCNS	43	0.87	0.10	0.97	3.1
3 M KCNS	14	0.43	0.03	0.46	1.6

[a] Source: Ref. (44).

of these experiments are summarized in Table 8.2. The enthalpy changes cited in Table 8.2 refer to changes per mole of peptide units present in the native fiber, rather than per mole of peptide units that are crystalline. Similar values for ΔH_u have been obtained by direct calorimetric measurement for a variety of different collagens.(46) Flory and Garrett (47), utilizing the diluent method, found that for the system collagen–ethylene glycol $\Delta H_u = 2.25$ kcal mol^{-1} crystalline units. The smaller value in water, determined by the method described above, can be attributed to an appreciable amorphous content of the native collagen fiber. Water entering the crystal lattice forms a hydrate with the polymer so that the melting behavior of identical species is not being compared. The reduction in the enthalpy of fusion as the KCNS concentration is increased may be more apparent than real, since any contribution to $\Delta \bar{\bar{H}}$ from the chemical reaction has not been taken into account. The results obtained for the collagen–water and collagen–water–KCNS systems give further evidence that we are dealing with a problem in phase equilibrium. Most important is the fact that a fundamental mechanism has been outlined wherein large tensions can be developed in protein fibers as a result of a chemical reaction.

The analysis of the aforementioned system is greatly simplified by its univariant behavior. The independence of the equilibrium force on the extent of the transformation implies a constancy of composition in the amorphous phase. For an open system containing a multicomponent supernatant phase, this result is not the one expected. More generally, as the transformation progresses at constant temperature, the composition of the amorphous phase will change, caused, for example, by an unequal partitioning of the monomeric components between the two phases. The equilibrium force must correspond to the composition of the mixed phase which in turn will depend on the total length of the specimen. When systems possess more than one degree of freedom at constant pressure, the two-phase region is no longer depicted by a horizontal straight line, as in Fig. 8.4, but by a curve with positive slope and curvature. The change in force with length (in the two-phase region) reflects this compositional change and is thus affected by the corresponding change in T_m^i and the ratio $\Delta \bar{\bar{H}}/\Delta \bar{\bar{L}}$. Since at constant temperature and pressure, $f = f(L, n)$,

$$df = \left(\frac{\partial f}{\partial L}\right)_{T,p,n} dL + \sum_i \left(\frac{\partial f}{\partial n_i}\right)_{T,p,L,n_i} dn_i \tag{8.76}$$

with the summation extending over all components. For equilibrium between the phases,

$$\left(\frac{\partial f}{\partial L}\right)_{T,p,\text{eq}} = \left(\frac{\partial f}{\partial L}\right)_{T,p,n} + \sum_i \left(\frac{\partial f}{\partial n_i}\right)_{T,p,L,n_i} \left(\frac{\partial n_i}{\partial L}\right)_{T,p,\text{eq}} \tag{8.77}$$

Only when the terms in the summation vanish is the idealized behavior of a pure one-component system realized.

Experiments have been carried out where the melting temperatures of swollen networks have been measured as a function of the elongation ratio when subject to a tensile force prior to the development of crystallinity.(48,49) These experiments are akin to those described in the previous section for unswollen networks. Surprisingly, for polyethylene networks immersed in p-xylene, the melting temperature only increases 2.5 °C in going from the undeformed state $\alpha = 1$ to an extension ratio of 4.(48) In contrast, for dry networks of comparable cross-linking density the increase in melting temperature would be about an order of magnitude larger. The reason for this relatively small increase in melting temperature is the result of two opposing factors. One is the melting point depression by diluent and the other the expected increase due to the deformation. For an open system the amorphous network will imbibe solvent upon elongation.

A detailed analysis, that accounts for these effects, results in the following expression,

$$\frac{1}{T_m} - \frac{1}{T_m^0} = \frac{R}{\Delta H_u} \left\{ \left(\frac{V_u}{V_1}\right) [v_1 - \chi_1 v_1^2] \right.$$
$$\left. - \left[\left(\frac{6}{\pi m}\right)^{1/2} v_2^{1/3} \alpha - \left(\frac{\alpha^2}{\alpha} + \frac{1}{\alpha}\right) v_2^{-2/3} \bigg/ m\right] \right\} \tag{8.78}$$

when Eq. (8.46) is used. Here T_m is the melting temperature of the swollen deformed network, while T_m^0 is that of the pure unrestrained polymer. The volume fraction of polymer is v_2, the elongation ratio is referred to the isotropic length of the swollen network. The other quantities have already been defined. For an open system the values of v_1 and v_2 are determined by the conditions of swelling equilibrium. The above analysis explains the small increase that is observed in melting temperatures of swollen networks.(48) Good quantitative agreement is found at the higher extension ratios. At the lower extension ratios the same shortcoming of the deformed system, which was previously discussed with the unswollen network, also manifests itself. In this range the observed values are slightly larger than predicted.

8.4 Oriented crystallization and contractility in the absence of tension

Axially oriented crystalline polymers of either synthetic or natural origin contract upon melting. Examples are shown in Figs. 8.10 and 8.11 for fibrous natural rubber (16) and for a collagen fiber immersed in water.(42) Here the change in length

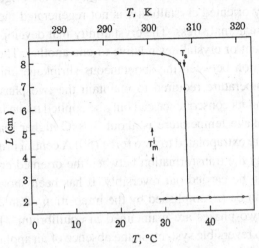

Fig. 8.10 Length, under zero force, as a function of temperature for fibrous natural rubber. (From Oth and Flory (16))

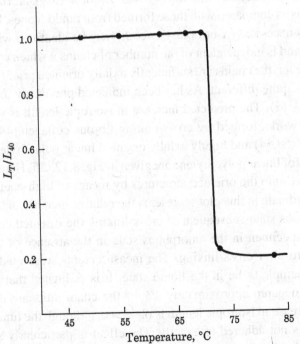

Fig. 8.11 Relative length as function of temperature for rat tail tendon collagen. (From Flory (42))

with temperature, under zero force, is illustrated for both fibers. In both cases a large axial contraction is observed over a narrow temperature interval. The shrinkage is accompanied by the disappearance of properties characteristic of the crystalline state such as discrete x-ray diffraction reflections and optical birefringence. Melting can therefore be deemed to have occurred. However, the original or native state, typified by axially oriented crystallinity, is not regenerated merely by cooling in the absence of an external stress. The crystallinity that develops is typified by the random arrangement of crystallites relative to one another. There is an important theoretical distinction between the spontaneous shrinkage under zero force and the stress and temperature required to maintain the two phases in equilibrium. Equation (8.17) and its consequence can only be applied to the latter situation. The spontaneous shrinkage temperature is about 7–8 °C higher than the equilibrium melting temperature extrapolated to zero force.(50) A central problem is to develop conditions whereby the transformation between the oriented crystalline state and the liquid state can be carried out reversibly. It has been shown in the previous section that this can be accomplished by the imposition of an appropriate tensile force so that the two phases are maintained in equilibrium. However, it is also possible to develop reversible systems in the absence of an applied tensile force by taking advantage of the increase in isotropic length, L_i, that ensues when axially oriented polymers are cross-linked.

There is a fundamental distinction between L_i for networks formed from highly oriented chains as compared with those formed from random ones. In the latter case the network is necessarily isotropic. Hence L_i may be identified with the length of the specimen and is independent of the number of chains ν which comprise the network. If, however, the chains are sufficiently axially oriented prior to cross-linking, the situation is quite different. As has been indicated previously, L_i is expected to increase as $\nu^{1/2}\langle\alpha\rangle$. The predicted increase in isotropic length is substantiated by studies on networks formed by cross-linking fibrous collagen,(44,51,52) fibrous natural rubber,(53,54) and highly axially oriented linear polyethylene.(55) The results obtained for linear polyethylene are given in Fig. 8.12.(55) Here the cross-links were introduced into the oriented structures by means of high-energy ionizing radiation. The ordinate in this plot represents the relative increase in length observed in the amorphous state subsequent to cross-linking the oriented chains. L_0 is the length of the specimen in the amorphous state in the absence of cross-links and L_i is the length after cross-linking. The measurements were made at 140 °C in order for the sample to be in the liquid state. It is estimated that for a radiation dose of 1000 megarep approximately 4% of the chain units are cross-linked. A substantial increase in isotropic length is observed, although the functional relation of Eq. (8.20) is not adhered to exactly. The effect is particularly striking in this highly oriented polyethylene where a 20-fold extension of length is developed in

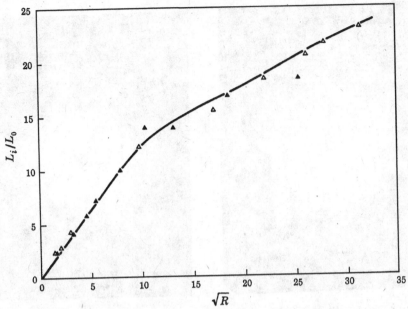

Fig. 8.12 Plot of L_i/L_0 at 140 °C against square root of radiation dose for highly oriented polyethylene fibers. ▲, ^{60}Co gamma irradiation; △, irradiation by high-energy electrons.(55)

the amorphous state without the application or maintenance of an external force. The extension ratios that can be developed when isotropic amorphous networks are mechanically deformed are severely restricted. Either crystallization intervenes to limit the extent of the deformation or the rupture of the network occurs. In either case, mechanical deformation does not result in extension ratios comparable to those depicted in Fig. 8.12.

The polyethylene networks are easily crystallized by reducing the temperature. Wide-angle diffraction patterns, characteristic of the recrystallized fibers are shown in Fig. 8.13. Cross-linking the original highly oriented fibers by ionizing radiation results in no sensible difference in the wide-angle x-ray pattern. However, after cross-linking, melting, and subsequent recrystallization, significant differences are exhibited, depending on the number of cross-links introduced. This becomes apparent in the patterns for the four samples illustrated. For the specimen into which no cross-links have been introduced, the pattern resulting after melting and recrystallization consists of a series of concentric rings. The crystalline state is thus characterized by a collection of randomly arranged crystallites. It is evident, however, from the other patterns that, as an increasing number of cross-links are introduced, a preferential orientation of the crystallites progressively develops. For example, the pattern in Fig. 8.13d, which is observed for a fiber characterized by ρ of approximately 2.65×10^{-2} and L_i/L_0 of 18.3, indicates that the *c*-axes

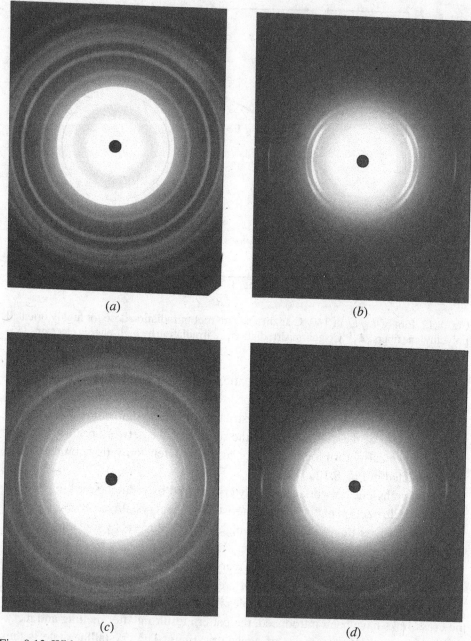

Fig. 8.13 Wide-angle x-ray diffraction patterns, taken at room temperature, of cross-linked melted and recrystallized polyethylene fibers for various radiation doses. (a) $R = 0$, $L_i/L_0 = 1$; (b) $R = 179$ megarep, $L_i/L_0 = 13.7$; (c) $R = 353$ megarep, $L_i/L_0 = 16.8$; (d) $R = 660$ megarep, $L_i/L_0 = 18.3$.(55)

of the crystallites are again preferentially oriented along the macroscopic fiber axis.

A collection of axially oriented crystallites can thus be developed without the necessity of a tensile force being applied during the crystallization process. These observations are intimately related to the extremely large values of L_i/L_0 that can be achieved in the amorphous state by the cross-linking process. The large extension ratios developed result in the establishment of a preferential axis for the subsequent transformation. Therefore, nuclei of the crystalline phase, which must form in order for the transformation to occur, are also preferentially directed. As a result, axially oriented crystallization occurs. The preferred orientation of the crystallites in the fibers described is now a built-in inherent part of the system. It should be present after any subsequent melting–recrystallization cycles, as long as the cross-linkages are maintained.

Appropriate dimensional changes must therefore accompany the melting and crystallization of such networks. Specifically, because of the axial orientation, contraction should occur on melting and spontaneous re-elongation on crystallization from the melt. Such dimensional changes are in fact observed as is illustrated in Fig. 8.14.(55) Here the relative length is plotted as a function of temperature for a fiber corresponding to the one illustrated in Fig. 8.13d. Starting with the crystalline

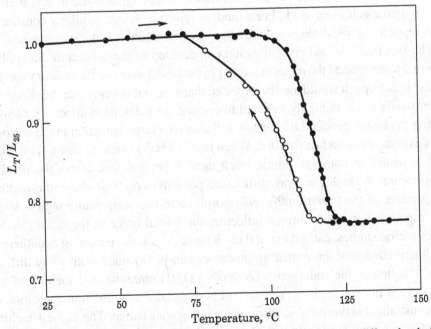

Fig. 8.14 Plot of relative length against temperature for a reversible contractile polyethylene fiber. ●, heating; ○, cooling.(55)

fiber, a slightly positive thermal expansion coefficient, typical of the crystalline state, is observed during the initial heating. Concomitant with melting, a 25% axial contraction occurs sharply over a narrow temperature interval. The observed shrinkage is consistent with the initial axial orientation of the sample in the crystalline state. Above the melting temperature the thermal expansion coefficient is slightly positive, as is expected in the liquid state. On cooling, crystallization of the specimen occurs, and the fiber regains its original dimensions. The heating process can then be repeated, and essentially the same melting temperature is obtained. Therefore, coupled with the crystal–liquid transformation, a reversible contractile system is obtained that is cyclic and does not require the imposition of an external stress for its operation. The shrinkage temperature can be identified with the equilibrium melting temperature in this case. When a slow cooling process is utilized after fusion, as is illustrated in Fig. 8.14, supercooling is observed. Supercooling is a characteristic of the crystallization of all polymeric systems. This effect, reflected in a dimensional lag, can be minimized by rapid cooling to low temperatures.

The sharpness of the observed contraction is a consequence of the melting of a homopolymer, in harmony with the view that the process is a first-order phase transition. The imposition of a stress on the system will raise the melting temperature, and the reversible contractility is still maintained. Consequently, fibers such as those described can serve as the working substance of an engine that converts thermal energy into mechanical work. For a random type copolymer, similarly constituted with respect to cross-linkages, the melting and contraction range is broadened.

The fact that oriented crystallization can develop in the absence of an applied external force reflects the molecular order in the liquid state and the concomitant decrease in entropy. It would then be expected that a similar development of molecular order would occur in highly oriented fibers that are maintained at constant length during the fusion process, since there will also be a large reduction in the entropy. This expectation is in fact fulfilled. When highly axially oriented fibrous polyethylene is heated at constant length, birefringence persists well above the melting temperature. Wide-angle x-ray diffraction patterns show that above the melting temperature of the orthorhombic polymorph there is a temperature region where a hexagonal structure is formed reflecting the lateral order of the chain.(56–58) Calorimetric studies, carried out at fixed length, exhibit three melting endotherms. An interpretation of these melting processes can be obtained from x-ray diffraction. The first of the endotherms (at about 141 °C) represents the melting of unconstrained fibrillar regions. The second is associated with the transformation of the constrained orthorhombic form into a hexagonal lattice. The highest melting endothermic peak is a result of the hexagonal structure being transformed into the melt.(58)

Since the crystal–liquid equilibrium can be governed by chemical processes, the transformation, and the concomitant dimensional changes, can occur isothermally. Once the principle has been established that the contractile process involves melting, or partial melting, it becomes important to distinguish between the actual contractile mechanism and the processes or chemical reactions that induce or regulate the phase transition. When these concepts are accepted it becomes possible to investigate many contractile systems from a unified point of view. Particularly important in this connection are macromolecules of biological interest. Contractility is widespread and known to be induced by a diversity of chemical reagents in this class of fibers. The principles of reversibility deduced for the polyethylene fibers serve as a useful model in investigating these more complex systems. The underlying basis of this contractile mechanism does not find its origin in a detailed crystallographic analysis of the fiber.

8.5 Contractility in the fibrous proteins

In a series of pioneering studies, Astbury and coworkers (21,59–61) established that fibrous proteins occur naturally in the crystalline state. In addition to being crystalline, these protein systems also possess a high degree of axial orientation. There are several different categories of fibrous proteins. These include the α- and β-keratins, collagen, elastodin and muscle fibers. The fibrous proteins as a class possess the basic initial structural requirements for contraction to accompany melting. In certain of these fibers, particularly the keratins, intermolecular covalent cross-links are also present. It can be presumed that in these cases the cross-links are formed subsequent to fiber formation, i.e. they are thus imposed on an initially axially oriented structure. Hence, based on the principles that have been developed, reversible contractility would be expected to accompany the crystal–liquid phase transition for these fibrous proteins. For the fibrous proteins that are not intermolecularly cross-linked, or for those in which the cross-linkages are not maintained during the melting process, only irreversible dimensional changes would be anticipated. The fact that axial contractions can be induced in different fibrous proteins by a variety of reagents and conditions does not vitiate the premise of a common underlying mechanism. It remains, therefore, to examine specific contractile systems to ascertain whether the principles that have been outlined are in fact obeyed.

There is a substantial body of evidence that demonstrates that the hydrothermal shrinkage of collagen, characterized by a contraction of about one-fifth the length of the native state, occurs directly as the result of melting.(44,61,62) However, neither the oriented crystalline state nor the original dimensions of the fiber are regenerated merely by cooling the specimen below the melting temperature. There is no indication in the amino acid composition of collagen that covalent intermolecular cross-links are present. Consequently, in harmony with the conclusion drawn from the studies of polyethylene fibers, regeneration of the oriented crystalline state in the absence of an external tensile force would not be expected. However, if collagen is cross-linked (tanned) with formaldehyde in its native state the hydrothermal melting–crystallization process is accompanied by a reversible anisotropic dimensional change.(63) Axially oriented crystallization develops from the molten state, as is evidenced by the wide-angle x-ray diffraction pattern and the simultaneous recovery of a significant portion of the low-angle x-ray diffraction pattern.(64,65) The recrystallized fiber contracts once again upon subsequent heating so that the process can be carried out cyclically.(44,51) An example of the latter observation is given in Fig. 8.15.(51) The initial melting of the native cross-linked fiber is extremely sharp. On cooling, a spontaneous re-elongation to about half the original length is

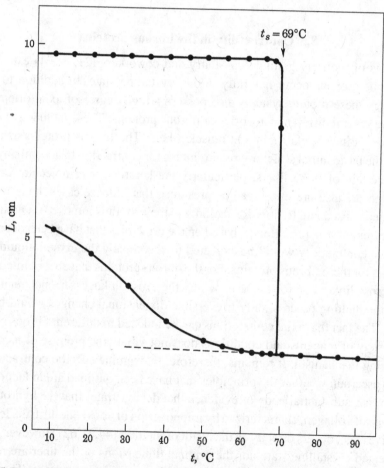

Fig. 8.15 Reversible contraction of cross-linked (tanned) collagen fibers. Upper curve, initial melting and shrinkage. Lower curve, melting after recrystallization.(51)

observed. On subsequent fusion the length diminishes more gradually with increasing temperature. The termination of the melting process is clearly defined, and a difference of only a few degrees exists between the two melting temperatures. The diffuse melting and slightly lower melting temperature observed during the second fusion can be attributed to hydrolysis at the higher temperatures and to kinetic difficulties that retard the development of the crystallinity typical of the native state. In this example, the reversible anisotropic dimensional changes accompany the phase transition in the absence of an external stress.

A similar example of contractility is demonstrated by the fibrous protein elastoidin. In the native state, the crystalline structure of elastoidin is similar to that of collagen. The amino acid compositions of the two proteins are also similar. However, elastoidin contains about 1 to 2% crystine residues, whose side groups can form stable intermolecular covalent cross-links. Consequently, it is not unexpected that elastoidin displays reversible contraction and relaxation concomitant with melting and crystallization.(66,67) When the native fiber is heated in water, a large axial contraction is observed at about 65°. On subsequent cooling to room temperature, about half of the initial length is regained without the application of any external force. After the initial shrinkage, the process can be carried out cyclically with contraction occurring on heating above 65° and relaxation occurring on cooling. The initial oriented collagen-type wide-angle x-ray diffraction pattern typical of elastoidin is completely converted to an amorphous pattern on shrinkage and is recovered on the subsequent relaxation.

In the previous examples the fibers were immersed in a liquid medium. This procedure serves to lower the melting temperature so that fusion occurs without degradation. Melting can also result from the interaction of groups on the polymer chain with specific species present in the supernatant liquid. In this case melting takes place at constant temperature. An example of such isothermal melting is shown in Fig. 8.16.(68) Here the effect of varying the pH of the supernatant aqueous phase on the melting, contraction temperature of cross-linked elastoidin is illustrated. The results summarized in Fig. 8.16 represent melting temperatures for a reversible process. The melting temperature remains invariant over a large pH range centered about neutrality. It then decreases sharply at the very high and low pH regions. In each instance contraction accompanies melting. Reversibility, in dimensions and in the return to the crystalline state, is obtained on cooling. Irrespective of the details of the chemical mechanism involved, Fig. 8.16 demonstrates the importance of the composition of the mixed phase (the amorphous polymer phase) in governing the melting temperature. In this figure, the equilibrium swelling ratio at the transformation temperature is also plotted as a function of the pH of the supernatant. A striking parallelism exists between the swelling ratio and melting temperature. The melting temperature remains constant with pH when the polymer concentration in

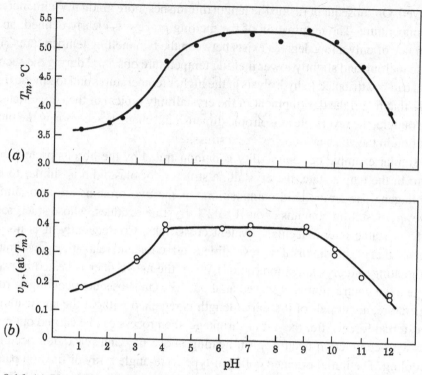

Fig. 8.16 (a) Plot of melting temperature T_m of elastoidin fibers as a function of the pH of the supernatant aqueous phase. (b) Plot of equilibrium swelling ratio of elastoidin, at its melting temperature, as a function of the pH of the supernatant aqueous phase.(68)

the molten phase is constant. When the polymer concentration decreases, the melting temperature does likewise. Hence it is clear that a major influence of pH on the isotropic melting is due to the changes that occur in the swelling of the amorphous protein.

The addition of certain monomeric reagents to the supernatant phase is known to affect the melting (contraction) temperature of protein fibers. Figure 8.17 illustrates the melting of elastoidin fibers when different monomeric reagents are added to the supernatant aqueous phase.(69) In each case, axial contraction accompanies melting. X-ray diffraction analysis indicates the complete disappearance of the ordered structure. Melting, and the accompanying contractility, can be induced by many different reagents. Substantial depressions in the melting temperature can be achieved. When the transformed fiber is cooled in the melting medium, two distinctly different results are obtained. In urea solutions, oriented recrystallization with spontaneous re-elongation occurs on cooling. However, in neutral salt solutions recrystallization does not occur. The fiber remains in the amorphous state after cooling. However, upon transferring the fiber to pure water, an almost instantaneous

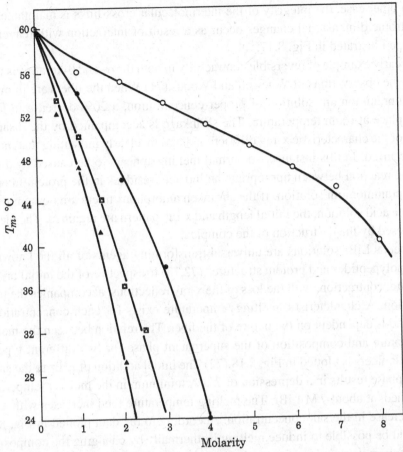

Fig. 8.17 Plot of melting temperature T_m of elastoidin fibers against concentration of monomeric reagent present in supernatant phase. ○ urea; □ CaCl$_2$; ▲ KCNS; △ KI; ● LiBr.(69)

regeneration of the oriented crystalline state and original length takes place. In contrast to the behaviors in urea solutions, and in pure water, recrystallization from the salt solution requires both cooling and dilution. These results indicate that, besides the usual disordering of the chain that occurs upon melting, additional structural alterations are imparted that prevent recrystallization. When the reagent is removed, crystallization and re-elongation ensue.

The α- and β-keratins exist in the oriented crystalline state and possess a high concentration of the cystine residues. They also undergo contraction when subjected to the action of a wide variety of reagents.(70) It is recognized that two distinctly different types of contractile processes can be observed in α-keratin fibers. One of these involves the interaction with reagents known to sever disulfide cross-links. As would be expected, in this case the observed dimensional changes are irreversible.

In the other case, the integrity of the intermolecular cross-links is maintained and anisotropic dimensional changes occur as a result of interaction with reagents of the type illustrated in Fig. 8.17.

An early example of reversible contractility in both the α- and β-keratins is inherent in the observations of Whewell and Woods.(71) When the fibers are immersed in a cuprammonium solution of proper concentration, a 20% decrease in length takes place at room temperature. The shrinkage is accompanied by the disappearance of the characteristic x-ray diffraction diagram, clearly indicating that melting has occurred. In this instance isothermal melting appears to be caused by a complexing reaction between appropriate amino acid residues in the protein fiber and the cuprammonium solution. If the shrunken amorphous fibers are now immersed in dilute acid solution, the initial length and x-ray pattern are regained. The melting is reversed by the destruction of the complex.

Aqueous LiBr solutions are universal transforming agents of all the known ordered polypeptides and protein structures.(72,73) Irrespective of the initial ordered structure, contraction, with the loss of the x-ray reflections, accompanies the transformation. A characteristic melting temperature exists for each concentration of LiBr and is dependent on the nature of the fiber. The relation between the melting temperature and composition of the supernatant phase for two different types of α-keratin fibers is plotted in Fig. 8.18.(72) The initial addition of LiBr to the supernatant phase results in a depression of T_m. A minimum in the melting temperature is reached, at about 7 M LiBr. The melting temperature then increases with a further increase in the salt concentration. According to the data plotted in Fig. 8.18, it should be possible to induce melting isothermally by changing the composition of the supernatant, in analogy to changing the pH of the supernatant of elastoidin fibers. Starting with a native fiber immersed in a high concentration of LiBr at 24 °C and following the pattern established in Fig. 8.18, contraction accompanies melting upon dilution at constant temperature. As the molten state is traversed, the length does not change with further dilution. However, when a concentration prescribed by the data of Fig. 8.18 is reached, recrystallization accompanied by re-elongation is observed.

The demonstration that the crystal–liquid phase transition can be conducted isothermally, by changing the concentration of the supernatant phase, portends the possibility of the utilization of fibrous macromolecules as the working substance of an engine that isothermally converts chemical energy to mechanical work.(3,54)

It is reasonable to inquire at this point whether the principles that have been set forth above have any applicability to natural functioning contractile systems. Muscles are very intricately constructed fibrous structures developed by nature to convert chemical energy into mechanical work. Detailed and sophisticated electron microscopic and x-ray diffraction studies have established the fine structure of muscle. The chemical processes and enzymatic activity that are intimately involved

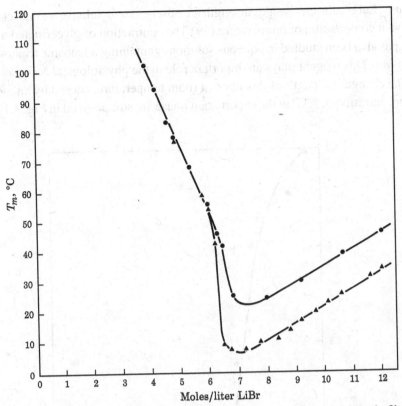

Fig. 8.18 Plot of melting temperature T_m (contraction temperature) of α-keratin fibers as a function of molarity of LiBr solution in the supernatant. ▲ Lincoln wool; ● horse hair.(72)

in controlling contractility and motility are complex. It is far beyond the scope of this work to discuss the detailed fiber structure and the chemical processes involved in muscular contraction. Irrespective of the intricacies involved, and the complex control mechanisms that are operative, the fact remains that muscle fibers are comprised of proteins that occur naturally in the axially oriented state. The primary transducing element in muscle fiber consists of macromolecules in a highly ordered conformation which under stimulus are transformed, at least in part, to a random conformation.(74,75)

In nonphysiological laboratory type experiments muscle fibers can be studied based on the concepts that have been developed for synthetic polymers. Because of their native structure, substantial shrinkage has been observed in muscle fibers by interaction with reagents known to cause melting and contraction in other fibrous proteins.(76–78) The underlying contractile mechanism can be presumed to be the same for muscle fibers. In a step closer to physiological conditions evidence for contraction accompanying melting has also been reported for glycerinated muscle fibers immersed in ATP–glycerol–water mixtures and in ATP–ethylene glycol–water mixtures. Large, abrupt changes in length are observed with relatively small changes in solvent composition or temperature.(79) The contraction of glycerinated muscle fibers has also been studied in aqueous solutions containing adenosine triphosphate (ATP).(80) This reagent plays an important role in the physiological action of muscle. The changes in length of this fiber, at room temperature, caused by an increasing concentration of ATP in the supernatant phase are summarized in Fig. 8.19.(80)

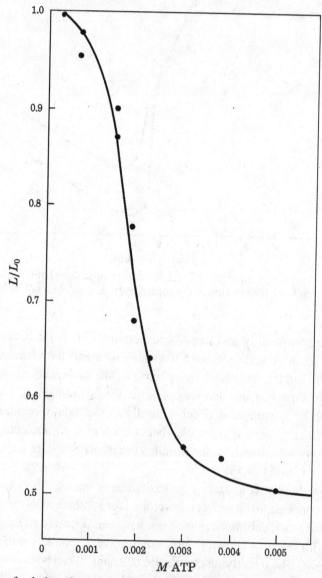

Fig. 8.19 Plot of relative change in length of glycerinated muscle fibers at 25 °C as a function of ATP concentration in the supernatant phase.(80)

The curve in Fig. 8.19 gives clear indication that a cooperative phase transition is taking place. Only a slight change in the ATP concentration is needed to induce the transformation at constant temperature. A melting process is clearly indicated, and the structural changes that are expected to accompany fusion can be demonstrated. The wide-angle x-ray pattern shows that concomitant with the completion of contraction, the native, α-keratin type, oriented crystalline structure has disappeared. The results just cited show that in nonphysiological experiments, muscle fibers behave as typical fibrous proteins with regard to melting and anisotropic dimensional changes. The development of tension, at constant length, must follow.

Although these results do not represent actual systems operating under physiological conditions, a basis for a possible mechanism for this complex process can be discerned. In naturally functioning systems the regulating processes are complex and involve a series of chemical reactions. Direct experimentation of the kind needed is thus made more difficult. In principle, however, if a phase transition is involved, its major characteristics should be discernible. It is not necessary that the complete transformation be involved. Contractility and tension development can also result when the system only operates over a portion of the transformation range, i.e. only partial melting and recrystallization is involved. The transformation range will be relatively broad for such multicomponent systems.

To summarize, the results that have been described demonstrate that in laboratory experiments anisotropic dimensional changes, or complementary tension development, can be produced in the fibrous proteins as a consequence of a phase transition between the oriented crystalline and amorphous states. This transition can be induced either thermally or isothermally by interaction with a diversity of chemical reagents. The same physical-chemical principles are followed that govern the structurally simpler fibrous synthetic polymers.

8.6 Mechanochemistry

The characteristic high deformability, coupled with the ability to regain initial dimensions allow long chain molecules to serve as converters of thermal or chemical energy into mechanical work. There are no *a priori* reasons for excluding biological processes from this generalization. There is always the possibility that other mechanisms may be operative in specific cases. However, it can be expected that nature will take advantage of the conformational versatility of macromolecules.(81)

In analyzing this problem it is necessary to distinguish between processes that are restricted to the amorphous phase and those that involve a crystal–liquid phase transition. Rubber elastic deformation involves an increase in the mean-square end-to-end distance of the chains in the liquid state, in compliance with the imposed macroscopic strain. Closely associated with this phenomenon is the deformation that results from changes in the degree of swelling of a network immersed in an excess of the supernatant phase. Swelling or deswelling of a network can be caused by changes in the intensity of polymer–solvent interactions or by various chemical reactions. W. Kuhn and collaborators pioneered studies of changes in the degree of swelling of polyelectrolyte networks with alterations in the pH of the surrounding medium.(82,83) Length changes can also be induced electrolytically in polyelectrolyte fibers.(84) Deformations restricted to the amorphous phase are usually isotropic unless either a large stress is applied or the network is dimensionally constrained by mechanical means. The anisotropic deformation and stress response involved in the crystal–liquid phase transition have already been discussed.

The utilization of macromolecules to convert thermal energy into mechanical work can be analyzed by referring to a Carnot cycle. There are no restrictions on the nature of the working substance in the operation of a Carnot cycle. However, it is required that all processes be conducted reversibly and that all heat received or rejected by the working substance be exchanged at constant temperature. Thus, processes for which temperature of the working substance changes are reversible adiabatics. By recalling the analogy between the intensive–extensive sets of variables, p, V and $-f$, $-L$, a schematic diagram for a reversible Carnot thermal engine, utilizing a pure amorphous polymer as the working substance, is given in Fig. 8.20a.(45) The isothermals are represented by curves AD and CB, and the polymer is in contact with large heat reservoirs at temperatures T_1 and T_2, respectively. AB and CD represent the reversible adiabatics, with the system being isolated from the surroundings. A reversible thermal engine can be constructed from a deformable substance if the tension at constant length is increased by a rise in temperature and the tension–length adiabatics possess a greater slope than the corresponding isothermals. These criteria are consistent with the previously discussed thermoelastic properties of amorphous polymers. An engine of this type is exemplified in the self-energizing pendulum described by Wiegand (85) which utilizes natural rubber as a working substance and is restricted to the noncrystalline state. The thermodynamic efficiency of the engine illustrated in Fig. 8.20a is directly given by Carnot and depends only on the two operating temperatures. The amount of work performed per cycle is represented by the area ABCD. A more detailed thermodynamic analysis of such an engine has been given.(86)

If instead of a rubberlike deformation a phase transition occurs, the isothermal processes are represented by horizontal lines (since the force is independent of the length), as is shown in Fig. 8.20b. If in each of the two cycles described the same adiabatics are involved, the net work performed is greater for the one with the phase transition. This is analogous to using a condensed vapor in the more conventional Carnot cycle. The thermodynamic efficiency remains the same since it depends only on the two temperatures at which the engine operates. The deliverance

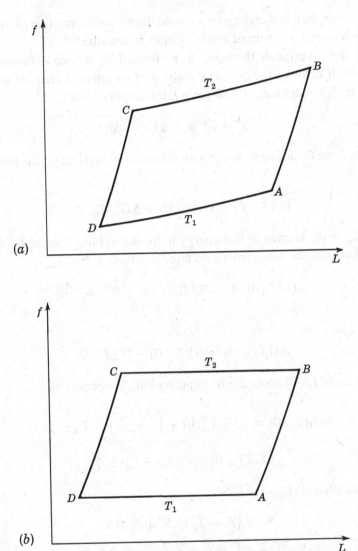

Fig. 8.20 Schematic diagram for a Carnot cycle utilizing a pure polymer as the working substance. (a) Polymer always in amorphous state; (b) intervention of an isothermal phase transition.

of more net useful work makes the intervention of a phase transition advantageous, irrespective of the molecular nature of the working substance. When fibrous macromolecules are involved, the oriented crystalline structure allows for such a transition. A transition of this type allows for the "razor-edge" character of the contraction displayed by natural systems and permits the working substance to be a more sensitive converter of thermal energy into mechanical work. A thermal engine that uses fibrous polyethylene that undergoes a crystal–liquid transformation as the working substance has been constructed and its operation described.(87)

It is possible to estimate the work, W, performed during one isothermal portion of the cycle. It involves the complete melting of an oriented fiber. At the melting temperature T_m, under the external force f, it is required that

$$G^a - G^c = f\,\Delta L = -W \qquad (8.79)$$

It is assumed that G^c does not change with deformation, and that G^a can be expressed as

$$G^a(T_m, f) = G^a(T_m, 0) + \Delta G_{el}^a(T_m, f) \qquad (8.80)$$

where ΔG_{el}^a is the change in free energy in the amorphous state at T_m due to the elastic deformation in going from zero force to a force f. Hence

$$\Delta G_f(T_m, 0) + \Delta G_{el}(T_m, f) = f\,\Delta L = -W \qquad (8.81)$$

where

$$\Delta G_f(T_m, 0) = G^a(T_m, 0) - G^c(T_m, 0) \qquad (8.82)$$

Expanding $\Delta G_f(T_m, 0)$ about the isotropic melting temperature T_m^0

$$\Delta G_f(T_m, 0) = \Delta G_f(T_m^0, 0) + \left(\frac{\partial \Delta G_f}{\partial T}\right)_{T_m^0} (T_m - T_m^0) \qquad (8.83)$$

$$\Delta G_f(T_m, 0) \cong -(T_m - T_m^0)\Delta S_f^0 \qquad (8.84)$$

For a system where $\Delta G_{el} = -T\,\Delta S_{el}$

$$W \cong (T_m - T_m^0)\Delta S_f^0 + T_m \Delta S_{el} \qquad (8.85)$$

$$W \cong -T_m^0 \Delta S_f^0 + T_m(\Delta S_{el} + \Delta S_f^0) \qquad (8.86)$$

Thus the work performed by the fiber on melting depends on two terms. One term is independent of the force, and the other depends on it through the terms T_m and ΔS_{el}. In the isotropic case, the work done must necessarily be zero, and, at $T = T_m^c$, $\Delta L = 0$ so that W is zero. Hence the work done in a single cycle passes through a maximum with increasing temperature and force. The latter two quantities are related by Eq. (8.30). For small forces, where T_m only slightly exceeds T_m^0, and ΔS_{el} is small, the work done is of the order of R cal mol^{-1}. This is comparable in magnitude to that observed in naturally occurring systems.

Of more general interest, particularly with respect to biological systems, are engines that operate isothermally and are based on chemical interactions and reactions.

For example, the naturally functioning muscle fiber system can be considered an engine that converts chemical energy into mechanical work. It operates by means of reversible strains induced in the working substance, i.e. the fibrous muscle proteins.(45) An isothermal chemical engine can be devised whose operation is similar to that of the thermal engine.(45,88,89) In place of the two heat reservoirs, the fiber is maintained in contact with two large baths of absorbing or reacting species, each being maintained at a constant chemical potential. Processes carried out at constant chemical potential have been termed isopotentials in analogy with the isotherms of a heat engine.(88) The transfer of the fiber from one chemical potential to another occurs as an isolated system whose composition remains fixed. These processes have been termed isophores and correspond to the adiabatics of the thermal engine. In the latter case, when the entropy is held constant, the thermal potential or temperature changes. A chemical engine, with a fibrous working substance, functions when the tension at constant length increases with increased concentration of reactants and the slope of the isopotential $(\partial f/\partial L)_{\mu_i}$ is less than the slope of the isophore $(\partial f/\partial L)_{n_i}$.

The consequences of the working substance crystallizing in a chemical engine are the same as for a heat engine. For operation between the same chemical potential levels, with common isophores, the net work delivered per cycle is greater when a phase transition occurs during the isopotential portions. An isopotential phase transition ensures that the required relations between the isophoric and isopotential force–length relations will be met. Several fibrous protein systems have already been referred to which can serve as the working substance and the reactants of a chemical engine with isopotential phase transitions.

For a simple engine, comprised of only a single component (besides the working substance), that is maintained in two reservoirs at chemical potentials μ^{I} and μ^{II}, respectively, the net work accomplished per cycle is given by (88)

$$W = (\mu^{I} - \mu^{II})\Delta n \tag{8.87}$$

where Δn is the quantity of reactant transferred from one reservoir to the other. The immediate source of the work obtained during the contraction can be attributed mainly to the increased entropy resulting from melting. However, for a complete cycle where the working substance returns to its original state, all the changes must be found in the surroundings. The ultimate source of the work performed comes from the free energy change involved in transferring the species from one reservoir to the other. More complex chemical engines, involving multicomponent systems, have been discussed in detail by Katchalsky and collaborators (88,90) along the principles outlined here. Chemical engines operating isothermally that are based on collagen fibers immersed in aqueous salt solutions known to induce melting, have been constructed.(89–91) The foregoing discussion of idealized cycles does not imply that real systems must rigidly adhere to them. It is meant only to serve as a basis for the molecular understanding of mechanochemical processes that involve macromolecules and the basic principles that are involved.

References

1. Li, T. T., R. J. Volungis and R. S. Stein, *J. Polym. Sci.*, **20**, 194 (1956); J. T. Judge and R. S. Stein, *J. Appl. Phys.*, **32**, 2357 (1961).
2. Flory, P., *J. Chem. Phys.*, **15**, 397 (1947).
3. Flory, P., *Science*, **124**, 53 (1956).
4. Flory, P., *J. Amer. Chem. Soc.*, **78**, 5222 (1956).
5. Mark, J. E., in *Physical Properties of Polymers*, J. E. Mark ed., American Chemical Society (1993) p. 3.
6. Treloar, L. R. G., *The Physics of Rubber Elasticity*, Oxford University Press (1949); P. J. Flory, *Trans. Faraday Soc.*, **57**, 829 (1961).
7. Flory, P. J., *Principles of Polymer Chemistry*, Cornell University Press (1953) Chapter XI.
8. Mark, J. E. and B. Erman, *Rubberlike Elasticity. A Molecular Primer*, Wiley-Interscience (1988).
9. Erman, B. and J. E. Mark, *Structure and Property of Rubber-Like Networks*, Oxford University Press (1999).
10. Flisi, U., G. Crespi and A. Valuassori, *Kautsch. Gummi Kunstst.*, **22**, 154 (1969).
11. Candie, F. D., G. Romano and R. Russo, *J. Polym. Sci.: Polym. Phys. Ed.*, **23**, 2109 (1985).
12. Hsu, Y. H. and J. E. Mark, *Polym. Eng. Sci.*, **27**, 1203 (1987).
13. Jiang, C. Y., J. E. Mark, V. S. C. Chang and J. P. Kennedy, *Polym. Bull.*, **11**, 319 (1987).
14. Roberts, D. E. and L. Mandelkern, *J. Res. Nat. Bur. Stand.*, **54**, 167 (1955).
15. Gee, G., *Quart. Rev. (London)*, **1**, 265 (1947).
16. Oth, J. F. M. and P. J. Flory, *J. Am. Chem. Soc.*, **80**, 1297 (1958).
17. Roberts, D. E. and L. Mandelkern, *J. Am. Chem. Soc.*, **77**, 781 (1955).
18. Smith, W. H. and C. P. Saylor, *J. Res. Nat. Bur. Stand.*, **21**, 257 (1938).
19. Tobolsky, A. V. and G. M. Brown, *J. Polym. Sci.*, **17**, 547 (1955).
20. Gent, A. N., *Trans. Farady Soc.*, **50**, 521 (1954).
21. Astbury, W. T., *Size and Shape Changes of Contractile Polymers*, A. Wasserman ed., Pergamon Press (1960) p. 78.
22. Natta, G. and P. Corradini, *Rubber Chem. Technol.*, **33**, 703 (1960).
23. Bull, H. B., *J. Am. Chem. Soc.*, **67**, 533 (1945).
24. Weber, A. and H. Weber, *Biophys. Acta*, **7**, 214 (1951).
25. Morales, M. and J. Botts, *Discuss. Farady Soc.*, **13**, 125 (1953).
26. Allegra, G., *Makromol. Chem.*, **181**, 1127 (1980).
27. Allegra, G. and M. Bruzzone, *Macromolecules*, **16**, 1167 (1983).
28. Sharaf, M. A., A. Kloczkouski, J. E. Mark and B. Erman, *Comp. Polym. Sci.*, **2**, 84 (1992).
29. Krigbaum, W. R. and R. J. Roe, *J. Polym. Sci. A*, **2**, 4394 (1964).
30. Kim, H. G. and L. Mandelkern, *J. Polym. Sci.*, *Pt A-2*, **6**, 181 (1968).
31. Flory, P. J., *Statistical Mechanics of Chain Molecules*, Wiley-Interscience (1969).
32. Krigbaum, W. R., J. W. Dawkins, G. H. Via and Y. J. Balto, *J. Polym. Sci.*, *Pt A-2*, **4**, 475 (1966).

33. Ronca, G. and G. Allegra, *J. Chem. Phys.*, **63**, 4990 (1975).
34. Cesari, M., G. Peraga, A. Zazzetta and L. Gargani, *Makromol. Chem.*, **180**, 1143 (1980).
35. Kloczkowski, A., J. E. Mark, M. A. Sharof and B. Erman, in *Synthesis, Characterization and Theory of Polymeric Networks and Gels*, S. M. Aharoni ed., Plenum Press (1992) p. 227.
36. Erman, B. and P. J. Flory, *J. Chem. Phys.*, **68**, 5363 (1978).
37. Flory, P. J. and B. Erman, *Macromolecules*, **15**, 800 (1982).
38. Erman, B. and P. J. Flory, *Macromolecules*, **16**, 1601 (1983).
39. Popli, R. and L. Mandelkern, unpublished results.
40. deCandia, F. and V. Vittoria, *Makromol. Chem.*, **155**, 17 (1972).
41. Oomo, R., K. Miyasaka and K. Ishikawa, *J. Polym. Sci.: Polym. Phys. Ed.*, **11**, 1477 (1973).
42. Flory, P. J., *J. Cellular Comp. Physiol.*, **49** (Suppl. 1) 175 (1957).
43. Scheraga, H. A., *J. Phys. Chem.*, **64**, 1917 (1960).
44. Flory, P. J. and O. K. Spurr, Jr., *J. Amer. Chem. Soc.*, **83**, 1308 (1961).
45. Pryor, M. G. M., in *Progress in Biophysics*, vol. I, J. A. V. Butler and J. I. Randall eds., Butterworth-Springer (1950) p. 216.
46. McLair, P. E. and E. R. Wiley, *J. Biol. Chem.*, **247**, 692 (1972).
47. Flory, P. J. and R. R. Garrett, *J. Amer. Chem. Soc.*, **80**, 4836 (1958).
48. Posthuma de Boer, A. and A. J. Pennings, *Faraday Discuss. Chem. Soc.*, **68**, 345 (1979).
49. Smith, K. J., Jr., A. Greene and A. Cifferi, *Kolloid Z. Z. Polym.*, **194**, 94 (1964).
50. Oth, J. F. M., E. T. Dumitra, O. K. Spurr, Jr. and P. J. Flory, *J. Amer. Chem. Soc.*, **79**, 3288 (1957).
51. Oth, J. F. M., *Kolloid.-Z.*, **162**, 124 (1959); *ibid.*, **171**, 1 (1966).
52. Gerngross, O. and L. R. Katz, *Kolloid-Beih.*, **23**, 368 (1926).
53. Roberts, D. E., L. Mandelkern and P. J. Flory, *J. Am. Chem. Soc.*, **79**, 1515 (1957).
54. Roberts, D. E. and L. Mandelkern, *J. Am. Chem. Soc.*, **80**, 1289 (1958).
55. Mandelkern, L., D. E. Roberts, A. F. Diorio and A. S. Posner, *J. Am. Chem. Soc.*, **81**, 4148 (1959).
56. Clough, S. B., *J. Macromol. Sci.*, **B4**, 199 (1970).
57. Clough, S. B., *J. Polym. Sci.: Polym. Lett.*, **8**, 519 (1970).
58. Pennings, A. J. and A. Zwijnenburg, *J. Polym. Sci.: Polym. Phys. Ed.*, **17**, 1011 (1979).
59. Bailey, K., W. T. Astbury and K. M. Rudall, *Nature*, **151**, 716 (1943).
60. Astbury, W. T., *Proc. R. Soc. (London), Ser. A*, **134**, 303 (1947).
61. Astbury, W. T., *Trans. Faraday Soc.*, **34**, 378 (1948).
62. Wright, B. A. and N. M. Wiederhorn, *J. Polym. Sci.*, **7**, 105 (1951).
63. Ewald, A., *Z. Physiol. Chem.*, **105**, 135 (1919).
64. Bear, R. S., *Adv. Protein Chem.*, **7**, 69 (1952).
65. Rice, R. V., *Proc. Nat. Acad. Sci. USA*, **46**, 1186 (1960).
66. Faure-Fremet, R., *J. Chim. Phys.*, **34**, 126 (1937).
67. Champetier, G. and E. Faure-Fremet, *J. Chim. Phys.*, **34**, 197 (1937).
68. Mandelkern, L. and W. T. Meyer, *Symposium on Microstructure of Proteins, J. Polym. Sci.*, **49**, 125 (1961).
69. Mandelkern, L., W. T. Myer and A. F. Diorio, *J. Phys. Chem.*, **66**, 375 (1962).
70. Alexander, P. and R. F. Hudson, *Wool, Its Chemistry and Physics*, Reinhold Publishing Corporation (1954) p. 55.
71. Whewell, C. S. and H. J. Woods, *Symposium on Fibrous Proteins*, Society of Dyers and Colourists (1946) p. 50.

72. Mandelkern, L., J. C. Halpin, A. F. Diorio and A. S. Posner, *J. Am. Chem. Soc.*, **84**, 1383 (1962).
73. Mandelkern, L., J. C. Halpin and A. F. Diorio, *J. Polym. Sci.*, **60**, 531 (1962).
74. Harrington, W. F., *Proc. Nat. Acad. Sci. USA*, **76**, 5066 (1979).
75. Sutoh, K., K. Sutoh, T. Karr and W. F. Harrington, *J. Mol. Biol.*, **126**, 1 (1978).
76. Bowen, W. J. and K. Laki, *Amer. J. Physiol.*, **185**, 91 (1956).
77. Mandelkern, L. and E. A. Villarico, *Macromolecules*, **2**, 394 (1969).
78. Bowen, W. J., *J. Cellular Comp. Physiol.*, **49** (Suppl. 1) 267 (1957).
79. Hoeve, C. A. J., Y. A. Willis and D. J. Martin, *Biochemistry*, **2**, 279 (1963).
80. Mandelkern, L., A. S. Posner, A. F. Diorio and K. Laki, *Proc. Nat. Acad. Sci. USA*, **45**, 814 (1959).
81. Flory, P. J., *Protein Structure and Function, Brookhaven Symp. Biol.*, No. 13 (1960) p. 89.
82. Kuhn, W., *Makromol. Chem.*, **35**, 54 (1960).
83. Kuhn, W., A. Ramel, D. H. Walters, G. Ebner and H. J. Kuhn, *Adv. Polym. Sci.*, **1**, 540 (1960).
84. Hamlen, R. P., C. E. Kent and S. N. Shafer, *Nature*, **206**, 1149 (1965).
85. Wiegand, W. B., *Trans. Inst. Rubber Ind.*, **1**, 141 (1925).
86. Farris, R. J., *Polym. Eng. Sci*, **17**, 737 (1972).
87. Mandelkern, L. and D. E. Roberts, U. S. Patent, 3,090,735 (1963).
88. Katchalsky, A., S. Lifson, I. Michaelis and H. Zwich, in *Size and Shape Changes of Contractile Polymers*, A Wassermann ed., Pergamon Press (1960) p. 1.
89. Pryor, M. G. M., *Nature*, **171**, 213 (1953).
90. Rubin, M. M., K. A. Piez and A. Katchalsky, *Biochemistry*, **8**, 3628 (1969).
91. Sussman, M. V. and A. Katchalsky, *Science*, **167**, 45 (1970).

Author index

411

412

Author index

Subject index

Page numbers in italics, e.g. *46*, indicate references to figures. Page numbers in bold, e.g. **41**, denote entries in tables.